Communications
in Computer and Information Science 305

T0238508

Jimson Mathew   Priyadarsan Patra
Dhiraj K. Pradhan   A. J. Kuttyamma (Eds.)

Eco-friendly Computing and Communication Systems

International Conference, ICECCS 2012
Kochi, India, August 9-11, 2012
Proceedings

 Springer

Volume Editors

Jimson Mathew
University of Bristol, UK
E-mail: jimson@cs.bris.ac.uk

Priyadarsan Patra
Intel Corporation, Hillsbro, OR, USA
E-mail: priyadarsan.patra@intel.com

Dhiraj K. Pradhan
University of Bristol, UK
E-mail: pradhan@cs.bris.ac.uk

A. J. Kuttyamma
Rajagiri School of Engineering and Technology
Kochi, Kerala, India
E-mail: kuttyamma_aj@rajagiritech.ac.in

ISSN 1865-0929   e-ISSN 1865-0937
ISBN 978-3-642-32111-5   e-ISBN 978-3-642-32112-2
DOI 10.1007/978-3-642-32112-2
Springer Heidelberg Dordrecht London New York

Library of Congress Control Number: 2012942857

CR Subject Classification (1998): I.4, I.2.4, C.2.1, E.3, I.6, J.3, H.3

Typesetting: Camera-ready by author, data conversion by Scientific Publishing Services, Chennai, India

Printed on acid-free paper

Springer is part of Springer Science+Business Media (www.springer.com)

Preface

The International Conference on Eco-friendly Computing and Communication Systems (ICECCS) was held in Kochi during August 9–11, 2012. ICECCS 2012 was organized by the Rajagiri School of Engineering and Technology (RSET) in association with the Oil and Natural Gas Corporation (ONGC) of India and Kerala State Council for Science, Technology and Environment (KSCSTE), and Computer Society of India Cochin Chapter. Established in 2001, RSET is a premier professional institution striving for holistic excellence in education to mould young, vibrant engineers.

ICECCS 2012 was a three-day conference that provided an opportunity to bring together students, researchers and practitioners from both academia and industry. ICECCS was focused on "Information Revolution Through Green and Eco-friendly Computing" and this theme was particularly relevant considering the fact that the conference was held in Cochin, a city in the Evergreen State of Kerala, India. This conference provided a forum for sharing insights, experiences and interaction on various aspects of evolving technologies and patterns related to computer science, information technology and electronics, and it was a platform not only for researchers from Asia but also from other continents across the globe, making this conference more international and attractive for participants. ICECCS 2012 received 133 research papers from different countries including Canada, France, Germany, India, Kuwait, Singapore, South Africa, Spain, and the UK. This clearly reflects the truly international stature of ICECCS. All papers were rigorously reviewed internationally by an expert technical review committee. The committee selected 50 papers (acceptance rate: 37.5%) for presentation at the conference. We would like to thank the authors for having revised their papers to address the comments and suggestions of the referees. The ICECCS 2012 conference program also included two workshops: the International Workshop on Mathematical Modeling and Scientific Computing (IWMS 2012) and the International Workshop on Green Energy Technologies (IWGET). We owe thanks to all workshop organizers and members of the Program Committee for their diligent work, which ensured a very high quality.

The establishment of a new conference with unique approaches and strengths reaching out globally to the technical community no doubt requires strong dedication, effective mindshare and good-old hard work from many volunteers. We would like to thank the Steering Committee and the entire Organizing Committee, for putting together a fantastic technical program. We also note the effort by our Publicity Chairs, the Publication Chair, and the Local Arrangements Chair. We received strong support from ONGC and KCSCTE. We are thankful for the help and friendship of several colleagues in academia and industry who volunteered to deliver a workshop or plenary talk despite many challenges. We would like to thank Springer for the fruitful collaboration during the preparation of the proceedings. The proceedings of ICECCS 2012 are organized in a single CCIS volume. We hope that the ICECCS 2012 proceedings will serve as an important intellectual resource for computing and communication systems researchers, pushing forward the boundaries of these two fields and enabling better collaboration and exchange of ideas.

August 2012

Jimson Mathew
Priyadarsan Patra
Dhiraj K. Pradhan
Kuttyamma A.J.

Organization

ICECCS 2012 was organized by the Rajagiri School of Engineering and Technology (RSET) in association with the ONGC and Kerala State Council for Science, Technology and Environment (KSCSTE).

Program Committee

Conference Chair	Dhiraj K. Pradhan, University of Bristol, UK
	J. Issac, RSET, India
Program Chair	Jimson Mathew, University of Bristol, UK
	Priyadarsan Patra, Intel Corporation, Oregon, USA
Organizing Chair	Kuttiyamma A.J., RSET, India
	Biju Paul J., RSET, India
Tutorials	Mahesh Panicker, GE
	K.S. Mathew, RSET
	Babita Roslind Jose, Cochin University of Science and Technology, India
Workshops	Rajan M.P., IISER Trivandrum, India
	Vinodkumar P.B., RSET, India
	S.K. Sinha, IISc Bangalore, India
	Chikku Abraham, RSET
	Rajesh Joseph Abraham, IIST, India
Publicity Chairs	Mahesh Panicker, GE, India
Publication	Raphael Guerra, Technical University of Kaiserslautern, Germany
	B.A. Jose, Intel Corporation, USA
Registration	Mahesh Poolakkaparambil, Oxford Brookes University, UK
	B.A. George, Technical University of Kaiserslautern, Germany

Referees

D. Samanta	S. Bhattacharyya	M. El-Hadedy Aly
T. Wooi Haw	F.W. Büsser	S. Bhattacharjee
E. Barrelet	J. Mathew	M.P. Rajan
H. Zarandi	S. Cherian	S. Manikandan
C.H. Ma	B. Paul	H.M. Kittur
U. Bhattacharya	F. Balasa	C. Abraham
V. Pangracious	S. Maity	V. Chandra

M.G. Mini

V. Panicker

D. Pradhan

P. Ooi Chee

G. Vincet

V. Mishra

V. Ojha

B.R. Jose

P. Chanak

D.P. Sudha

B.A. Jose

S. Sivaguru

R. George

D. Jayaraman

S. Kasarian

J. Mulerikkal

D. Sankar

L. Jones

S.D. Kolya

N. Murthy

P. Ghosal

Q. Dong

R. Islam

J. Mandal

G. Pramod

R. Logeswaran

A. Beldachi

I. Banerjee

K.P. Rao

R. Diptendu

C. Feng

D. Mahapatra

A. Sarkar

S. Ahuja

K.G. Smitha

M. Poolakkaparambil

A. Patanaik

N. Pandya

J. Chiverton

R.A. Shafik

S. Rana

V. Ambat

M. Andrei

R. Thottupuram

B. Palayyan

P. Kharat

T. Samanta

B. Neethu

M.V. Chilukuri

D. Mavridis

P. Prasad

R. Paily

A. Binu

M.T. Sebastian

A.S. Nair

N. Malmurugan

S.P. Maity

R. Govindan

S. Banerjee

M. Bouache

D. Das

L. Sun

P. Patra

B. Joshi

H. Patil

H. Jim

A.M. Jabir

A. Manuzzato

C. Thomas

A.K. Singh

C. Argyrides

P.S. Deshpande

Table of Contents

Image and Signal Processing

Bioinformatics and Emerging Technologies

Secure and Reliable Systems

Mathematical Modeling and Scientific Computing

Pervasive Computing and Applications

Energy-Aware Mobile Application Development by Optimizing GCC for the ARM Architecture

Mahalingam P.R. and Shimmi Asokan

Department of Computer Science,
Rajagiri School of Engineering & Technology, Rajagiri Valley, Cochin, India
{prmahalingam,shimmideepak}@gmail.com

Abstract. Nowadays, mobile domain is growing much faster than the desktop domain. This is since people are obsessed with the concept of "portable devices", which is catered to by handheld devices. Even if small in size, portable devices are responsible for a considerable share of power consumption, primarily due to their abundance. So, ample attention has to be provided in optimizing the applications in this field also. Some of the power consumption is reduced by the ARM core, which is well-known for its power-efficient working. If the applications running on the mobile can also be optimized, the end result will be a considerably efficient and low-power mobile device. We tweak the existing optimizations in such a way that they provide the best possible performance. To increase the level of optimization, we perform optimization reordering, instruction selection, and so on. The end result is projected to give around 32% improvement, which translates to a considerable change in power consumption (up to 35,000 MW per year), and ultimately forms a big step forward in green computing. We take the instance of GCC, which is one of the major contributors to application development.

Keywords: ARM core, GNU Compiler Collection, RISC architectures, Machine Idioms, Instruction selection, Optimization, Architecture dependency, Green applications.

1   Introduction

Mobile domain is one of the most booming fields in technology. When we depend heavily on mobiles for our daily processes, their performance also matters a lot. To satisfy this need, the programs that create the applications have to be tuned for performance. Almost all mobile devices (phones, PDAs, etc.) that arrive in the market rely on the processing power provided by the ARM core[7]. ARM is a RISC-based processor core that is famous for its low power usage. So, any performance tuning should be adapted to the ARM core so that they can be used directly on the applications.

Importance of Optimization: Optimization[8] in compilers[9] refers to improving the efficiency of the code such that the semantics remains unchanged. This phase performs processes like unwanted code elimination, redundancy removal, code reordering, etc. We can also perform pipeline-dependent operations that will be of advantage in the ARM point of view.

J. Mathew et al. (Eds.): ICECCS 2012, CCIS 305, pp. 1–8, 2012.
© Springer-Verlag Berlin Heidelberg 2012

Even in a non-pipelined system, the optimizer can bring about a lot of improvement. Consider a program to sort an array of 100,000 elements using selection sort. The GNU Compiler Collection provides 3 levels of optimization, in addition to an unoptimized level. The number of cycles taken up by the program will be as below.

Table 1. Comparison of optimization levels

Level	Number of CPU Cycles
No optimization	71,280,000
Level 1 (-O1)	18,910,000
Level 2 (-O2)	26,410,000
Level 3 (-O3)	27,160,000

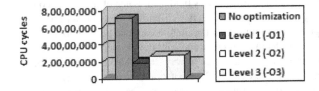

Fig. 1. Performance of different optimization levels

2   GNU Compiler Collection

GNU Compiler Collection (GCC) consists of a group of compiler packages, and is designed to be a generic, flexible compilation unit. Originally called *GNU C Compiler*, it was later expanded to accommodate other languages. Right now, 17 language front-ends are supported by GCC. Also, it has the back-ends for 63 architectures, divided into 3 categories: Standard release (20 architectures), Additional release (23 architectures), FSF releases (20 architectures)

General Features: GCC is a complete language processing system[12] made up of 4 components: Preprocessor (cpp), Compiler (cc), Assembler (as), Linker & Loader (ld).

It uses a variety of intermediate formats during the translation from the high-level source program to the assembly code. They are GENERIC, GIMPLE and RTL[11][18]. When we improve the performance at the architecture level, we have to work on RTL, which is attached to the code generation phase. GCC performs a variety of optimizations[3] on the intermediate code.

It is also a portable compiler package[3], allows cross-compilation and is modular. It is distributed as a free software under GNU GPL. The presence of a multitude of front-end and back-ends has made it nearly impossible to make GCC specific for a particular architecture[14]. The only option is to selectively invoke certain optimizations. But if that is stressed upon too much, it might indirectly cause issues to other architectures. All this has led GCC to be built as a "generic" package. So, there is a lot of scope for improving GCC from the ARM point of view.

3   RISC Architectures and ARM Core

ARM stands for Advanced RISC Machine. It is currently used in a variety of embedded system applications like mobile phones, PDAs, handheld organizers, etc. ARM is not a single core, but a whole family of designs sharing similar design principles and a common instruction set. ARM is not a 100% RISC architecture[4], due to the limitations posed by the embedded systems[7] themselves.

RISC Features: ARM core uses RISC architecture[7]. RISC is a design philosophy aimed at delivering simple, but powerful instructions that execute within a single cycle. The complexity of instructions is reduced. The main RISC design rules are given as: Instructions, Pipelines, Registers and Load-store architecture.

Application Domains of ARM Core: ARM cores were primarily aimed at the mobile domain. Nowadays, every handheld device (including smartphone breeds) is powered by an ARM-based processor. The attractive feature that pulls mobile devices to use this core is the power-efficient operation. ARM consumes very little power compared to the desktop processors. This core is now being tested on server systems, for which, a more robust core has been developed. Once that is completely successful, any ARM-based enhancement will have far-reaching applications.

4   Machine-Dependent Operations in GCC

GCC supports some machine-dependent operations, mainly in the code generation phase (and in the code optimization phase, as selective optimizations).

Code Support for Architectures: The modularity of GCC ensures that all the architecture-level operations are clearly laid out as individual modules, allowing addition of new architectures, and modification of existing ones. They are put in the configuration folders of each architecture, and they have their own set of *Machine Descriptions*[16] that describe the appropriate RTL processes[11][17] required. The RTL supports two operations[13]: Direct translation[13] and Conversion[13]. The RTL conversion adds more architecture-specific code generation.

Adding Specific Operations: To modify the architecture-level features, we can work at different levels. Initially, we can modify the command line options for the architecture. These options are used while building GCC from the source, to add or remove features from the final compiler. Another option is to modify the machine descriptions[13], which are defined in the file arm.md. The machine descriptions govern the code generation module by outputting assembly segments specific to the architecture. If any code template can improve the performance over some other, they can be explicitly defined in the RTL conversion options, as an expand option. In that case, the operation will be automatically cast to the required template, and we can perform any optional further processing on them.

5   Proposed Methodology

The proposed methodology enhances the existing performance by adding pipeline-friendly instruction templates, and removing non-performing features. We have to first analyze the performance of the existing optimization functions and determine whether they can improve the code in any way. Since the GCC structure is generic, an optimizer module might be placed, keeping a specific architecture in view. But the same operation might worsen the performance in a different architecture. We can also add new optimizations in the machine level, as *postpass optimizer*[2]. Here, we stress on some basic modifications that can be done.

Reordering of Optimizations: Many optimizations remove unwanted code from the program. But at the same time, they might render some additional code useless. To remove that, we have to repeat some optimizations later in the process, and remove the code rendered useless by other optimizers. The same is done in GCC by using multiple passes for an optimization[1].

The default ordering of optimizations and their configuration is suited for most architecture. But in the case of ARM, some changes can be made in the optimizers, both in order and configuration.

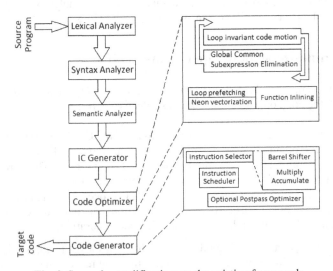

Fig. 2. Scope for modifications to the existing framework

Some of the configurational are:

- Swap the usages of loop-invariant code motion and global common subexpression elimination[1][8]
- Configure loop prefetching[1] to avoid excessive cache misses, and at the same time, avoid unnecessary re-fetching during branch statements.
- Selectively invoke neon vectoriation[1] so that any underperforming scenario is avoided.

Machine Idioms: They refer to the machine-specific operations[2] that can replace the conventional instruction templates and bring an improvement in performance. They are commonly combined with *instruction combining*[8] so that a complete template is replaced. It has been observed that some idioms can replace entire loops.

Here, we consider two main idioms that can improve the performance to a good degree – Barrel Shifter and Multiply Accumulate.

Barrel Shifter: It performs the *shift* operation[10] in ARM. By default, barrel shifter shifts the bits to the left or right, adding zeroes as necessary. This operation can benefit in some cases. Multiplications can be done indirectly by the barrel shifter. Consider the example:

$$5 \times 6 = 30$$

In binary, $101 \times 110 = 11110$
We can perform the process indirectly as: $5 \times (2+4) = 30$
It can now be distributed to separate operations in the following way.

$$5 \times 2 + 5 \times 4 = 30$$

Now, we can split the multiplication in such a way that it is the sum of multiplications, with powers of 2. ie, we now split the operation as:

$5 \times 2^1 = t1$	(shift one bit left)
$5 \times 2^2 = t2$	(shift two bits left)
$t1 + t2 = 30$	(conventional addition)

So, the composite multiplication is now converted to a set of two shifts, followed by an addition. This will be beneficial in cases of small decompositions. But if it is a large number, the summations will outweigh the improvement provided by the shifter.

Multiply Accumulate: Multiply accumulate[7] is an option that can eliminate a lot of memory transfers that occur during RISC operations[10], by maintaining intermediate results in the accumulator. This is an option that makes extensive advantage out of pipelining, by maintaining the results, and directly executing subsequent instructions. This is a heavily used process in RISC architectures, especially in fields like Image Processing. They use special architecture that implements integer multiplication with early termination.

Instruction Selection and Ordering: Even if we use idioms, there are still other instructions that can be manipulated for efficiency. The pipelined architecture enables the instructions to be ordered in such a way that the dependencies are minimized. Also, proper scheduling ensures that loops and sequences are selected and executed efficiently.

6   Challenges in Implementation

Even if the methodology seems straightforward, there are a lot of issues that cause difficulties while implementing. Some of them are as follows.

- Size of code[1] is one of the main factors that affect the implementation.
- ARM-specific optimization is a less explored domain[15]. Many efforts have been limited to simply tuning the performance of the compiler and optimizers, but not much effort had been introduced in adding features.
- Performance estimate of existing optimizations has to be analyzed in depth to decide what all to remove. An example is in the case of vectoriation, where we have to identify the cases in which the optimization should be enabled.
- ARM provides no support for threads[6]. It uses a method called Simultaneous MultiThreading that uses rapid switching. So, even separate, and independently executable functions should be scheduled sequentially.

7   Evaluation

Considering machine idioms, each idiom takes a different number of cycles to execute compared to the original. On average, if we consider the possible templates which get replaced, we can see the improvements as follows.

Table 2. Improvement by machine idioms

Instruction	Improvement
Barrel shifter	30%
Multiply accumulate	25%
Both	32%

When we consider the reordering of optimizations, we have varying improvements, since they depend on the entire program to adjust the code. The same applies for instruction selection and reordering, since the performance depends on the dependencies imposed by all the instructions.

But when we consider the performance as a whole, considering typical embedded system programs[2], we can expect a performance improvement of around 20%-30%.

Consider a program running in ARM that takes 1,000,000 cycles to execute. This scenario is normally encountered during the execution of simple applications. An improvement of 32% will translate to savings of 320,000 cycles. If we consider an average clock rate of 300 MHz, the difference in 320,000 cycles is shown as 0.1% improvement in the execution time. So, if we take a complex application, or even the case of mobile operating systems, the performance improvement can be safely scaled up to around 10%.

---

[1] GCC is more than 70,000 files in size, and reaches over 500 MB in size. There are nearly 2 million lines of code to be analyzed.

[2] We consider cases like image rendering and manipulation, which involves a lot of complex multiplications and other mathematical operations.

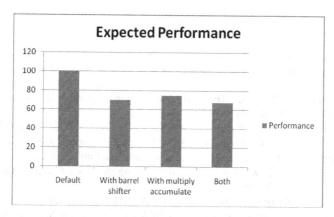

Fig. 3. Expected performance of proposed optimizations (on a scale of 1-100)

If we estimate the power consumption at 0.1milliwatts per MHz, a difference of 10% will translate to 0.01 milliwatts (or 10 microwatts) saved per second. An improvement of 10 microwatts per second may sound very small, but over a day, this will translate to 864,000 microwatts, which is equivalent to 0.864 watts (or can be approximated to 0.1 watts). Now, with nearly 5 billion mobile devices in the world, this improvement translates to atleast 5 million watts, or 5 MW.

So, even the smallest improvement has translated to 5 MW of reduced consumption per day. Once Calxeda project, which uses ARM cores in its servers, takes off, the same improvement will get multiplied by a factor of nearly 20, which is nearly 100 MW. So, per year, we save up to 36,500 MW.

8   Conclusion

With the advent of the mobile domain, more stress has been placed on application development for mobile users. But the stress on application optimization hasn't increased to that degree. When we consider the ARM core, which is at the heart of mobile devices, and try to optimize the applications, we are getting a considerable improvement (more than 35,000 MW per year).

This shows the power of optimization, which has an impact even in the mobile domain. Taking the case of GCC, some optimizations were proposed, and taking typical examples, the improvement is estimated at 32%. If we consider other packages, it might get better. But even if we don't achieve this large an improvement, we would have a minimum of 5% improvement, translating to around 6000 MW per year.

9   Future Work

The current work has been done on the mobile domain. But the same optimization can apply to desktop computing itself. It will further improve by nearly 50%, and in the end, the savings will be quite considerable, which is a great step in green computing. If we can add some level of improvement in hardware, we can exploit all possible opportunities in green computing at the application level.

References

1. Melnik, D., Belevantsev, A., Plotnikov, D., Lee, S.: A case study: optimizing GCC on ARM for performance of libevas rasterization library. In: Proceedings of GROW 2010 (2010)
2. Wang, L., Lu, B., Zhang, L.: The Study and Implementation of Architecture-dependent Optimization in GCC. IEEE (2010)
3. Stallman, R.M.: Using and Porting the GNU Compiler Collection. GNU Press (2001)
4. Sloss, A.N., Symes, D., Wright, C.: ARM System Developer's Guide–Designing and Optimizing System Software, pp. 1–5. Morgan Kaufmann Publishers (2006)
5. Stallman, R.M.: GCC Developer Community, Using the GNU Compiler Collection for GCC version 4.4.2. GNU Press (2008)
6. Dong, L., Ji, Z., Suixiufeng, Hu, M., Cui, G.: A SMT-ARM Simulator and Performance Evaluation. In: Proceedings of the 5th WSEAS Int. Conf. on Software Engineering, Parallel and Distributed Systems, pp. 208–210 (2006)
7. Furber, S.: ARM System-on-Chip Architecture, pp. 47–52. Pearson Education (2007)
8. Muchnik, S.S.: Advanced Compiler Design and Implmentation, pp. 319–704. Morgan Kaufmann Publishers
9. Aho, A.V., Lam, M.S., Sethi, R., Ullman, J.D.: Compilers–Principles, Techniques and Tools, pp. 769–902. Pearson Education (2008)
10. ARM DDI 0100 E, ARM Architecture Reference Manual, ARM Limited
11. Novillo, D.: GCC Internals-Internal Representations. In: GCC IR-2 (2007)
12. Gough, G.: An Introduction to GCC for the GNU compilers gcc and g++. Network Theory Limited (2004)
13. Novillo, D.: GCC Internals–Code Generation. In: GCC IR-2 (2007)
14. Wirth, N.: Compiler Construction. Addison-Wesley (2005)
15. Den, W.: ARM7TDMI Optimization Based on GCC, pp. 639–642. IEEE (2010)
16. Khedker, U.: GCC Translation Sequence and Gimple IR. GCC Resource Center, Department of Computer Science and Engineering. Indian Institute of Technology, Bombay (2010)
17. Stallman, R.: GCC Internals–GCC 4.7.0 pre-release. GNU Press (2010)
18. Merrill, J.: GENERIC and GIMPLE: A New Tree Representation for Entire Functions. GCC Developers Summit (2007)

Compiler Efficient and Power Aware Instruction Level Parallelism for Multicore Architecture

D.C. Kiran, S. Gurunarayanan, Faizan Khaliq, and Abhijeet Nawal

Department of Computer Science and Information Systems,
Birla Institute of Technology and Science-Pilani,
Pilani, 333031, Rajasthan, India
{dck,sguru,h2010137,h2010198}@bits-pilani.ac.in

Abstract. The paradigm shift to multicore processors for better performance has added a new dimension for research in compiler driven instruction level parallelism. The work in this paper proposes an algorithm to group dependent instructions of a basic block of control flow graph into disjoint sub-blocks during the SSA form translation. Following this an algorithm is presented which constructs a graph tracking dependencies among the sub-blocks spread all over the program. A global scheduler of the compiler is presented which selectively maps sub-blocks in the dependency graph on to multiple cores, taking care of the dependencies among them. The proposed approach conforms to spatial locality, aims for minimized cache coherence problems, communication latency among the cores and overhead of hardware level instruction re-ordering while extracting parallelism and saving power. The results observed are indicative of better and balanced speedup per watt consumed.

Keywords: Control Flow Graph (CFG), Static Single Assignment (SSA), Data Dependency, Multicore Processors, Instruction Level Parallelism, Global Scheduler, Local Scheduler, Energy Efficiency, Performance per watt, Sub-block Dependency Graph (SDG).

1 Introduction

Increasing demand for speed up in processors has resulted in many revolutions over the past years. Multicore processors are one of them in which a single chip has two or more independent cores which read and execute program instructions. Finding an effective way to exploit the parallelism or concurrency inherent in an application is one of the most daunting challenges in multicore environment. This parallelism can be achieved in two ways, first is scheduling different tasks on different cores. Second is scheduling portions of same tasks on different cores, Intra process parallelism or Instruction level parallelism. In first approach the scheduler of the operating system picks up a task and schedules it on the free core. In second approach compiler has to analyze the program for the possibilities of parallelism and suitable scheduler should

J. Mathew et al. (Eds.): ICECCS 2012, CCIS 305, pp. 9–17, 2012.

map the parallel constructs on to multiple cores. By analyzing the dependencies between instructions, the compiler can identify the independent instructions that can run in parallel. There are three types of dependencies to look for [1], Write after Write dependency (WAW), Write after Read dependency (WAR), and Read after Write dependency (RAW). Converting program to static single assignment (SSA) form will remove WAW and WAR dependencies and hence compiler has just to look for RAW dependencies. SSA form is an intermediate representation of a program in which each variable is defined only once [2].

ILP can achieved by existing scheduling techniques, where one instruction at a time is scheduled on to multiple execution units [3][4]. To perform this, critical path of instructions is created by analyzing the dependencies. The instruction with longest critical path was scheduled first to enable other instruction to get scheduled. This approach was well suited for parallel architectures which were available before multicore era. But this technique cannot be applied on multi-core environments because two dependent instructions may get scheduled on different cores (may not be in the same time) resulting in increased communication latency.

A technique was proposed to overcome these limitations [5], in which a set of dependent instructions within a basic block which are in SSA form were grouped. This set is disjoint and is referred as sub-block. The intra block scheduler is used to schedule these sub-blocks on to multiple cores.

The creation of sub-blocks can be done at two levels. In the first approach [5] a SSA form program is taken as input. These SSA form instructions are then analyzed for dependencies in a separate pass so as to form the sub-blocks. In the second approach [6], dependency analysis and sub-block creation is done along with variable renaming step in the generation of SSA form program. This avoids the need for an extra pass done in the first approach. The proposed work is an extension of the second approach. In this work a technique is proposed which considers extracting parallelism among the sub-blocks created across the program referred thereafter as inter-block parallelism. This technique facilitates global scheduling. This technique also conforms to the principal of spatial locality as the closely related or dependent instructions are grouped together in sub blocks. This also minimizes the cache coherence problems as the instruction stream of a sub block scheduled to a core is not dependent on what is scheduled on the other cores at a time. This in turn reduces the communication latency among the cores. The task of instruction stream reordering in the processor hardware consumes power. Since as per this technique the dependency analysis in the instruction stream is done in the compilation phase itself, the overhead of hardware level reordering is reduced. This makes the technique power aware. The scheduler being presented takes the granularity of instruction stream to be presented to a core as sub block and so it is possible to selectively utilize the cores. This makes it more energy efficient.

The rest of the paper is organized as follows. Section 2 gives a brief overview of the architecture and compiler used in the work. Section 3 gives details of creating sub-block dependency graph (SDG). Section 4 gives the detail of the inter block scheduling technique. Analysis and discussion of results are given in section 5. Finally, the paper concludes with Section 6.

2   Realm

The framework for the proposed work has two parts, the target architecture and the compiler. The target architecture will expose the low level details of hardware to compiler. They implement minimal set of mechanisms in the hardware and these mechanisms are fully exposed to software, where software includes both runtime system and compiler. Here runtime system manages mechanisms historically managed by hardware, and compiler has responsibility of managing issues like resource allocation, extracting parallel constructs for different cores, configurable logic, and scheduling. These types of architectures can be seen in some network processors and RAW architecture [7][8].

The multicore environment has multiple interconnected tiles and on each tile there can be one RISC like processor or core. Each core has instruction memory, data memory, PC, functional units, register files, and source clock. FIFO is used for communication. Here the register files are distributed, eliminating the small register name space problem, thus allows exploiting ILP at greater level.

The proposed work uses Jackcc Compiler [9]. This is an open source compiler developed at university of Virginia. The basic block in CFG of Jackcc is called *Arena,* and instruction inside the block is called *Quad.* The DAG generated by front end of the compiler is converted into *quad* intermediate representation, and then these quads are used to construct the basic blocks of CFG. Instructions are in SSA form. The process of converting a Non-SSA form program to SSA form program has two steps.

Step 1: Placing Phi (φ) statements by computing iterated dominance frontier.

Step2: Renaming variables in original program and Phi (φ) functions, using dominator tree and rename stack to keep track of the current names.

3   Inter Block Parallelism

The proposed work is an extension of intra block technique where sub-blocks within the basic blocks of CFG were considered for scheduling [5]. Extracting inter block parallelism will facilitate global scheduling. In this approach the sub-blocks across the program are considered for scheduling. The sub-blocks are disjoint within a basic block of CFG, but the sub-blocks across the basic blocks need not be disjoint. The non-disjoint sub-blocks should be executed one after the other.

Inter block parallelism is a process of finding the non-disjoint sub-blocks across the basic blocks. The process of extracting inter block parallelism involves two steps, one creating sub-block and second creating sub-block dependency graph.

The technique discussed here to create sub-block and sub-block dependency graph is compiler efficient, because it is done during SSA translation itself as shown in Figure1.

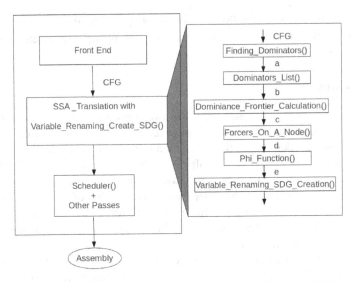

Fig. 1. Modified flow or Jackcc to create disjoint sub-blocks and SDG

Disjoint sub-blocks were created by modifying variable renaming step of SSA translation function [6]. There are two tasks which need to be done simultaneously during translation. First is variable renaming and second is creating disjoint sub-blocks.

Its first role is accomplished by traversing each node of the CFG which has now Phi (ϕ) functions inserted and rename every variable in such a way that each use corresponds to exactly one definition. Each definition is renamed with a new version of that variable. Second job is accomplished by updating the use of the renamed variable, there by anti dependency (RAW) of the current statement is gathered to forms the disjoint sub- blocks.

3.1   Sub Block Dependency Graph (SDG)

This section explains the process of creating of a sub block dependency graph along with variable renaming and creating disjoint sub-blocks during SSA translation. The dependencies among the sub blocks from different basic blocks will be analyzed by the compiler to form the sub block dependency graph.

The SDG is graph G (V,E), where vertex vi $\in$ V is sub-block B_pSB_i of basic block B_p and the edge e $\in$ E, is drawn between vertex $SB_i \in B_p$ and $SB_j \in B_q$ where p≠q, and instruction in the sub-block SB_j is dependent on one or more instructions in SB_i.

SDG is represented as dependency matrix. In dependency matrix all sub-blocks are arranged in first column. If the sub-block B_pSBi is dependent on sub-block B_qSBj, then B_qSBj is added in the dependency list of B_pSBi, meaning B_pSBi should be scheduled only after B_qSBj completes its execution. The sub-block B_pSBi can be scheduled only if the list is empty otherwise it should wait till the list becomes empty.

3.2   Algorithm for Sub Block Dependency Graph Creation

Algorithm for creation of Sub block Dependency Graph is given below

```
Scan the current statement
if(variables on right hand side==0)
        if statement has phi function
                for each predecessor of this block do
                        find_sub_block_and_block
                        add_sub_block_to_dependency_list
                endfor
        endif
endif
else if(number of variables on right hand side==1)
        find where this variable is defined
        if(variable is not defined in the current block)
                find_sub_block_and_block
                add_sub_block_to_dependency_list
        endif
endelseif
else if(number of variables on right hand side==2)
        find where these variables are defined
        if(one variable is not defined in current block)
                find_sub_block_and_block
                add_sub_block_to_dependency_list
        endif
        else if(no variable is defined here)
                find_sub_block_and_block_for_both_variables
                add_sub_blocks_to_dependency_list
        endelseif
endelseif
```

Here each statement is scanned for inter block dependencies. If statement has got phi function as it's R value then there would be inter dependencies coming from the predecessors of given basic block as per definition of phi function. While observing every predecessor, dependency edges are created between the sub blocks and those sub blocks are added to the dependency matrix. Second case could be presence of one variable in R value of statement such that this variable is not defined in current basic block. Hence inter block dependency exist and the graph is traversed upwards covering its predecessor to look for the dependency. Hence one edge would be added and the sub block is added to the dependency matrix. Similar action would be taken for the case when R value consists of two variables, both of which are defined in predecessors of the given basic block.

3.3   Complexity Analysis

Let number of nodes in CFG=N
Let average number of statements per block=S
Let total number of variables definitions in program (before conversion to SSA) =D
Let average number of variables defined per block (after conversion to SSA) =d
Let average number of sub blocks formed in a basic block after processing s statements = B
Let average number of statements in each sub block = S'
Let average number of predecessors of a basic block= P
Then complexity of the algorithm turns out to be

$$O(max(N*S*(max(D,d,B*S',N))),(N*S*(max(D,P*max(N,d))))) \qquad (1)$$

This complexity is of variable renaming plus intra block plus sub block dependency graph creation since all are being done in the same pass. Two conditions contributing to the worst scenario can be observed. First one is when statement doesn't have a phi function and there are two variables in the R value and both of them cause intra block dependency. In that case complexity would be given by $N*S*max(D,B*S',N,d)$. In second condition, there is a phi function in the statement and complexity is given by $N*S*max(D,P*max(N,d))$. So the final complexity would be maximum of these two values.

If the formation of sub blocks and formation of sub block dependency graph takes place in the separate pass then complexity would be given by

$$O(N*S*max(D,d))+O(max(N*S*max(B*S',N,d)),N*S*(max(D,P*max(N,d)))) \qquad (2)$$

In equation the Big O notation gives the worst case complexity of variable renaming phase and Big O notation on the right gives the worst case complexity of creating sub block and forming sub block dependency graph. It can be observed that compared to equation 1 at least $O(N*S)$ time is more in equation 2 and hence complexity in equation 2 is greater than equation 1.

4   Scheduler

In the proposed compiler framework, the scheduling phase follows the SSA translation phase. The scheduler takes SDG as input and maps the sub-blocks on to multiple cores. To schedule sub-blocks on multiple cores scheduler needs information like height of each sub-block, number of instructions in each sub-block, schedule time of each core, and ready-time of sub-block. This information is stored in a structure along with respective sub-block. This structure is an element in priority queue.

The information required by Scheduler is calculated as:
  I.   Scheduling time of core

Schedule time of core I = current schedule time of core I + no. of instructions in currently scheduled sub-block. $\qquad (3)$

II.   Height of sub-block

Height of sub-block $i=$ max(height of all immediate successors) + 1)   (4)

III.   Ready Time of sub-block

Ready-time of sub-block j = max (finish time of all immediate predecessors).   (5)

Finish time of sub-block i = number of cycles in i + schedule time of i.   (6)

The algorithm for scheduling is as given below. This is a power aware scheduler which uses optimum number of cores to balance speed up and power.

```
For the core having minimum schedule time.
Select the sub-block from the queue
Find all the nodes with highest height.
Among them find node with max number of instructions.
if ready time of node is less then or  equal to current
 core execution time, then select it.
Else
Repeat from step 1 to find best node which can be
scheduled.
```

5   Experimental Results

This section assesses the efficiency of the proposed technique for detecting and scheduling inter-block disjoint sub-blocks on to multiple cores. The outcome of the work can be viewed at two levels, the compilation time and the performance gain when scheduled by the proposed scheduler in terms of speedup and balanced power consumption.

In section 4, algorithm complexity for creating disjoint sub-block and sub-block dependence graph is analyzed. It is shown that the proposed technique is compilation time efficient and can be done in linear time O(N).

To check the gain in terms of speed up, experiments are done on systems, with one core, 2 cores (Dual core) and 4 cores (Quad core). The results of 4 different extreme test cases are shown in Table 1. All the test cases are mapped on to multiple cores using the proposed scheduler and their execution time in terms of cycles are captured.

Amdahl's law for multicore architecture proposed by Hill-Marty [10] is used to analyze the performance in terms of speedup. The result is normalized to the performance metric to that of a basic "unit core", which is equivalent to the "Base Core Equivalents (BCE)" in the Hill-Marty model. The results obtained with intra block scheduler are used to compare the results of the proposed technique. In Table 1, it is observed that the proposed technique attain better performance in terms of speedup with inter-block parallelism than intra-block.

Another grand challenge faced by computer architecture beyond performance is energy efficiency. It can be observed in Table 1 and 2, that speedup is gained as the number of cores is increased, but at the same time the power consumption also is increased. To balance power and speed up, the proposed scheduler will try to use less number of cores where ever is possible by applying some optimizing techniques like merging the sub-blocks based on register requirement. The speedup gain when the test cases were mapped to 3 active cores and its performance per power is also shown in Table 1 and Table 3.

To check the energy efficiency of the approach, performance per power model proposed by Woo-Lee model [11] is used. The results obtained by the inter block scheduler on dual core and quad core is compared with result obtained by intra block scheduler.

Performance per power on quad core machine is decreased in Table 3. It is obvious, because power consumed by all four cores is considered in the calculation. To balance this, the same scheduler is used to map the sub-blocks on 3 active cores. Though the speed up obtained is slightly less, performance per power is increased.

Table 1. Speedup Statistics

System	Test Case 1	Test Case 2	Test Case 3	Test Case 4
Single Core	1	1	1	1
Dual Core (Intra block)	1.72	1.72	1.65	1.77
Dual Core (Inter block)	1.93	1.91	1.93	1.89
Quad Core (Intra block)	2.25	2.69	2.14	1.84
Quad Core (Inter block)	3.11	3.44	2.93	2.44
3–Active Cores (Intra block)	2.25	2.61	2.14	1.84
3–Active Cores (Inter block)	2.79	2.87	2.82	2.59

Table 2. Power (Watt) Consumed by each core to execute the test cases

System	Test Case 1	Test Case 2	Test Case 3	Test Case 4
Single Core	1	1	1	1
Dual Core (Intra block)	1.86	1.86	1.82	1.88
Dual Core (Inter block)	1.96	1.96	1.96	1.94
Quad Core (Intra block)	3.13	3.43	3.07	2.92
Quad Core (Inter block)	3.56	3.72	3.46	3.22
3–Active Cores (Intra block)	2.63	2.80	2.57	2.42
3–Active Cores (Inter block)	2.90	2.93	2.91	2.80

Table 3. Performance Per Power (Watt)

System	Test Case 1	Test Case 2	Test Case 3	Test Case 4
Dual Core (Intra block)	0.93	0.92	0.90	0.94
Dual Core (Inter block)	0.99	0.98	0.98	0.99
Quad Core (Intra block)	0.72	0.80	0.70	0.63
Quad Core (Inter block)	0.88	0.95	0.92	0.85
3–Active Cores (Intra block)	0.86	0.93	0.83	0.76
3–Active Cores (Inter block)	0.96	0.98	0.97	0.93

6   Conclusion

Exploiting the parallelism in an application is a challenge in multicore environment. The work proposed in this paper discusses an efficient parallelism technique to create and schedule disjoint sub-blocks across the program. The idea of grouping the dependent instructions has cheering side effects, like minimized cache coherence problems, minimized communication latency among the cores, and reduced overhead of hardware level instruction re-ordering while extracting parallelism. The scheduler proposed in this paper balances both speed up and power consumption. Outcome of the proposed work is compared with the intra block scheduler. The assessment shows that the inter block scheduler will perform better in terms of speedup by balancing the power consumption.

Acknowledgments. We thank Nick Johnson of University of Virginia, for providing the compiler and its assembler.

References

1. Tyson, G., Farrens, M.: Techniques for extracting instruction level parallelism on MIMD architectures. In: The 26th International Symposium on Microarchitecture, pp. 128–137 (1993)
2. Cytron, R., Ferrante, J., Rosen, B.K., Wegman, M.N., Zadeck, F.K.: Efficient computing static single assignment form and the control dependence graph. ACM Transaction on Programming Languages and Systems 13(4), 451–490 (1991)
3. Colwell, R.P., Nix, R.P., O'Donnell, J.J., Papworth, D.B., Rodman, P.K.: A VLIW Architecture for a Trace Scheduling Computer. In: ASPLOS-II Proceedings of the Second International Conference on Architectural Support for Programming Languages and Operating Systems. IEEE Computer Society Press, Los Alamitos (1987)
4. Adam, T.L., Chandy, K.M., Dickson, J.R.: A Comparison of List Schedules for Parallel Processing Systems. Communications of the ACM 17(12) (December 1974)
5. Kiran, D.C., Gurunarayanan, S., Misra, J.P.: Taming Compiler to Work with Multicore Processors. In: IEEE Conference on Process Automation, Control and Computing (2011)
6. Kiran, D.C., Gurunarayanan, S., Misra, J.P., Khaliq, F.: An Efficient Method to Compute Static Single Assignment Form for Multicore Architecture. In: 1st International Conference on Recent Advances in Information Technology (2012)
7. Waingold, E., Taylor, M., Srikrishna, D., Sarkar, V., Lee, W., Lee, V., Kim, J., Frank, M., Finch, P., Barua, R., Babb, J., Amarasinghe, S., Agarwal, A.: Baring it all to software: Raw machines. Computers 30(9), 86–93 (1997)
8. Gebhart, M., Maher, B.A., et al.: An evaluation of the TRIPS computer system. In: ACM SIGPLAN Notices–ASPLOS 2009, pp. 1–12 (March 2009)
9. The Jack Compiler, http://jackcc.sourceforge.net
10. Hill, M.D., Marty, M.R.: Amdahl's law in the multicore era. IEEE Computer, 33–38 (2008)
11. Woo, D.H., Lee, H.-H.S.: Extending Amdahl's Law for Energy-Efficient Computing in the Many-Core Era. IEEE Computer, 24–31 (2008)

MotherOnt: A Global Grid Environment for Interconnecting the Ontologies Semantically to Enable the Effective Semantic Search

Rohit Rathore[1], Rayan Goudar[1], Sreenivasa Rao[2], Priyamvada Singh[1], and Rashmi Chauhan[1]

[1] Dept. of Computer Science & Information Technology,
Bell Road, Graphic Era University, Dehradun, India
{rohitrathor,rhgoudar,priyakip,rashmi06cs}@gmail.com
[2] CIHL, MSIT Division, IIIT Hyderabad, Gachbowli, Hyderabad, India
prof.srmeda@gmail.com

Abstract. Due to the problems (poor precision, poor recall etc) associated with current keyword based search engines, a number of semantic search architectures have been proposed in near past. These semantic approaches though effective in returning better and more relevant search results, suffer from a number of drawbacks when it comes to scaling their implementation to a full web search. We have identified a number of limitations associated with semantic search; like lack of semantics for the web documents, domain limited search, search limited to the semantic parts of the web, lack of a unified ontology knowledge base etc. Through our work, we put forward a search-architecture for today's web (partly semantic and mostly non semantic information). It addresses the fundamental issues of ontology standardization by introducing the notion of semantic annotations for ontologies using a global ontology. The proposed model combines the content based (keyword) and context based (semantic) search approaches thus taking into account the present (mostly non semantic data) and the future (semantic information).

Keywords: Semantic Web, Ontology, Semantic Search, Semantic Annotation, Ontology Mapping, Knowledge Base, Global Ontology, MotherOnt, CrawlOnt, Ontology based Automatic Classification, Ontology Standardization, Semantic Search Architecture.

1   Introduction

Semantic Search techniques for the web works on the basic assumption that the underlying information has been modeled or represented in a structured way which allows it to be given a semantics . This semantic structure can be used by a context based search mechanism (semantic search) to return results that are not only lesser in number but are more accurate and more relevant to the user query (in comparison to the results returned by the keyword based approaches)

Various research works like [1] in past few years have identified a number of problems associated with the current generation web search engines which are based upon

J. Mathew et al. (Eds.): ICECCS 2012, CCIS 305, pp. 18–29, 2012.

the keyword based index construction and Boolean logic based query matching. The major problems with keyword based approaches are poor precision and recall. i.e. out of the total number of results possible for a search query only a fraction is retrieved (poor precision) and out of all the results that are returned only a fraction of them is relevant (poor recall). The search results that are returned are so large in number that it is impossible for the user to segregate the meaningful results manually (information explosion) Also, It has been established that the task of effective search query building cannot be entrusted to the user as is the case with current search engines; in other words it is impossible for the user to pre-determine what query terms will return more relevant results.

Semantic Web as defined by Tim Berners Lee [2] is a vision for the foreseeable future when all the information present on the web will have a structure or meaning (semantics) so as to make it machine understandable as well as machine process-able. Various software agents will be able to process this information automatically in different ways depending on the context and the user requirements. The search for the web documents will take into account the contents as well as the search-context and thus more accurate, meaningful results could be returned by a search engine which can utilize the underlying semantic structure of the web information.

A semantic search architecture applies the various tools and techniques of semantic web to the Web search for returning the results to the user. In recent past various Semantic search architectures have been put forward [1] [3][11][12][14][15] which provides the foundation for semantic search and try to resolve the problems associated with the keyword based approaches. The fundamental difference between the two search paradigms is that the keyword based search engine view the web as a collection of web documents (web resources in general).Each web document in turn is viewed as a collection of keywords by the search engine. On the other hand, a semantic search engine views the Web as a collection of domains. Each domain is defined by a number of concepts which are connected to each other through relationships and are defined by a number of attributes.

An Ontology is often used to model domain-specific knowledge for a particular domain. According to their formal definition: Ontologies are explicit specifications of the conceptualization and corresponding vocabulary used to describe a domain [Gruber 1993].The four key terms can be identified from the above definition. "Conceptualization" refers to identifying the key components of the domain in terms of which the domain can be fully defined. "Explicit" means that each concept or entity should be uniquely, completely and unambiguously defined in such a way that software could understand it. "Specification" refers to identifying the various attributes or properties that are used to define a particular concept and defining the relationships between them. A number of axioms are also specified for the domain, which constitute of the rules written in description logic form that helps in making inferences on the asserted knowledge thus extending the semantic knowledge representation for the domain.

We have been able to identify a number of issues that needs to be addressed while designing and implementing a Semantic Search Engine which can work for the World Wide Web as a whole.

1. **Semantic Search is domain specific:** Various semantic search architectures that have been proposed so far are limited to particular domains. The reason behind this domain restriction is that these architectures use Ontologies as the underlying foundation/structure to provide semantic markup to the unstructured and semi-structured web documents. Ontologies as a knowledge representation tool, in turn, are highly domain specific. That is, an Ontology is created for a particular domain and is used to model the knowledge for that particular domain in terms of: Concepts, Attributes or properties of these concepts, Relationships between these concepts, Properties of these relationships, and Rules for the domain on the basis of which inference can be drawn and implicit knowledge can be extracted (by the use of a reasoning tool / Description Logic).So, an effective Semantic Search Architecture provided for the whole Web should be able to work with different domains and thus different Ontologies.

2. **Heterogeneity in Domains:** Traditional keyword search which is content-based has an advantage in the sense that all the web documents can be treated in the same (or similar) manner to identify the keywords and creating the index. A semantic search engine for the Web, on the other hand, needs to work with different domains (and thus different domain Ontologies) to semantically annotate the web documents and creating a semantic index or an information knowledge base on the basis of which user queries can be answered. So once the web crawler provides a web document, a semantic search engine should provide a mechanism for finding out the domain to which that document belongs to, and to determine what ontology should be applied to that document for marking it semantically (during the document processing phase of semantic search).

3. **Standardization of Ontologies:** This is another important issue which needs to be addressed and resolved before any semantic search engine architecture can be implemented for the whole web (without being domain specific). Today, many ontologies exist for multiple domains. Same or similar domains may have different, more than one ontologies published for them by the different set of users or organizations which have been created keeping in view their very specific data modeling requirements. Now, a semantic search engine should be able to find out all such ontologies and should be able to map, merge and align the various overlapping concepts and relationships occurring in these ontologies to provide some sort of uniform ground for semantically marking up the documents for these domains. Ontology mapping is a challenging task and an area of research in itself.

Also, a standard mechanism is required for providing a standardization of ontologies themselves in terms of:

- **Uniform Ontology Representation:** Today, Ontologies on the web are available in a number of languages like OWL, DAML+OIL and RDFS etc. From a semantic search engine point of view all the Ontologies, it is working with should have a common underlying representation so there arise a need to express or convert the Ontology in a standard language. W3C has recommended the use of RDF based Web Ontology Language (OWL) to construct the ontologies which are stored as .owl files and provides a standard way for Ontology modeling.

- Ontologies themselves should be semantically marked so as to facilitate easy discovery on the web and to build a common Ontology Knowledge Base which is required for providing a complete Web search.

2   Related Work and Motivation

Ahmad Maziz Esa et al [12] have used an open source search engine Nutch [13] to add semantic capabilities to it through a plugin module which they called Zenith. In their architecture, the original design of data flow is intercepted and modified before being re-channeled to the indexer. This plug-in based approach though provides certain semantic capability to the search, does not address the problem from the ground up.

Junaidah Mohamed Kassim et al [3] have made a comparison between keyword based and semantic search approaches. In their work they have listed the advantages that semantic search offers over traditional keyword search. They have proposed a generalized architecture of a semantic search engine.

Jiang Huiping [15] has proposed a semantic web search model to enhance efficiency and accuracy of information retrieval for unstructured and semi-structured documents. The unique features of his proposed architecture are Search Arbiter and Ranking Evaluator modules. Search Arbiter determines whether enough ontologies are available to perform the semantic search or whether the query should be directed to the keyword search. The idea though useful does not elaborate upon the mechanism which needs to be applied to take such a decision. In our work, we have given a process on the basis of which such decision could actually be taken through the use of an Ontology Knowledge Base.

Many researchers ([4] [5] [6] [7] [9] etc.) have given a number of different approaches for ontology mapping which is an important component in our proposed architecture. We have used the concept of Ontology Mapping to achieve the unification of ontologies so as to provide a standard and unified view of a domain.

Several paper explains the different methods for Ontology based Automatic Classification of web documents for respective domains ([17] [18] etc). Again, Ontology based automatic classification of web documents is required in order to semantically mark the unstructured documents on the web and has been used as a module in our proposal.

Zeng Dan [10] presented a conceptual model for representing a generic Ontology which can be used to represent any Ontology in terms of Classes, Relationships, Enumeration Types, Axioms and Instances.

In our architecture, we have proposed the use of a Mother Ontology which is used to combine all the existing ontologies and to provide semantic markup to them (ontologies themselves) so that a knowledge base for ontologies can be created. This approach offers the following advantages:

- An ontology knowledge base makes it easy for an Ontology Crawler to find out the ontologies that exist for various domains
- The information about the overlapping of various domains i.e. how various domains and their concepts are related to each other can be captured in a global Ontology knowledge base.

- The information that exists today on the Web is mostly semi-structured or unstructured in nature. Semantic web is a vision for future, for the time when almost all the available knowledge on the web will have a semantic structure. So any effective, modern day web search architecture should combine both the traditional keyword based search technique with semantic search approach [16]. In order to decide what search to perform on a particular user query, the search engine needs to determine whether enough ontologies are available for the key terms or not. This can be established by the use of OKB.

3   Proposed Idea

3.1   Need for Creating a Global Ontology or MotherOnt

Ontologies are used to provide semantic structuring or meaning to the web documents in their particular domains so as to offer a number of advantages including enabling the context-based or semantic search for web documents. We realize that there is a need to add semantic markup or annotations to the ontologies of various domains so as to offer the same advantages in searching for existing ontologies on web, as are offered by semantically annotating the documents while performing the document search.

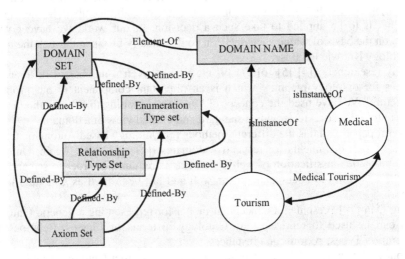

Fig. 1. Conceptual Model of Global Ontology "MotherOnt"

Some sorts of standardization efforts are required for interconnecting all the ontologies semantically to produce a global grid environment for enabling the effective Semantic Search. For achieving such an interconnection, a global ontology, we call it MotherOnt in our model, can be created and maintained by a central governing body like W3C which should identify and define the various available information domains, their properties, and how they are related or connected. An overlapping of domains which is inevitable should also be defined by this MotherOnt ontology.

In addition to making the search for ontologies easier, this approach also provides a common unifying point for various existing ontologies currently available on the web by creating a common Ontology Knowledge Base (or OKB), thus taking a step ahead in standardizing the ontologies which is a huge issue and a roadblock in making the effective Web wide (across all the domains) semantic search a reality.

We have extended the conceptual model proposed by [10] for representing Ontology so as to accommodate the expression for various domains and interrelationships between them. Fig. 1 gives our extended model diagram which can be used as a basis for defining the MotherOnt.

This MotherOnt can be used to provide the required semantic markup to the existing heterogeneous ontologies and to facilitate the creation of a common Ontology Knowledge Base (OKB). "Medical" and "Tourism" which are two different domains, for example, can be the instances of the class DOMAIN NAME and a relationship between them can be defined to depict the concept of "Medical Tourism".

3.2   Ontology Knowledge Base Creation (OKB)

As stated earlier, the main use of MotherOnt is to annotate the existing ontologies and to combine their representation to create an Ontology Knowledge Base (OKB). This section explains in some detail how, for our proposed model, the OKB is created starting from the creation of Ontologies by Domain Experts followed by a number of steps like discovering these Ontologies on the web by an ontology crawler, combining the ontologies that represent the same domain through ontology mapping, and finally creating an OKB by semantically annotating these aligned (or mapped) ontologies with the help of MotherOnt. Various steps in the Ontology Knowledge Base creation can be summarized as follows: (Refer to Fig. 2)

1) Ontology Creation: Ontologies are created by ontology engineers with the help of domain experts. These ontologies are then made available on the Web and can be used in a number of ways by the users and the software agents.

Ontology construction is an iterative process and involves the following steps:

- **Design:** Specifies the scope and purpose of the ontology. Also reveals the relationship among classes and subclasses.
- **Develop:** Decides whether construction of ontology has to be done from scratch or to reuse an existing ontology.
- **Integrate:** Combine the developed ontology with the already existing one.
- **Validate and Feedback:** The completeness of the constructed ontology is verified with the help of automated tools or by seeking the opinion of the experts.
- **Iterate:** Repeat the process and incorporate the changes given by the expert. Ontologies are commonly encoded using **ontology languages**.

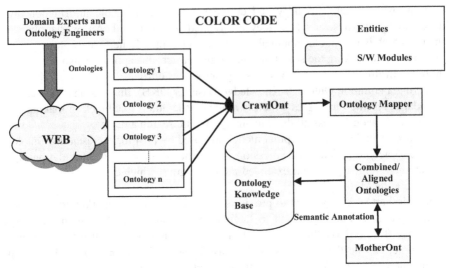

Fig. 2. Ontology Knowledge Base Creation

Protégé is the tool of choice for constructing ontologies which is open source and can be used to build frame based as well as OWL ontologies. The OWL plug-in for Protégé allows the Ontology Engineer to load, save, edit and visualize ontologies in OWL and RDF [21].

Domain experts are responsible for identifying the set of concepts for their specific domain. These concepts are defined by their attributes (called data properties in Protégé) and relationships between various concepts of the domain (called object properties in Protégé). A set of rules called axioms are defined which allows a reasoner (Protégé provides a number of integrated reasoning tools like fact ++ and pallet etc) to draw inferences for extending the ontologies (future needs) and for extracting and modeling the non explicit domain knowledge.

2) Discovering the Ontologies on the Web: The ontologies published on the Web can be discovered or searched by the use of an Ontology Crawler (CrawlOnt). An Ontology- Crawler is a distributed crawler which is used to search for existing (published) ontologies on the web. CrawlOnt returns the ontologies as .owl files.

3) Ontology Mapping: If more than one ontologies are present for a domain, they must be merged or aligned using Ontology Mapping so that the domain information can be represented and viewed uniquely.

Same or similar domains may have different, more than one ontologies published for them by the different set of users or organizations which have been created keeping in view their very specific data modeling requirements. This leads to a heterogeneity in ontologies which is an obstacle in the interoperability and makes it difficult to take a consistent, unique view of a particular domain. To resolve the mentioned heterogeneity and consistency issues these various different ontologies representing the same domain needs to be combined or merged in some way. The technique of Ontology mapping is often used to find out the overlapping between ontologies and to map them. The purpose of ontology mapping is to first find the semantic relation between concepts in ontologies and establishing the various mapping rules, and then integrating the available ontologies into one using these rules [5].

Ehrig, Marc and Staab, Steffen [4] defines Ontology mapping as: Given two ontologies O1 and O2, mapping one ontology onto another means that for each entity (concept C, relation R, or instance I) in ontology O1, we try to find a corresponding entity, which has the same intended meaning, in ontology O2.

The process of ontology mapping has five steps: information ontology, obtaining similarity, semantic mapping execution and mapping post-processing [8]. The key of ontology mapping is the computation of conceptual similarity.

4) Semantic Annotation of Domain Ontology: The step 3) above (Ontology Mapping is optional). Once a unique representation for a particular domain is received in the form of domain ontology, either directly (only one standard ontology available for a domain) or through an extra step via ontology mapping (when more than one ontologies are available for the same domain), the next step is to semantically annotate the Ontology by using MotherOnt as the reference and thus populating the Ontology Knowledge Base.

3.3   Architecture of Proposed Semantic Search Engine

Fig. 3 gives an outline of the proposed architecture for Semantic Search. There are two aspects of providing the semantics to the web documents:

(a) All the new documents that are being published on the web can be semantically marked or annotated via the use of existing Ontologies for their particular domain. User or the web publisher before posting or publishing the web content can be made to select the Ontology and the contents can be automatically annotated on the basis of that. This involves the use of an automatic annotation tool that given an Ontology and web documents as input, produces the required annotations. This semantically annotated information is then published and becomes available on the web.

(b) Most of the already existing information on the web is unstructured or non-semantic in nature. Our proposed architecture combines a number of semantic web tools and artificial intelligence techniques (like Automatic Document Classification, Information Extraction etc) in an attempt to provide these documents with the desired semantic markup and to create a knowledge base of information.

A Web crawler is used to fetch the documents from the web so as to build an information knowledge base. If the document fetched is already semantically annotated it is added to the Information knowledge base directly after indexing. On the other hand if the web page found by the crawler is non-semantic, ontology based automatic classification is applied so as to attach meta- information to it. Following steps are taken:

(i)   Given a particular web document, an information extraction tool is applied to the document in order to identify the key concepts contained in the document.

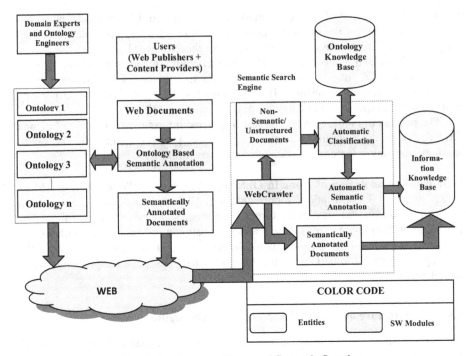

Fig. 3. Architecture of Proposed Semantic Search

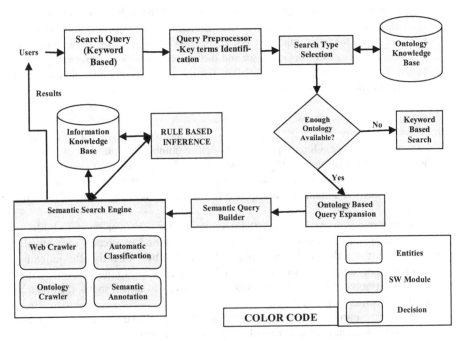

Fig. 4. Search Mechanism and Information Flow

[Mallet is a collection of tools in Java for statistical NLP: text classification, clustering and IE. It was created by Andrew Mccallum's information extraction lab at UMass. Mallet is one of the leading academic tools for text classification, topic modeling, and sequential tagging using Conditional Random Fields (CRFs).][19]

(ii) The ontology knowledge base (OKB) is searched for finding out the nearest match of these identified key concepts with the domain concepts contained in the ontologies. This process takes into account the semantic similarity between the ontology concepts present in the OKB and the key words identified in the document. As a result of this step an Ontology is found which automatically determines the domain to which this particular document belongs.

(iii) Once the ontology is identified the document can be semantically annotated using the same process as applied in step 1 above i.e. using the technique of automatic ontology based semantic annotation.

3.4   Search Mechanism and Information Flow

The proposed search engine architecture defines a number of steps to perform the search (Fig. 4); these can be listed as follows:

1) User issues a keyword based search query through the user interface of the search engine. Here we have taken the basic assumption that a user can not be entrusted in formulating an effective search query.

2) The normal query pre-processing is applied to the entered query words which includes removing the stop words and identifying the key-terms of the query

3) A semantic matching function of the search engine is called upon which uses the semantic similarity comparison of the identified query terms with the ontologies available in the Ontology Knowledge Base. This step decisively determines whether some Ontology exist for the domain terms being searched for or not. On the basis of it a decision is taken whether to go for a keyword-based search or for the semantic search.

4) If, for most of the query terms, the domain ontologies cannot be found in the OKB, keyword based search approach is taken, and the search is redirected to the keyword based search engine component of the architecture.

5) If enough ontologies are found in the OKB, the semantic search is implied and we move to the next step

6) Ontology based Query Expansion is applied on the identified search query terms. The ontologies that are used for Query Expansion are the same as found in the step 3 above.

7) Semantic Query Builder translates the expanded query (i.e. the final set of query terms) to a query in a semantic query language; SPARQL which is a W3C standard [20] could be used as the semantic query language for RDF database.

8) The Semantic Search Engine directs the query to the knowledge base of information. With the help of the Inference engine the query is resolved and the required RDF triples are returned.
[The use of a rule based inference engine allows the domain experts to define a number of axioms for the domain on the basis of which inferences can be drawn to extend the asserted or represented knowledge]

9) A suitable ranking function is used to re-order the results on the basis of their semantic importance to the context and results so re-ordered are returned to the user.

4   Future Scope

The proposed architecture is a conceptual one at present but various modules which constitute it have been implemented and are being used in different application areas. These components can be identified, re-tuned for the Search Engine and can be integrated together to actually implement the proposed search model. Some work has been done in the areas of Ontology Mapping and Ontology based Automatic Classification of Web Documents. The results and implementation from these areas can be incorporated in the proposed system with little difficulty. Standardization and Unification of Ontologies though is a larger problem which needs to be addressed at a higher level so as to make the effective semantic search for the whole web a reality.

5   Conclusion

In this paper we have identified a number of obstacles in getting the semantic search right, across all the domains i.e. for the whole web. Much of the information available on the web is unstructured or semi-structure so by the time the vision of semantic web is realized any search approach needs to combine the keyword search architecture with the semantic search. Also, for integrating the number of domains for which ontologies exist, there is a need for ontology unification so as to present a consistent, single view to the semantic search engine. We have introduced the idea of a Global Ontology which can be used to combine all the existing ontologies across various domains and providing the semantic structure to ontologies themselves. We have combined a number of semantic web techniques like Ontology Mapping and Ontology based Automatic Classification of web documents to make the search scalable for semantic as well as the non-semantic web.

References

1. Wang, L., Hou, J., Xie, Z., Wang, X., Qu, C., Li, H.: Problems and Solutions of Web Search Engines. IEEE (2011)
2. Berners-Lee, T., Hendler, J., Lissila, O.: The Semantic Web (2001),
   http://sciam.com/article.cfm?articleID=00048144
   -10D2-1C70-84A9809EC588EF21

3. Kassim, J.M., Rahmany, M.: Introduction to Semantic Search Engine. In: 2009 International Conference on Electrical Engineering and Informatics, Selangor, Malaysia (August 2009)
4. Ehrig, M., Staab, S.: QOM – Quick Ontology Mapping. In: McIlraith, S.A., Plexousakis, D., van Harmelen, F. (eds.) ISWC 2004. LNCS, vol. 3298, pp. 683–697. Springer, Heidelberg (2004)
5. Li, W.-J., Xia, Q.-X.: Study of Semantic-based Ontology Mapping Technology (2010)
6. Singh, S., Cheah, Y.-N.: Hybrid Approach Towards Ontology Mapping (2010)
7. Zheng, C., Shen, Y.-P., Mei, L.: Ontology Mapping based on Structures and Instances. In: Proceedings of the Ninth International Conference on Machine Learning and Cybernetics, Qingdao, July 11-14 (2010)
8. Li, A.W., Zhang, X., Wei, X.: Semantic Web-Oriented Intelligent Information Retrieval System. In: Proceedings of the First International Conference on BioMedical Engineering and Informatics (BMEI 2008), Sanya, China, vol. 1, pp. 357–361 (2008)
9. Zheng, L., Li, G., Shajing, Liang, Y.: Design of Ontology Mapping Framework. In: International Conference on Computational Intelligence for Modelling Control and Automation and International Conference on Intelligent Agents, Web Technologies and Internet Commerce (CIMCA-IAWTIC 2006) (2006)
10. Dan, Z.: Research of Semantic Information Retrieval Based on Grid (2010)
11. Zhai, J., Zhou, K.: Semantic Retrieval for Sports Information Based on Ontology and SPARQL. In: 2010 International Conference of Information Science and Management Engineering (2010)
12. Esa, A.M., Taib, S.M., Hong, N.T.: Prototype of Semantic Search Engine Using Ontology. In: 2010 IEEE Conference on Open Systems (ICOS 2010), Kuala Lumpur, Malaysia, December 5-7 (2010)
13. Nutch - Open-source Web-search software, built on Lucene Java,
http://nutch.apache.org/
14. Qi, H., Zhang, L., Gao, Y.: Semantic Retrieval System Based on Corn Ontology. In: 2010 Fifth International Conference on Frontier of Computer Science and Technology. IEEE (2010)
15. Jiang, H.: Information Retrieval and the Semantic Web. In: 2010 International Conference on Educational and Information Technology, ICEIT 2010 (2010)
16. Mayfield, J., Finin, T.: Information retrieval on the Semantic Web: Integrating inference and retrieval. In: Workshop on the Semantic Web at the 26th International ACM SIGIR Conference on Research and Development in Information Retrieval (SIGIR 2003), Toronto, Canada (2003)
17. Prabowo, R., Jackson, M., Burden, P., Knoell, H.-D.: Ontology-Based Automatic Classification for the Web Pages: Design, Implementation and Evaluation. In: Proceedings of the 3rd International Conference on Web Information Systems Engineering, WISE 2002 (2002)
18. Alani, H., Kim, S., Millard, D.E., Weal, M.J., Hall, W., Lewis, P.H., Shadbolt, N.R.: Automatic Ontology-based Knowledge Extraction and Tailored Biography Generation from the Web
19. http://mallet.cs.umass.edu/
20. http://www.w3.org/TR/rdf-sparql-query/
21. http://protege.stanford.edu/

FFMS: Fuzzy Based Fault Management Scheme in Wireless Sensor Networks[*]

Prasenjit Chanak, Indrajit Banerjee, Tuhina Samanta, and Hafizur Rahaman

Department of Information Technology
Bengal Engineering and Science University, Shibpur,
Howrah, India
prasenjit.chanak@gmail.com,
{ibanerjee,t_samanta,rahaman_h}@it.becs.ac.in

Abstract. Fault detection and fault management in wireless sensor network is a major challenge in its performance analysis. In this paper we propose a fuzzy logic based fault detection and fault management scheme (FFMS). Fuzzy rules are proposed for efficient detection of different types of nodes named as normal node, traffic node, end node, and dead node. Fuzzy interface engine categorizes the different nodes according to the chosen membership function and the defuzzifier generates a non-fuzzy control to retrieve the various types of nodes. Next, active nodes' performance analyses are done using five traditional metrics. Experimental results manifest that the proposed FFMS scheme is more effective for fault management in WSN.

Keywords: Wireless sensor networks (WSNs), fault detection, fuzzy logic, power efficiency.

1   Introduction

Wireless sensor network is a collection of hundreds and thousands of low-cost, low power miniaturized electronic programmable devices, which are deployed in a monitoring area [1], [2]. These devices are capable for data sensing of monitoring environment, data processing and communicating between each other. Low cost and effortlessly deployable nodes makes sensor network an attractive solution for a plethora of applications in various filed, such as military tracking, fire monitoring etc. Low power, low computation capability of the node is the main obstacle in augmentation of sensor networks application [3]. Due to distant deployment of sensor nods, the probability of fault in sensor network is very high [4]. Consequently, node failure degrades the life time and quality of service of the networks [2]. In addition, many WSN applications requires harsh node deployment, which makes difficult to control and monitor individual nodes manually [5]. Such networks demand automatic fault management technique in WSN.

---

[*] This work is partially supported by the grant from the Council of Scientific and Industrial Research (CSIR), human resource development group (Extramural research division), Govt. of India.

J. Mathew et al. (Eds.): ICECCS 2012, CCIS 305, pp. 30–38, 2012.
© Springer-Verlag Berlin Heidelberg 2012

In WSN, the node fault can be broadly classified into two groups (i) hardware fault and (ii) software fault. In hardware fault different hardware components of the nodes are damaged [6]. On the other hand in software fault, the system software of the node is malfunctioning [7]. The data routing fault is another obstacle in WSN [8, 9, 10]. In this paper we propose the fuzzy rule based distributed fault detection technique. The proposed technique analyzes nodes' hardware condition by fuzzy logic. The fuzzy representation of hardware circuit condition reduces dead nodes' percentage and improves the network coverage. This technique also categorizes nodes according to different hardware conditions. These different categories of nodes perform different jobs that improve the network life time.

The remainder of this paper is organized as follows. The FFM Scheme is presented in Section 2. Section 3 provides the fuzzy logic interface system for fault detection and management in the node. Simulation results are detailed and analyzed in Section 4. Finally, we summarize our work and draw conclusion in Section 5.

2   Proposed Fault Detection and Scheme

The FFMS scheme is divided into two parts, fault detection and fault management. In fault detection phase the hardware fault detection is carried out either in the cluster head or in the node itself. The transmitter circuit or microcontroller circuit fault detection is carried out in cluster head and the receiver circuit fault, sensor circuit fault or battery fault is detected by the node itself with the help of fuzzy logic system. The FFMS technique represents sensor node hardware condition with the help of fuzzy logic linguistic variable. The fault detection is conceded out on the basis of fuzzy logic rules. On the basis of the hardware condition the node declare itself as *traffic node*, *end node*, *normal node* or *dead node* and then inform it to the cluster head [6]. The cluster head after collecting that information from its member nodes organizes them in such way that there is no overlapping of the sensing region. The cluster head also selects effective routing path via *traffic/normal node* for efficient data transmission to base station.

This section describes different fault detection mechanism.

Transmitter circuit/ Microcontroller fault: Transmitter circuit/ microcontroller condition is checked by the cluster head. Every sensor node sends a *heartbeat message* to cluster head in precise time interval. When cluster head receives *heartbeat message* from the cluster member node, it also sends a *heartbeat-ok message* with respect to *heartbeat message*. The nodes transmitter circuit/ microcontroller circuit condition is represent by the function,

$$f(TM) = \psi/L_{time} \qquad (1)$$

Where ψ is the number of *heartbeat message* received by the cluster head and L_{time} is the total time spends by the WSN. Depending on the function value cluster head takes design about the member node's *transmitter circuit / microcontroller* fault.

Sensor circuit fault detection: the sensor circuit fault of a node is detected by the node itself. The difference between sensed information of a node and its received information from the neighboring node is represented in fuzzy logic. The sensor circuit condition of every node is represented by the linguistic variables. If the linguistic variable value is low then the sensor circuit of the node is faulty. However, if the value is medium then sensor circuit fault occurs after some time interval. Otherwise, if the value is high then the sensor circuit of the node is non-faulty.

Battery power fault: The battery power fault of the sensor node is detected by node itself. Remaining battery power of the sensor node is represented by the fuzzy logic linguistic variables. If the linguistic variable's value is low the battery fault occur, if the value is medium then sensor node battery fault occur after some time, and if the value is high then sensor node battery condition is good.

Table 1. Fuzzy Rule Based Decision Making in Node

Rule No.	Sensor circuit condition	Battery condition	Receiver circuit condition	Node Decision
0	Low	Low	Low	*Dead Node*
1	Low	Low	Medium	*Dead Node*
2	Low	Low	High	*Dead Node*
3	Low	Medium	Low	*Dead Node*
4	Low	Medium	Medium	*Traffic Node*
5	Low	Medium	High	*Traffic Node*
6	Low	High	Low	*Dead Node*
7	Low	High	Medium	*Traffic Node*
8	Low	High	High	*Traffic Node*
9	Medium	Low	Low	*Dead Node*
10	Medium	Low	Medium	*Dead Node*
11	Medium	Low	High	*Dead Node*
12	Medium	Medium	Low	*End Node*
13	Medium	Medium	Medium	*Normal Node*
14	Medium	Medium	High	*Normal Node*
15	Medium	High	Low	*Traffic Node*
16	Medium	High	Medium	*Normal Node*
17	Medium	High	High	*Normal Node*
18	High	Low	Low	*Dead Node*
19	High	Low	Medium	*Dead Node*
20	High	Low	High	*Dead Node*
21	High	Medium	Low	*End Node*
22	High	Medium	Medium	*Normal Node*
23	High	Medium	High	*Normal Node*
24	High	High	Low	*End Node*
25	High	High	Medium	*Normal Node*
26	High	High	High	*Normal Node*

Receiver Circuit fault: The receiver circuit condition is checked by the node itself. The receiver circuit condition is represented by the function.

$$f(R)=\beta)/L_{time} \tag{2}$$

Where β is the number of *heartbeat-ok message* received by a node. The function values are represented by the linguistic variables. If the value of the linguistic variable is low then receiver circuit is faulty. If the value of the linguistic variables is medium then receiver circuit faulty occur after some time, and if the value is high then receiver circuit condition is good.

According to hardware condition, sensor nodes are taking decision about its type through fuzzy logic rules. Then the node informs its cluster head about its type (*normal node, traffic node, end node*). The cluster head then sends few nodes to energy saving standby state, if two or more node are covering same region.

3   Fuzzy Logic Interface System for Fault Detection and Management

Fuzzy logic system is used in each node for fault analysis. Figure 1 shows the fuzzy logic system with three input variable, the sensor circuit condition, battery condition, and receiver circuit condition.

The input variables in the fuzzy system are BATTERY_CONDITION, SENSOR_CONDITION and RECEIVER_CONDITION. The output of the fuzzy system classifies the nodes into four types, *normal node, traffic node, end node, dead node* (Table 1). The labels of input variables are as follows:

- SENSOR_CONDITION= {Low, Medium, High}.
- BATTERY_CONDITION={Low, Medium, High}.
- RECEIVER_CONDITION={Low, Medium, High}.

The output labels are as follow:

- NODE_DECISION = {*normal node, traffic node, end node, faulty node*}.

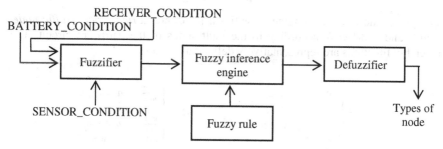

Fig. 1. Fuzzy based system with three input variable

The fuzzy logic system is divided into four part a) Fuzzifier b) Inference engine c) Fuzzy rule d) Defuzzifier. The fuzzifier phase converts three inputs into linguistic variable through the membership functions. The output of fuzzifier goes to inference engine. The fuzzy rule is the collection of linguistic control rules as is described in Table 1. The inference engine makes decision on the basis of fuzzy control rules and output of the fuzzifier phase (Table 1). The defuzzifier collects aggregated value from inference engine and generates a non-fuzzy control which presents the types of node, *normal node, traffic node, end node* and *dead node*.

We have used triangle or trapezoid shape membership function for reducing computational complexity which is the major issue in WSN. The domain of the input variables has been selected according to our simulation environment, but can be easily modified for other environments. Boundary condition a, b, c, d are used for member ship function representation. The membership function of battery conditions are represented as

$$\mu_{BH} = \begin{cases} 0, & x \le a \\ \frac{x-a}{b-a}, & a < x < b \\ 1, & x \ge b \end{cases} \quad \mu_{BM} = \begin{cases} 0, & x \le a \\ \frac{x-a}{b-a}, & a < x \le b \\ 1, & b \le x \le c \\ \frac{d-x}{d-c}, & c \le x < d \\ 0, & x \ge d \end{cases} \quad \mu_{BL} = \begin{cases} 1, & x \le a \\ 1 + \frac{x-a}{b-a}, & a < x < b \\ 0, & x \ge b \end{cases}$$

Sensor circuit condition represented by linguistic term set "Low", "Medium" and "High". Membership function of sensor circuit condition is represented as follows:

$$\mu_{SDH} = \begin{cases} 0, & x \le a \\ \frac{x-a}{b-a}, & a < x < b \\ 1, & x \ge b \end{cases} \quad \mu_{SDM} = \begin{cases} 0, & x < a \\ \frac{x-a}{b-a}, & a \le x < b \\ 1, & b \le x < c \\ \frac{d-x}{d-c}, & c \le x \le d \\ 0, & x > d \end{cases} \quad \mu_{SDL} = \begin{cases} 1, & x < a \\ 1 + \frac{x-a}{b-a}, & a \le x \le b \\ 0, & x > b \end{cases}$$

Receiver circuit condition represented by linguistic term set "Low", "Medium" and "High". Membership function of receiver circuit condition is represented as follows:

$$\mu_{BH} = \begin{cases} 0, & x < a \\ \frac{x-a}{b-a}, & a \le x \le b \\ 1, & x > b \end{cases} \quad \mu_{BM} = \begin{cases} 0, & x < a \\ \frac{x-a}{b-a}, & a \le x \le b \\ 1, & b \le x \le c \\ \frac{d-x}{d-c}, & c \le x \le d \\ 0, & x > d \end{cases} \quad \mu_{BL} = \begin{cases} 1, & x < a \\ \frac{x-a}{b-a}, & a \le x \le b \\ 0, & x > b \end{cases}$$

The output of the inference engine classifies a node into *normal node, traffic node, end node* and *dead node* according to the fault status of the hardware circuit. The membership functions are representatively defined as follows:

$$\mu_{NN} = \begin{cases} 0, & x = a \\ \frac{x-a}{b-a}, & a \le x \le b \\ 1, & x > b \end{cases} \quad \mu_{NT} = \begin{cases} 0, & x < a \\ \frac{x-a}{b-a}, & a \le x \le b \\ 1, & x = b \\ \frac{c-x}{c-b}, & b < x \le c \\ 0, & x > c \end{cases}$$

$$\mu_{NE} = \begin{cases} 0, & x < a \\ \frac{x-a}{b-a}, & a \le x \le b \\ 1, & x = b \\ \frac{c-x}{c-b}, & b < x \le c \\ 0 & x > c \end{cases} \quad \mu_{ND} = \begin{cases} 1, & x < a \\ \frac{x-a}{b-a}, & a \le x \le b \\ 0, & x > b \end{cases}$$

Use of membership function deliberately increases the percentage of active node and improves network coverage.

4   Experimental Result

A series of simulations are conducted to evaluate the performance of the proposed technique by using MATLAB. Each sensor node collects data from sensing environment and sent the data to the base station. 8000 to 10000 nodes are deployed randomly across an area of $100 \times 100m^2$. The nodes are functionally equivalent and had fixed transmission radius. The simulation parameters are used as we have used in [6].

In order to evaluate the performance of FFMS, described in Section 2, five traditional metrics of WSN have been considered.

Average packet delivery ratio: The ratio of number of distinct packet received at destination node and original number of sent packet from source node.

Global energy of network: This is the sum of residual energy of each active node in the network. We calculate this value at each time interval.

Network lifetime: The time elapsed from the start of simulation until all the nodes run out of energy.

Active nodes: The total numbers of nodes which are in active state either as a normal node or as traffic node or as end node.

Coverage: The percentage of the area that is monitored by the sensing range of all active nodes in a specific period of time excluding the overlapping area.

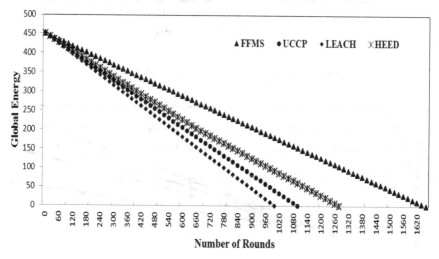

Fig. 2. Global energy loss comparison between FFMS and other popular clustering techniques

The chosen membership functions efficiently discriminate the active nodes from the faulty node and then the Equation 1 and Equation 2 are used for transmitter circuit fault and receiver circuit fault calculation respectively.

Figure 2 shows the impact of using FFMS technique on a networks lifetime. The FFMS increases the network life time compared to UCCP [9], LEACH [3] HEED [8] energy saving techniques. The UCCP enhances network life time by optimizing different communication costs involved in transmission of data to base station. The algorithm LEACH forms cluster by using a distributed algorithm, where nodes make the decision by itself without any central control. Therefore, LEACH suffers uneven energy drainage. On the other hand HEED prolongs the network life time by certifying balanced energy dissipation as well as uniform circulation of the cluster head nodes. In FFMS technique traffic nodes are mainly responsible for reduction of transmission energy loss.

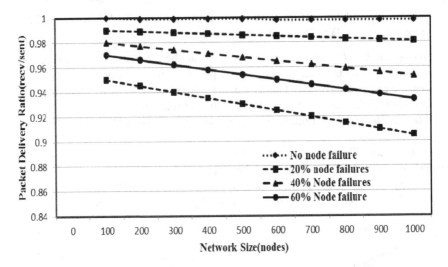

Fig. 3. Average packet delivery ratio with node fault

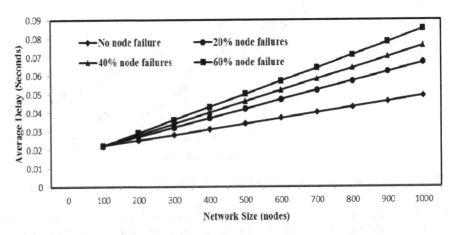

Fig. 4. Average delay with node failures

Average packet delivery ratio depending on the percentage of node failure is shown in Figure 3. In FFMS number of dead node is moderately low because the node with specific hardware failure like sensor circuit fault or receiver circuit fault are reused as traffic node or end node. Therefore, packet delivery ration very high at fairly high percentage of different hard ware fault. If percentage of node fault is more the 60% then the packet delivery ratios decreases very rapidly, as the percentage of dead node increases in the network.

Average data transmission delay in FFMS technique with different percentage of node failure is shown in Figure 4. If transmitter circuit, receiver circuit, microcontroller and battery fault occur at very high rate then data transmission delay increases in the network. At low percentage of node fault the probability occurrence of dead node is low. Therefore, average delay of data transmission is reasonable. On the other hand with more than 60% of node fault the probability occurrence of dead node increase. Therefore, average delay of data transmission is increased.

5   Conclusion

In the proposed fuzzy based fault management scheme (FFMS) schemer we analyses the node's hardware condition through fuzzy logic. According to the hardware condition, the nodes are categorized in to four categories. FFMS scheme detect hardware fault in distributed manner and reduces rapidly increment of faulty node in the network. The microcontroller and transmitter circuit condition check in cluster head, on the other hand receiver circuit, sensor circuit and battery condition check by node itself. Therefore, large number of energy saves in FFMS scheme comparison to other fault detection technique. The FFMS scheme allocates different work according to nodes hardware condition. Therefore, energy utilization in FFMS scheme is better compare to other existing energy saving techniques.

References

[1] Tarique, M., Tepe, K.E., Adibi, S., Erfani, S.: Survey of multipath routing protocols for mobile ad hoc networks. Journal of Network and Computer Applications, 1125–1143 (2009)

[2] Banerjee, I., Chanak, P., Sikdar, B.K., Rahaman, H.: EERIH: Energy efficient routing via information highway in sensor network. In: IEEE International Conference on Emerging Trends in Electrical and Computer Technology, Kanyakumari, India, March 23-24, pp. 1057–1062 (2011)

[3] Heinzelman, W.R., Chandrakasan, A., Balakrishnan, H.: Energy-efficient communication protocol for wireless microsensor networks. In: Proceedings of the 33rd Annual Hawaii International Conference on System Sciences, pp. 3005–3014 (2000)

[4] Akyildiz, I., Su, W., Sankarasubramaniam, Y., Cayirci, E.: A survey on sensor networks. IEEE Communication Magazin 40(8), 102–114 (2002)

[5] Ruiz, L.B., Siqueira, I.G., Oliveira, L.B., Wong, H.C., Nogueira, J.M.S., Loureiro, A.A.F.: Fault management in event-driven wireless sensor networks. In: Proceedings of the 7th ACM Internationl Symposium on Modeling, Analysis and Simulation of Wireless and Mobile Systems, New York, pp. 149–156 (2004)

[6] Banerjee, I., Chanak, P., Sikdar, B.K., Rahaman, H.: DFDNM: A Distributed Fault Detection and Node Management Scheme for Wireless Sensor Network. In: Abraham, A., Mauri, J.L., Buford, J.F., Suzuki, J., Thampi, S.M. (eds.) ACC 2011, Part III. CCIS, vol. 192, pp. 68–81. Springer, Heidelberg (2011)

[7] Chessa, S., Santi, P.: Comparison-based system level fault diagnosis in ad hoc network. In: Proceeding of 20th IEEE Symposium on Reliable Distributed Systems, pp. 257–266 (2001)

[8] Younis, O., Fahmy, S.: Distributed clustering in ad hoc sensor network: a hybrit energy-efficient approach. IEEE Transitions on Mobile Computing 3(4) (2004)

[9] Aslam, N., Phillips, W., Rebertson, W.: A unified clustering and communication protocol for wireless sensor network. IAENG International Journal of Computer Science 35(3), 1–10 (2008)

Evaluation of WSN Protocols on a Novel PSoC-Based Sensor Network

Rakhee[1], P. Sai Phaneendra[2], and M.B. Srinivas[2]

[1] Department of Computer Science and Information Systems
[2] Department of Electrical Engineering
Birla Institute of Technology and Science-Pilani, Hyderabad Campus, India

Abstract. Sensor networks have found widespread use in a variety of applications such as structure monitoring, environment monitoring, etc... Commercially available sensor nodes (that integrate sensors, micro-processor/controller, memory and peripherals) are often used to design and deploy sensor networks. In this work, authors describe a sensor network built using PSoC (Programmable System-on-Chip) from Cypress Semiconductor. The advantage of PSoC is that it integrates analog, digital and controller components (all required to process sensor data) on a single chip, thus resulting in a smaller footprint for sensor nodes. Communication with other nodes is achieved through CyFi low power RF module operating in the 2.4GHz ISM band. Two specific WSN protocols namely, LEACH (Low Energy Adaptive Clustering Hierarchy) and SPIN (Sensor Protocol for Information via Negotiation) have been implemented on this network and their performance studied using NS-2 simulator.

Keywords: PSoC, Programmable System on Chip, LEACH, Sensor network.

1   Introduction

Sensor networks have attracted wide attention over the years due to their utility in monitoring a wide variety of physical phenomena, often remotely. A wireless sensor network consists of spatially distributed sensor nodes to monitor physical or environmental conditions such as temperature, humidity etc... The nodes communicate among themselves using a specified wireless routing protocol.

In this paper, authors examine the suitability of PSoC as a sensor node to set up a wireless sensor network. After a brief description of PSoC architecture [1], it is shown how CYFISPI user module (a part of PSoC Designer software) along with CYFI transceiver (CYRF7936) RF module can be configured to act as a sensor node to both transmit and receive the data. A few such nodes have been used to build a sensor network and functionality has been tested with LEACH [2-3] and SPIN protocols [4-5].

2   Description of PSoC

2.1   PSoC Architecture

PSoC is a reconfigurable, embedded, mixed signal system-on-chip from Cypress Semiconductor [1]. In addition to all the standard elements of 8-bit microcontrollers,

J. Mathew et al. (Eds.): ICECCS 2012, CCIS 305, pp. 39–46, 2012.

PSoC features analog and digital programmable blocks, which themselves allow implementation of a large number of peripherals. PSoC can replace tradition fixed function ICs (Integrated Circuits), ASICs (Application Specific Integrated Circuits) and microcontrollers in many applications today [1]. PSoC architecture is described in [8].

2.2   RF Module in PSoC

The CyFi (Radio Frequency) module is a low cost device targeted and designed for low-power embedded wireless applications [4]. It can operate in unlicensed ISM band (2.4 GHz – 2.483 GHz). Internally CyFi has Direct Sequence Spread Spectrum (DSSS) radio transceiver. The advantage of using DSSS is its short latency time, 80 discrete 1 MHz selective channels and high dynamic data rates of 250 Kbps and 1 Mbps. The internal architecture of CyFi RF (CYRF7936) is as shown in Figure 1.

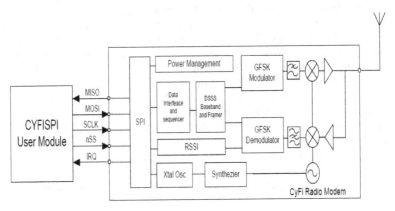

Fig. 1. CyFi RF Transceiver

2.3   CyFiSPI User Module and Its Application Programming Interfaces (APIs)

In PSoC to develop any application, pre-characterized analog and digital peripherals are dragged-and-dropped within the PSoC designer software development environment. In order to use the RF module, the CyFiSPI user module is placed. Then, API libraries are used to customize the design [5].

CYFISPI user module takes input as MISO (Master in Slave output) and provides MOSI (Master out Slave input) output which acts as input to RF SPI. These MOSI and MISO are used for data transfer (serial single byte or multiple bytes) between an application microcontroller unit (MCU) acting as a master and RF acting as a slave. The other inputs such as nSS (negative slave select) is an active low pin used to select the slave device and SCLK is synchronous clock to RF which runs at half the crystal frequency.

After receiving the data from SPI the baseband performs DSSS spreading/dispreading, adds headers like SOP (Start of Packets), length and 16-bit CRC (Cyclic Redundancy Check) which are then transmitted by radio modem. The baseband can be configured to automatically transmit Acknowledge (ACK) packets whenever a valid packet is received indicating successful transmission and reception of data. In an application, if a data rate of 1Mbps is required then GFSK (Gaussian Frequency-Shift Keying) is used. Communication is successful if channel is free from noise and has good signal strength. CYFI RF has RSSI (Receive Signal Strength Indicator) which automatically measures and stores (as five bit value) the relative signal strength when an SOP is detected in receive mode. Channel is changed if RSSI level falls below the threshold value.

2.4   PSoC as a Sensor Node

A conventional microcontroller needs to be interfaced with external analog circuits and sensors to build a sensor node. But as described earlier, PSoC consists of built-in analog and digital blocks and thus no external components would be required excepting for sensors.

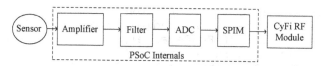

Fig. 2. Implementation of Sensor Node using PSoC

For example, as shown in figure 2 above, all signal processing related to sensor output can be done within the PSoC. Also, Serial Peripheral Interconnect Master (SPIM) can be used as the master block to control CyFiSPI.

3   Implementation Using PSoC

In order to implement LEACH protocol, two basic functions of transmitter and receiver are used. The frequency of transmission or reception and the data which is to be transmitted or received are passed as arguments to these functions. Each node will calculate their respective energy levels.

In the following, pseudo-code for communication between PSoC controller [6] (CY27443 PVXI) and CYFI RF module is presented.

```
// Transmit() is the packet transmission handler.
Transmit(BYTE freq, BYTE pkt)
  CYFISPI_SetPtr((BYTE *)&AppBuf);
  // Provide the radio data buffer parameters
  AppBuf [0] = pkt;
  CYFISPI_SetSopPnCode(0);   //Pn CODE
  CYFISPI_SetChannel(CHANNEL);   //Set the channel
  CYFISPI_SetFrequency(freq);   //Setting the frequency
  CYFISPI_StartTransmit(No. of Retry, Packet size);
  // Initiate transmission.
  while(1)
```

```
    if(CYFISPI_State& CYFISPI_COMPLETE)
     if(!(CYFISPI_State& CYFISPI_ERROR))//Check for er-
ror.
       //Handle successful packet transmission
     else
       //Handle failed packet transmission
     break;
   CYFISPI_EndTransmit(); // Ends the radio transaction.
```

// Receive() is the packet reception handler.
```
Receive(BYTE Freq, BYTE *pkt)
   CYFISPI_SetPtr((BYTE *)&pkt);
  // Provide the radio data buffer parameters
   CYFISPI_SetSopPnCode(0);   //Pn CODE
   CYFISPI_SetChannel(CHANNEL); //set the channel
   CYFISPI_SetFrequency(Freq); //Setting the frequency
   CYFISPI_StartReceive();// Start listening for a packet.
   while(1)
     if(CYFISPI_State& CYFISPI_COMPLETE)
     // Check to see if the radio stopped listening.
       RxLen = CYFISPI_EndReceive();
     // Complete the radio transaction.
       if(!(CYFISPI_State& CYFISPI_ERROR))
       //Check for error.
          // Handle a successful packet reception
       else
         //Handle failed packet transmission
       break;
```

3.1   LEACH Protocol Implementation:

The stepwise implementation of LEACH protocol is as follows:

3.1.1   Energy Acquisition

Initially default node requests energy from all other nodes in range. It transmits a Request packet (i.e. ReqPkt) and receives energies (i.e. EnergyPkt) from all the nodes using the function given below:

```
EnergyAcquisition()
  for(i=1; i<=N; i++) //N= Number of Nodes
    Transmit(Node_Frequency[i], ReqPkt);
    Receive(Node_Frequency[i], EnergyPkt);
```

3.1.2   Cluster Head Selection

Received energy values of all the nodes are stored in buffer array (i.e. energy []). To find the node with the highest energy level in this function all the energy values are passed as argument. The node having the highest energy value is assigned as the new cluster head.

```
ClusterHeadSelection(BYTE energy[])
  int j;
  BYTE temp;
  // energy[j] consists of energy value of jth nodes
  for(j=0;j<=n; j++)
    if (energy[j]>energy[j+1])
      temp=energy[j+1];
      energy[j+1]=energy[j];
      energy[j]=temp;
  NewClusterHead = NodeID(energy[n]);
```

3.1.3   Cluster Head Information

After selecting the cluster head, the old cluster head informs to that node that it is the cluster head.

```
ClusterHeadInformation()
  if (oldClusterHead!=NewClusterHead)
  Trans-
mit(Frequency[NewClusterHead],NodeID[NewClusterHead]);
```

3.1.4   Cluster Head Broadcast

The new cluster head broadcasts its node ID to all the nodes using this function. All the nodes which are in the range now bind to the new cluster head.

```
ClusterHeadBroadcast(int var)
  for(i=1; i<=N; i++)
    Transmit(Frequency[i], NodeID[NewClusterHead]);
```

3.1.5   Data Acquisition

Cluster head will send advertisement (i.e. Advt) for data to all the nodes and receive acknowledgement (i.e. Ack). The nodes then transmit the data packet (i.e. pkt).

```
DataAcquisition()
  for(i=1; i<=N; i++)
    Transmit(Frequency[i], Advt);
    //Send the Advertisement to all nodes
    Receive(Frequency[i], Ack);
    // Receive the Acknowledgement from all nodes
    Receive(Frequency[i], pkt);
    //Receive the data from all nodes
```

This whole process will be continued in the second round after certain timeout.

3.2   SPIN Protocol Implementation

The stepwise implementation of SPIN protocol is as follows:

3.2.1   Send Advertisement

SendAdvt function call is used to broadcast the advertisement (i.e. Advt) to all the neighbor nodes that are binded at different frequencies (i.e. Frequency[1], Frequency[2], ..., Frequency[N]).

```
SendAdvt(void)
  int i;
  for(i=1; i<=N ; i++)
    Transmit(Frequency[i], Advt);
```

3.2.2   Receive Advertisement

RcvAdvt function will receive the advertisement from the source nodes, which matches with the transmitted frequency of Node_Frequency. AdvtRecv is a pointer which stores the advertisement packet received from source node.

```
RcvAdvt(void)
  Receive( Node_Frequency, AdvtRecv);
```

3.2.3   Send Request

The interested nodes will call SendReq function to request the source node to send data through Request_Packet.

```
SendReq(void)
  Transmit (Node_Frequency, Request_packet);
```

3.2.4   Waiting for Request

At source node, Wait4Req function will look for the request from every node for a specific time. If it receives request from a node, a flag bit is set to indicate for sending the data.

```
void Wait4Req(void)
  int i;
  for(i=1; i<=N ; i++)
    Receive (Frequency[i], Request_packet);
```

3.2.5   Sending Data

SendData function will send the data to the requested nodes.

```
SendData(void)
  int i;
  for(i=1; i<=N ; i++)
    if(FlagBit[NODEid] == TRUE)
  Transmit(Frequency[i] , Data);
```

3.2.6   Receive Data

RcvData function will receive data from the source node. The DataPtr points to the data buffer to store the data.

```
RcvData()
  Receive(Node_Frequency, DataPtr);
```

Figure 3 shows the sensor nodes implemented using PSoC and CYFI module.

Fig. 3. Sensor Network implementation using PSoC based Sensor Nodes

4   Results

A sensor node is built using PSoC microcontroller and CYFI wireless module. Operation of the network has been tested by implementing two protocols, LEACH and SPIN protocol.

Table 1 shows various parameters of CYFI wireless module while Table 2 illustrates the power consumed by PSoC based sensor node.

The power results obtained in the hardware are used to model in NS-2 simulator. This simulator is used to scale the network for large number of nodes. Table 3 shows the results from NS-2 simulator for different clusters at different simulation times.

Table 1. CYFI Wireless module parameters

Receive Sensitivity	*-97dBm*
Max Transmitter Power	+4dBm
Peak Transmit Current at 0dBm	26.2mA
Peak Transmit Current at +4 dBm	34.1 mA
Peak Receive Current at 0dBm	23 mA
Sleep Current	0.8 µA

Table 2. Power Consumed by PSoC Sensor node

Mode	*Power*
Transmitter	0.57mW
Receiver	0.56mW
Sleep	0.498mW

Table 3. NS-2 Simulation for different clusters (Number of nodes = 100)

No. of Clusters	*Simulation Time (Sec)*	*No. of Nodes Alive*	*Energy Dissipated (J)*
2	200	95	116.702
3	300	12	193.828
4	400	16	196.546

5   Conclusion and Future Work

In this paper, authors explored the utility of PSoC to build a sensor network. PSoC, with its programmable analog, digital and microcontroller blocks, coupled with CYFI RF module, can function as a sensor node. Integration of all signals processing functionality in a single chip results in a small footprint for the node several of which have been combined to set up a sensor network. Well known WSN protocols such as LEACH and SPIN have been tested on the network and found to be functioning correctly. A model is created in NS-2 simulator to scale the network for larger nodes. More experiments are under way to scale up the network as well as test other routing protocols, to quantify the performance of the network in terms of latency, energy consumption and other important parameters.

References

1. Cypress Semiconductor website, http://www.cypress.com
2. Wang, G., Wang, Y., Tao, X.: An Ant colony Clustering Routing Algorithm for Wireless Sensor Networks. In: Third International Conference on Genetic and Evolutionary Computing (2009)
3. Li, L., Wu, H., Chen, P.: Discuss in Round Rotation Policy of Hierarchical Route in Wireless Sensor Networks. Digital Engineering Research Center, China
4. Abd-El-Barr, M.I., Youssef, M.A.M., Al-Otaibi, M.M.: Wireless sensor networks - part I: topology and design issues. In: Canadian Conference on Electrical and Computer Engineering, May 1-4, pp. 1165–1168 (2005)
5. Rehena, Z., Roy, S.: A Modified SPIN for Wireless Sensor Networks. In: International Conference on Communication Systems and Networks, pp. 1–4 (January 2011)
6. SPI-based CyFi^{TM} Transceiver Data Sheet. Cypress (2009)
7. CYRF7936 2.4 GHz CyFi^{TM} Transceiver Data Sheet. Cypress (2008)
8. CY8C27443 Datasheet. Cypress (2007)

Improving Lifetime of Structured Deployed Wireless Sensor Network Using Sleepy Algorithm

Jis Mary Jacob and Anita John

Department of Computer Science
Rajagiri School of Engineering & Technology
Rajagiri Valley, Kochi-39, Kerala, India
jismaryjacob@gmail.com, anitaj@rajagiritech.ac.in

Abstract. A wireless sensor network is a distributed system interacting with physical environment. It comprises of motes equipped with task–specific sensors to measure the surrounding environment Wireless sensor networks are designed with energy constraint. More often Wireless Sensor Networks are deployed in harsh environments where failure of sensor nodes and disruption of connectivity are regular phenomena. Energy efficiency remains the main concern to achieve a longer network lifetime. One of the main design issues in wireless sensor networks is to obtain long system lifetime, as well as maintain sufficient sensing reliability. Communication amongst sensing nodes consumes largest part of energy. Communication and sensing consume energy, therefore efficient power management can extend network lifetime. The overall energy consumption of system can be reduced by turning off some redundant nodes. The objective is to cover maximum of network with the minimum active sensors.

Keywords: Wireless sensor network (WSN), cluster head (CH), network lifetime, active sensors, multihop.

1   Introduction

Wireless sensor networks (WSNs) constitute the foundation of a broad range of applications. These have been created with a set of sensors with the aim of collecting their surrounding data and sending them to the sink node. Sensors should be able to interact with each other for exchanging information. Each sensor in the wireless sensor networks should cover all sensors in its neighbourhood in order to communicate by radio communication. Wireless sensor motes often run unattended on battery power for long periods. In most applications sensor node are immobile and left unattended once deployed. Clustering is an effective approach for self organizing large WSNs into a connected hierarchy. A WSN is partitioned into small groups called clusters. Each cluster has a coordinator called the cluster head (CH) to whom other member nodes (MNs) of the cluster communicate. CHs perform data aggregation and send results to the sink. Data relay may be single hop or multihop. Thereby CHs constitute the virtual backbone of network.

J. Mathew et al. (Eds.): ICECCS 2012, CCIS 305, pp. 47–53, 2012.
© Springer-Verlag Berlin Heidelberg 2012

The recent developments in the Wireless and electronic communications have caused lower prices, lower power consumption and smaller sizes in all sensor points. Each sensor contains a microprocessor capable of processing millions of instructions in few seconds, a limited storage as much as one kb, short degree of radio communication and a limited power supply. The sensors have various applications such as battle field control, displaying the ocean and atmosphere environments, detecting fire in the forest. The coverage of surrounding by the sensors is very important in the applications of sensor network. We need to compensate for the large difference in power consumption between a CH and a MN to prolong lifetime.

2   Motivation

Minimizing energy consumption and maximizing the system lifetime has been a major design goal for wireless sensor networks. In the last few years, researchers are actively exploring advanced power conservation approaches for wireless sensor networks. On the other hand, device manufacturers have been arriving for low power consumption in their products. In [1], an energy scavenging technique which enables self-powered nodes using energy extracted from the environment is presented. In [2] a probing-based density control algorithm is proposed to ensure long-lived, robust sensing coverage by leveraging unconstrained network scale. In this protocol, only a subset of nodes are maintained in working mode to ensure desired sensing coverage, and other redundant nodes are allowed to fall asleep most of the time. However, long system lifetime is expected by many monitoring applications. The system lifetime, which is measured by the time until all nodes have been drained out of their battery power or the network no longer provides an acceptable event detection ratio, directly affects network usefulness. Therefore, energy efficient design for extending system lifetime without sacrificing system reliability is one important challenge to the design of a large wireless sensor network

In [3], Chen et al. proposed an algorithm to turn off nodes based on the necessity for neighbor connectivity. They intend to reduce the consumption of system energy without significantly diminishing the connectivity of network. A node density based clustering algorithm is suggested in [3] for increased stability in dynamic wireless sensor networks. Self-stabilization is attained by minimizing the maximum number of nodes that need to change their topology information as a result of node mobility.

In [8], Tian et al. have proposed a coverage-preserving node scheduling scheme which guarantees sensing coverage when the redundant nodes are turned off. In their approach, each node decides on whether to turn itself on or off based on the off-duty fitness rule: a node is eligible to turn off itself if its neighbors can cover its sensing area. The node scheduling operation is divided into phases. Each phase includes a self-scheduling phase, followed by a sensing phase. In the self-scheduling phase, nodes exchange position information (and sensing range, if nodes have different sensing ranges) among neighbours. Each node computes its neighbours' covering sensing area (the sensing area that its neighbors can help it monitor), and based on this approach decides whether it is eligible for off-duty.

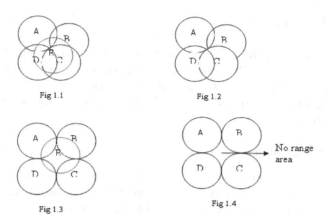

Fig. 1. Illustration of blind point occurrence

In [8] after completing the collection of neighbour information, each node eva-luates its eligibility for turning off by calculating the sponsored coverage, as described in the earlier section. However, if all nodes make decisions concurrently, blind points may appear, as shown in Figure 1. Node E finds its sensing area can be covered by node A, B, C and D. According to the off-duty eligibility rule, the node 'E' turns itself off as shown in Fig 1.2. While in Fig 1.3, believing that its sensing area can be cov-ered by node A, B, C and D, node 'E' turns itself off. Thus a blind point occurs after turning off node 'E', as in Figure 1.4. To avoid such a crisis, back-off scheme is in-troduced. Let each node start its determination after a random back-off time period T_d and broadcast a Status Advertisement Message (SAM) to announce its status, if it is eligible for turning off. Neighbouring nodes receiving a SAM will delete the sender's information from their neighbour lists. Thus, the nodes that have a longer back-off delay will not consider the nodes that have decided to be turned off before. In certain applications like temperature sensing, occurrence of blind points is not hazardous but highly sensitive applications cannot bear the occurrence of blind points.

A node density based clustering algorithm is suggested in [9] for increased stability in dynamic wireless sensor networks. Self-stabilization is attained by minimizing the maximum number of nodes that need to change their topology information as a result of node mobility. Related studies can be classified into two categories: those that deal with general robustness issue in WSNs without concerning clustering protocols and those that integrate robustness with clustering schemes. In [10], an easy-to-implement method named DED (distributed, energy-efficient, and dual homed clustering) which provides robustness for WSNs without relying on the redundancy of dedicated sen-sors, that is, without depending on node density is presented. DED uses the informa-tion already gathered during the clustering process to determine backup routes from sources to observers, thus incurring low message overhead. It does not make any as-sumptions about network dimension, node capacity, or location awareness and termi-nates in a constant number of iterations.

3   Methodology

WSN is application specific so it is not practical to seek a one size-fits-all solution in the WSN design. Some WSN applications require continuous monitoring capabilities that generate a constant stream of data. Nodes within a particular area report to a single data sink, which is assumed to possess a more stable energy supply and extra computing power

3.1   Sleepy Algorithm

In this paper, we consider a wireless sensor network whose nodes sleep periodically. We try to identify the manner in which the optimal sleep schedule varies with the length of the sleep period and the statistics of lifetime of the nearby nodes.

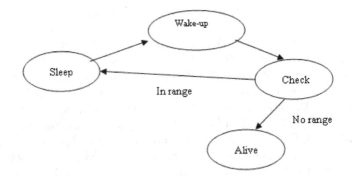

Fig. 2. State transition in sleepy algorithm

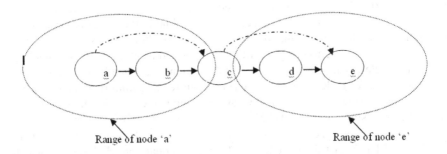

Fig. 3. Illustration of sleepy algorithm

Optimized routing (from source to sink) in wireless sensor networks constitutes one of the key design issues in prolonging the lifetime of battery-limited sensor nodes. In Fig.3, all the nodes a, b, c, d and e are involved in routing towards sink in WSN. Here in this case if there is data from node a it will be forwarded through b, c, d and then to e. As shown in the figure the nodes b and c comes in the range of node a (node b can be avoided in this case). Similarly node d can also be avoided. Then the nodes a, c, e is only involved, b and d can sleep and thereby save energy. This is represented as dashed line. In the first case five nodes are involved and in second case only 3 nodes are involved. In the scenario where there is thousands of operating nodes this is very critical. Sleeping method helps to improve network lifetime

3.2   Challenges Faced: How Long One Node Can Sleep

A standard WSN node is inoperative nearly all of the life time when it is not communicating. The energy utilization can be abridged radically by putting the radio in deep sleep during this state. The duty cycle for sleep state have to be optimized. More over as discussed in this paper, the nodes can also go to sleep not only because of no data to communicate; it can go to sleep if there are more nodes to sense in the same area. But how long it can sleep have to be optimized. This is very important since the energy expenditure of WSN node is considered in different operational states, e.g., Idle, Listen, Transmit and Sleep. In sleep state, the energy consumption is almost zero. Before going into details of the algorithm, let us see the existing scenario. To improve the life time the radio-section will be in sleeping mode. When there is a sensed data, it is passed to the processor normally. If it requires any processing, it is performed. Then the processor sends signals to make the radio-section to turn itself on from sleepy mode. Thus saves the energy and in that way improve lifetime.

In our new algorithm we have added a timer which is connected to all blocks in the mote like sensor, processor and radio. First the sensor senses the surroundings if the situation permits, the sensor, processor and radio go to sleeping mode. Before going to sleeping mode, the processor sets the timer. Once the set time elapse, the timer sent signal to all the blocks and turns it on. Again the sensor senses the surrounding and decides whether to go to sleep or to the listen mode.

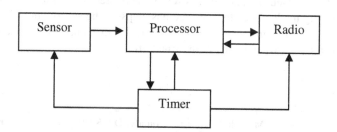

Fig. 4. Blocks comprising a node

4   Experimental Results

Analysis of improvement in network lifetime of WSN while using sleepy algorithm is shown in graph in Fig.5. The graph depicts three different scenarios where 1 indicates 50 nodes deployed in a specific area, 2 shows 100 nodes deployed in the same region

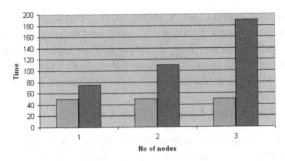

Fig. 5. Comparison

and 3 indicates 1000 nodes deployed in the same area. The Y-axis in the graph indicates network lifetime in WSN. The bar in blue colour implies the network lifetime when all deployed nodes are active whereas the bar in red colour implies the network lifetime while using sleepy algorithm. The graph demonstrates the comparison in lifetime when sleepy algorithm is not used and when it is employed in WSN.

5   Conclusion

Designing a sensor network with low energy consumption is one of the most fundamental issues. In this paper, with respect to the restriction of energy of each sensor, the maximum coverage of the network should be ensured through the least active sensors, and to achieve this goal, we have proposed a node scheduling scheme.

In this paper, we proposed a node-scheduling scheme, which can reduce energy consumption, and thereby increasing system lifetime, by turning off some redundant nodes. A basic model for off-duty eligibility among sensors was discussed. The off-duty eligibility guarantees that the original sensing coverage. A timer is attached to the blocks within the sensor. The timer is set by the processor within the sensor before it goes to sleeping mode. Once the set time has elapsed, the sensors in sleeping mode are turned on by the signals sent by the timer. Experiments show that sufficient redundancy is retained when nodes were turned off.

References

1. Rabaey, J.M., Ammer, M.J., da Silva, J.L., Roundy, D.P.S.: PicoRadio Supports Ad Hoc Ultra-Low Power Wireless Networking. IEEE Computer Magazine (July 2000)
2. Ye, F., Zhong, G., Lu, S., Zhang, L.: Energy Efficient Robust Sensing Coverage in Large Sensor Networks. Technical Report (2002)
3. Chen, B., Jamieson, K., Balakrishnana, H., Morris, R.: Span: An Energy-Efficient Coordination Algorithm for Topology Maintenance in Ad Hoc Wireless Networks. Wireless Networks 8(5), 481–494 (2002), doi:10.1023/A:1016542229220
4. Gupta, H., Das, S.R., Gu, Q.: Connected sensor cover: self-organization of sensor networks for efficient query execution. In: Proceedings of the ACM MobiHoc, pp. 189–200 (2003)

5. Polastre, J.P., Szewczyk, R., Mainwaring, A., Culler, D., Anderson, J.: Analysis of wireless sensor networks for habitat monitoring. In: Raghavendra, C.S., Sivalingam, K.M., Znati, T. (eds.) Wireless Sensor Networks. Kluwer Academic Publishers (2004)
6. Azim, M.A., Kibria, M.R., Jamalipour, A.: Wireless Communications & Mobile Computing Archive 9(1) (January 2009)
7. Pawar, M.S., Manore, J.A., Kuber, M.M.: Life Time Prediction of Battery Operated Node for Energy Efficient WSN Applications. IJCST 2(4) (October-December 2011)
8. Tian, D., Georganas, N.D.: A coveragepreserving node scheduling scheme for large wireless sensor networks. In: Proceedings WSNA, USA, (2002)
9. Mitton, N., Fleury, E., Lassous, I., Tixeuil, S.: Selfstabilization in self-organized multihop wireless networks. In: Proceedings of the IEEE International Workshop on Wireless Ad-Hoc Networking (ICDCSW 2005), vol. 9, pp. 909–915 (2005)
10. Hasan, M.M., Jue, J.P.: Survivable Self-Organization for Prolonged Lifetime in Wireless Sensor Network. International Journal of Distributed Sensor Networks 2011, Article ID 257156 (2011)

Cellular Automata Approach for Spectrum Sensing in Energy Efficient Sensor Network Aided Cognitive Radio

Jaison Jacob[1], Babita R. Jose[2], and Jimson Mathew[1]

[1] Dept. of ECE.,Rajagiri School of Engineering and Technology, Kochi, India
jaison_jacob@rajagiritech.ac.in
[2] Division of Electronics Engineering, SOE, Cochin University of Science
and Technology, Kochi, India
babitajose@cusat.ac.in

Abstract. Efficient spectrum sensing is an important requirement for
the success of the cognitive radio (CR) system. A novel spectrum sensing
approach through external sensing is proposed here. In external sensing,
an external agent performs the sensing and broadcasts the channel occupancy information to SUs. A large number of low cost and energy
efficient wireless sensors can be deployed in the field, to sense the spectrum continuously or periodically. Individual sensing result can be send
to the central node (CN) for final decision making through the sensor
network. Since we are looking for energy efficient and low cost network
installation, individual sensing results are expected to get affected by
noise, fading, and shadowing. In this paper we employed a Cellular Automata (CA) based approach at the CN to obtain spectrum status and
the correct coverage region of a Primary User (PU). This method requires
less number of computations compared to existing fusion rules that are
proposed for cooperative spectrum sensing. Impact of sensor density on
the percentage false positive and false negative are been carried out. It
was also found that with CA based approach the CN can calculate the
coverage region of a PU at any point of time accurately with minimum
computational effort.

Keywords: Cellular automata, wireless sensor networks, central node.

1 Introduction

Future mobile terminals will be able to communicate with various heterogeneous
systems which are different by means of the algorithms used to implement base
band processing and channel coding [1]. Current research is investigating different techniques of using cognitive radio (CR) to reuse locally unused spectrum
to increase the total system capacity. The biggest challenge related to spectrum sensing is in developing sensing techniques which are able to detect very
weak primary user signals while being sufficiently fast and low cost to implement [2-4]. Local spectrum sensing has some limitations and it is hard to detect

J. Mathew et al. (Eds.): ICECCS 2012, CCIS 305, pp. 54–61, 2012.

signals of low SNR for desired performance. Cooperative spectrum sensing is proposed in literatures as a solution to issues that arise in spectrum sensing due to noise, fading, shadowing and hidden terminals [5],[8]. Even though the distributed sensing reduces the errors in the individual sensing, extra hardware is required by the SU to fully exploit the spectrum opportunity. Additional management cost and communication overheads will be there as different SUs work under different organizations. Information security and sharing are also become challenges[11]. Another technique for obtaining spectrum information is external sensing. In external sensing, an external agent performs the sensing and broadcasts the channel occupancy information to SUs[5]. A large number of low cost and well designed wireless sensors can be deployed in the field to sense the spectrum continuously or periodically. The sensor results are communicated to a CN and it will be processed. Now instead of spending time to sense the spectrum hole SUs can obtain it from the agent. In this paper we explore the possibility of using CA in processing the sensor information at the CN, to obtain the spectrum status and the coverage area of a PU. The coverage area of a PU need not be perfect circle in practical situations. Contours of a constant received power based on a fixed transmit power at the base station form an amoeba like shape due to the random shadowing variations about the path loss [6]. The organization of the paper is as follows: In Section 2, the system model is described. Section 3 describes the proposed information combining strategy for the CN. The simulation results are presented in Section 4 with conclusions in Section 5.

2   System Model

We consider a scenario that an external agency is providing the spectrum hole information to the SUs. The external agency obtains this information through the wireless sensors deployed in the field. We assume that proper architecture is formed to deploy sensors in the field and necessary networks and protocols are available to transfer all the sensed data to a CN. After processing the data, CN will have the in-formation about the channel availability and coverage region of a particular PU. We also assume that these low cost and energy efficient sensors will undergo fading and random shadowing. Hence the sensor output may not be correct always. A simulation set up was formed in Matlab to model and evaluate the system. We consider the sensors are arranged in a 2-D grid within an area of 100 sq-km. And the transmitter is located at the center. Transmit power is chosen such that sensors will be present within and outside the coverage area of the transmitter. All the sensing results are transferred to a central node through the network. At the CN the data is processed with the proposed method to obtain channel availability and coverage region. Now this can be broadcasted or it can be provided on demand. Our estimations are based on the practical link budget design using path loss models [7]. The cell coverage area in a cellular system is defined as the expected percentage of locations within a cell where the received power at these locations is above a given minimum. The transmit

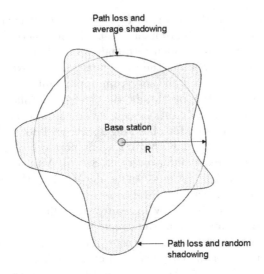

Fig. 1. Cell coverage area

power at the base station is designed for an average received power PR at the cell boundary. However shadowing will cause some locations within the cell to have received power below P_R, and others will have received power exceeding P_R [6]. This is illustrated in Fig.1.

Propagation path loss according to distance from the transmitter was defined according to the equation (1) as mentioned in [8].

$$PL_{dB} = 10 n log_{10} d + 20 log_{10}\left(\frac{4\pi}{\lambda}\right) + X_{dB} \qquad (1)$$

Where $PL_{(dB)}$ = Path loss in dB, n = Path loss exponent , d = Distance from transmitter in meters, λ = Wavelength of transmitted signal in meters, $X_{(dB)}$ = Shadowing factor. Thus the received power Pr (in dBW) at a receiver at a distance d meters from a transmitter with transmit power Pt (in dBW) will be:

$$P_{r(dBW)} = P_{t(dBW)} - PL_{(dB)} + G_{r(dBi)} \qquad (2)$$

Where Gr represents the receiver antenna gain (typically 2dBi was used). This value of received power for each receiver is used in the sensing calculations. The parameter X in equation 1 is a factor which models the shadow fading effects. The value of X depends on its standard deviation σ (quoted in dB, since X is log-normal in distribution), which typically varies between 6 and 10dB across different environments. Since we have considered the dense urban scenario the values of n and σ considered are 4.5 and 10 respectively[7][8]. This approach is verified in [12].We have considered the standard mobile communication systems (GSM 900MHz band) used in India.

3   Information Combining

In the case of external sensing the CN should have the clear information about the spectrum holes in the time -frequency space and geographical space. Spectrum space in the time frequency space can be obtained through simple fusion rules at the central node. Since we consider a larger area in the case of external sensing and the transmit power of PUs may vary from low to high and there will be spectrum holes in the geographical space. Information fusion process at the central node should have the capability to obtain the presence of PU and the coverage area of each PU. We expect same channel is used by different PUs at different geographical space.

From the literatures it is seen that weighted combining of the single node results gives acceptable results [8]. Also the number of neighboring nodes considered will have impact on the result. As the number of neighbors increases the result is also become better. Hence we need more computational power to get a better result. In this paper we propose a Cellular automata (CA) approach for the information combining which gives a better result with less computation.

3.1   Basic Definitions of CA

Let I denote the set of integers. A cellular space (I x I, V, v0, f) includes the set I x I, V is the set of cellular states, v0 is a distinguished element of V called the quiescent state, and f is the local transition function from n-tuples of elements of V into V. [9] [10]. Here each cell changes it state based on the transition function f based on the state of its neighbors.

3.2   CA Based Combining

Here we assume that the sensors are arranged in a 2-D grid and the sensing result will form the cellular space and the possible states are either Yes or No. In the process of information combining, each sensor result is considered as the cell state. And each cell will make a transition from the present state to the next state according the state of its neighbors based on the local transition function. There will not be any transition if the condition is not satisfied. Some of the rules for the decision making is given in the figure 2. The central pixel will go to its state 1 if each cell and its neighbor hood is same as the mask. Here white indicates state 1(sensed) and black indicates the state 0(not sensed). This can be applied repeatedly until no further change is happens to the Cellular space or to a specific number of times.

4   Simulation Results

In this section, we present the simulation results demonstrating the performance of the proposed CA based system for information combining to obtain the proper coverage region of a PU. We have considered the standard mobile communication

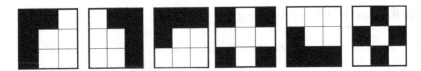

Fig. 2. Some of the CA-Based fusion Rules

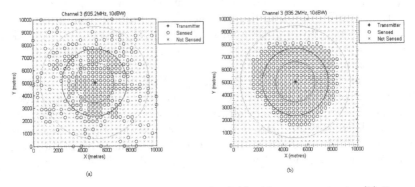

Fig. 3. (a) Wireless sensor distribution in the field with sensor outputs, (b) Processed data after information combining process at central node

systems (GSM 900MHz band) used in India. Transmitters with various power levels have been considered for simulation. Wireless sensors are considered to have installed in a 2-D grid within an area of 100 sq-km. as shown in the figure 3(a). Sensors with positive results(sensed) are marked 'o' and not sensed is marked as 'x'.

Since we are looking for energy efficient and low cost network installation we expect that these sensors will undergo the fading and random shadowing. Hence the coverage area of a Primary User(PU) need not be perfect circle. Contours of a constant received power based on a fixed transmit power at the base station form an amoeba like shape due to the random shadowing variations about the path loss [6].

Fig.3(b) shows that the result after information combining is in line with the expectation in [6]. The central circle indicates the coverage region based on path loss and average shadowing against the minimum received power requirement of noise floor. The other two circles are named as the outer fade boundary and the inner fade boundary from equation (1). Outer and Inner fade regions are calculated based on standard deviation σ of the shadowing which typically varies between 6dB and 10dB across different environments.

Sensing results are evaluated based on the sensors located outside the outer fade region and the sensors located inside the average coverage region. Positive sensing result of the sensors which are located outside the outer fade region is considered as false positive. Negative sensing result of the sensors which are located inside the average coverage region is considered as false negative. Percentage error is calculated based on the false positive or false negative and the total number of sensors located in the respective region.

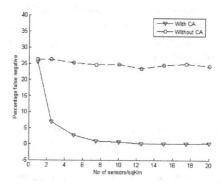

Fig. 4. Impact of Number of sensors/sq-km on the percentage false negative

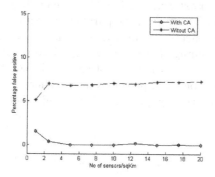

Fig. 5. Impact of Number of sensors/sq-km on the false positive

Analysis on the percentage error verses number of receivers are shown in fig.4. It has been formed by taking the average of 25 readings in each sensor density. It is carried out for three different transmit power and the result has no major change with the change in power. Fig.4 shows that false negative at the range of 25% is observed in the sensor result because of the shadowing. And sensors located inside the inner fade boundary are also giving false negative. After the CA based information combining false negative is almost zero within the inner fade boundary. Since we are considering sensors located inside the coverage region for calculating the false negative, we have some amount of false negative within this range. As per fig.1 these false negatives may be neglected. On an average 8-12 sensors are required within a square kilometer to have an acceptable performance.

Fig 5 shows that about 7% of the sensors located outside the outer fade boundary are giving false positive. After the CA based information combining false positives are totally get eliminated. On an average 2-5 sensors are required within a square kilometer to have an acceptable performance. It is better to have more number of sensors to obtain accurate coverage region. In the case of cooperative spectrum sensing weighted combining of the neighbor's results is performed to have acceptable results in a dense urban scenario. If we compare the computational requirement of this algorithm number of neighbors considered will be less in the case of CA. Since there is no weighted combining we have a saving of N (No. of neighbors considered (N)) multiplications. CA can be implemented using simple additions and comparisons.

5   Conclusion

In this paper an external sensing using wireless sensor networks for cognitive radio is being considered. We look for an energy efficient and low cost system for spectrum sensing. Since the number of SUs expected in the field is high the task of spectrum sensing may be given to an external agency. It will result in large savings in energy and cost. Also the battery of such mobile SUs may get a longer life. Here a simple and computationally efficient method for information combining is presented for external sensing. CA based information combining can be implemented very effectively with the parallel processing approach. Coverage region of a PU can be effectively formed even in the scenario of fading and random shadowing. Percentage false positive and false negative can be reduced to the minimum even with less number of sensors in the field. Also we have significant saving in the processing power required. There is a scope for exploring better CA rules to reduce the computational cost and number of sensors require in the field.

References

1. Menouni Hayar, A., Knopp, R., Pacalet, R.: Cognitive radio Research and Implementation Challenges, Mobile Communications Laboratory Institute, Eurécom, Sophia Antipolis, France
2. Kang, X., Liang, Y.C., Nallanathan, A.: Optimal power allocation for fading channels in cognitive radio networks under transmit and interference power constraints. In: Proc. IEEE International Conf. on Communications (ICC), Beijing, China (May 2008)
3. Zhang, L., Xin, Y., Liang, Y.C.: Weighted Sum Rate Optimization for Cognitve Radio MIMO Broadcast Channels. IEEE Trans. on Wireless Communications, 2950–2959 (June 2009)
4. Hoang, A.T., Liang, Y.C., Zeng, Y.H.: Adaptive Joint Scheduling of Spectrum Sensing and Data Transmission in Cognitive Radio Networks. IEEE Trans. on Communications, 235–246 (January 2010)
5. Yücek, T., Arslan, H.: A Survey of Spectrum Sensing Algorithms for Cognitive Radio Applications. IEEE Communications Surveys and Tutorials 11(1) (first quarter 2009)
6. Goldsmith, A.: Wireless Communications, ch. 2, p. 53. Cambridge University press (2005)
7. Rappaport, T.S.: Mobile Radio Propagation: Large Scale Path Loss. In: Wireless Communications: Principles and Practice. Prentice-Hall (1999)
8. Harrold, T.J., Faris, P.C., Beach, M.A.: Distributed Spectrum Detection Algorithms For Cognitive Radio, Centre for Communications Research, University of Bristol, United Kingdom

9. Cattell, K., Zhang, S., Serra, M., Muzio, J.C.: 2-by-n Hybrid Cellular Automata with Regular Configuration: Theory and Application. IEEE Transaction on Computers 48(3) (March 1999)
10. Jacob, J., Abraham Chandy, D.: Image edge detection using cellular automata. In: Proceedings of the ISCO 2006, Coimbatore, India, August 9-11 (2006)
11. Gao, M., Cheng, L., Liu, Y., Ni, L.: SCAS: Sensing Channel ASsignment for Spectrum Sensing Using Dedicated Wireless Sensor Networks. In: 16th International Conference on Parallel and Distributed Systems (2010)
12. Jacob, J., Panicker, A., Mathew, J., Vinod, A.P.: Exploration of a Distributed Approach for Simulating Spectrum Sensing in Cognitive Radio". In: Proc. of International Conference on Communication and Signal Processing, ICCSP 2011, Calicut, India (February 2011)

Economic Analysis of a Biomass/PV/Diesel Autonomous Power Plant

Subhadeep Bhattacharjee and Anindita Dey

Department of Electrical Engineering
National Institute of Technology (NIT), Agartala India-799055
subhadeep_bhattacharjee@yahoo.co.in, anindita.dey7@gmail.com

Abstract. This paper presents the design and optimization process of a stand-alone autonomous power system, for supplying continuous power in typical rice-mills of Tripura, together with an economic analysis and environmental considerations for the project life cycle. The simulation results indicate that the renewable fraction is 0.62 for the hybrid system composed of 15 kW PV, 7 kW biomass gasifier, 8 kW diesel system with battery storage. The cost of energy is found to be 0.803$/kWh.

Keywords: Hybrid energy technology, net present cost, levelized cost of energy, capital cost, renewable fraction, sensitivity analysis.

1 Introduction

Providing energy for a community, in a sustainable manner, nowadays has become a more and more important issue as we face global warming and climate change realities. Harnessing renewable energy sources which are abundantly available in nature provides an opportunity to produce energy in an environmentally friendly way [1]. These alternative sources of energy include wind, solar, biomass, geothermal, tidal and wave. The applications of these sources include the very small to large isolated, grid connected and hybrid power systems. Moreover the hybrid power systems exhibit higher reliability and lower cost of generation than those that use only one source of energy. A power generating system which combines two or more different sources of energy is called a hybrid system [2]. Of the renewable, clean and inexhaustible sources of energy, biomass power is catching the attention of engineers, environmentalists and financiers these days [3]. Biomass is organic, plant derived material that may be converted into other forms of energy [4]. This paper presents a design process, economic analysis and environmental considerations of a stand-alone hybrid power system based on solar and biomass energy for providing electricity in a typical rice mills of Tripura (one of the north-east states of India). National Renewable Energy Laboratory (NREL)'s, Hybrid Optimization Model for Electrical Renewable has been used as the sizing and optimization software tool. Sensitivity analysis with biomass price and diesel fuel price has been done. The economic issues analysed are the initial capital cost needed, the fuel consumption and operating cost, the total net present cost(NPC), the cost of electricity(COE) generated by the system per kWh for the project. The environmental considerations discussed are the amount of gas emissions, such as CO_2 and NO_x as well as particulate matter released into the atmosphere.

J. Mathew et al. (Eds.): ICECCS 2012, CCIS 305, pp. 62–68, 2012.

2   System Description

2.1   Methodology

The methodology includes simulation, optimization and sensitivity analysis process. The simulation process determines how a particular system configuration, a combination of system components of specific sizes, and an operating strategy that defines how those components work together, would behave in a given setting over a long period of time. In the optimization process, it simulates many different system configurations, discards the infeasible ones (those that do not satisfy the user-specified constraints), ranks the feasible ones according to total net present cost, and presents the feasible one with the lowest total net present cost as the optimal system configuration.

2.2   Resources Utilized

Biomass Resource
Rice husk is considered as biomass resource for gasifier. In Tripura, there are many scattered rice mills which are operated with grid electricity. There are no prominent applications of rice husk in rice mills of the state. In a typical rice mill of Tripura, available paddy in a day is 440 kg. So the rice husk production is 22% of the paddy= 0.22*440=96.8 kg. Immature paddy production is 3% of the paddy=0.03*440=13.2 kg. Total rice husk and immature paddy= 96.8+13.2=110 kg. So the Total biomass input- 0.11 tonne/day.

Solar Resource
The monthly average solar radiation data, has been collected from TREDA (Tripura Renewable Energy Development Agency). The annual average solar radiation is 5.025 kWh/m^2/day.

2.3   Electrical Load Input

Generally the rice mills of Tripura are operated with the help of 10-15 HP motor. There are also some other electric appliances. A rice mill of Tripura may consume around 91 kWh/day with a peak demand of nearly 25 kW over a year.

2.4   An Optimal Design Approach

The Fig. 1 below shows the propose scheme as implemented in the simulation tool. In this system, the renewable solar and biomass are taken as the primary sources and the diesel generator as a backup. Since the performance of the hybrid energy system is highly dependent on the environmental conditions, the battery is used as storage system. The above system operates under cycle charging strategy, means whenever a generator has to operate, it operates at full capacity with surplus power going to charge the battery bank. Excess power is used to charge the battery bank, so as to minimize the dumped energy.

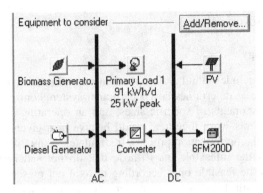

Fig. 1. The proposed scheme as implemented in simulation tool

3 Results and Discussions

Optimization analysis was carried out to arrive at the best possible sizing configurations. After simulating all of the possible system configurations, it displays a list of feasible systems, sorted by lifecycle cost. The optimization result is shown in Table 1.

This most optimized system includes 15 kW PV array, one 10 kW diesel generator, total 152 number of battery and a 20 kW converter. The initial capital cost, total net present cost (NPC), and cost of energy (COE) for such a system are $164,928, $335,559 and 0.793$/kWh, respectively for one year. The operating cost of the system is 13,348$/yr. Total diesel consumption of the system is 5,361L. Renewable fraction is 0.56, and the diesel generator will operate for total 1,648 hour over one year period.

Table 1. Optimization result

Double click on a system below for simulation results.   ● Categorized ○ Overall   Export...   Details...

PV (kW)	Gen 1 (kW)	Gen 2 (kW)	6FM200D	Conv. (kW)	Initial Capital	Operating Cost ($/yr)	Total NPC	COE ($/kWh)	Ren. Frac.	Capacity Shortage	Diesel (L)
15		10	152	20	$ 164,928	13,348	$ 335,559	0.793	0.56	0.00	5,361
15	7	8	152	20	$ 171,348	13,182	$ 339,862	0.803	0.62	0.00	4,534
15		8	152	20	$ 168,408	15,005	$ 360,228	0.851	1.00	0.00	
15	7	7		20	$ 43,320	26,320	$ 379,781	0.903	0.62	0.02	8,771
	15		152	20	$ 149,428	22,710	$ 439,741	1.039	0.00	0.00	12,054
	7	15	152	20	$ 156,848	22,736	$ 447,487	1.057	0.15	0.00	10,083
	10	7		20	$ 28,500	34,376	$ 467,940	1.118	0.48	0.03	10,182
15		14		20	$ 39,400	40,793	$ 560,866	1.333	0.34	0.02	18,297

The second optimized system utilizing biomass resource is consists of 15 kW PV array, 7 kW biomass generator, 8 kW diesel generator, total 152 number of battery and a 20 kW converter. The initial capital cost, Total net present cost (NPC), and cost of energy (COE) for such a system are $171,348, $339,862 and 0.803$/kWh. The operating cost of the system is 13,182$/yr. In this case diesel replacement is 62%, the diesel consumed is 4,534L, the biomass usage is 3 tonne. The biomass generator operating hour is 308 and the diesel generator will operate for total 1,732 hour over one year period.

The third optimized model utilizing only renewable energy sources includes 15 kW PV array, 8 kW biomass generator, total 152 number of battery and a 20 kW converter. The operating cost of the system is 15,005$/yr. In this case the renewable fraction is 1.

In the second most optimized model (i.e. PV/biomass/diesel), 56% of the electricity demand is produced from solar panels with 20,491 kWh/yr, while 6% of the energy requirement is supplied from biomass with 2,156 kWh/yr. 38% of energy requirement is met from diesel generator with 13,704 kWh/yr. The total annual electricity production is 36,352 kWh/yr. Fig.2 shows the contribution of electricity production from different sources.

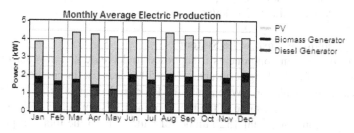

Fig. 2. Monthly average electrical production rates of the stand alone system

The total annual electric energy consumption in PV/biomass/diesel system is 33,105 kWh/yr. Excess electricity production is 157 kWh/yr. Unmet load is very less, there is no capacity shortage.The cash flow summary of the system is shown in Fig. 3. Each bar in the graph represents either a total inflow or total outflow of cash for a single year.

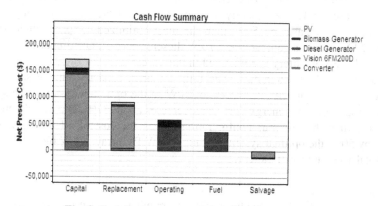

Fig. 3. Cash flow summary of the stand alone system

The rated capacity of the PV array is 15 kW. Mean output is 2.34 kW, capacity factor is 15.6%, PV penetration is 61.9%, hours of operation is 4380hr/yr as calculated by it.

Capacity factor of biomass generator is 3.52%, hours of operation is 308 hr/yr, operational life is 48.7 yr, fixed generation cost is 3.90 $/hr, minimum electrical output is 7 kW, biomass feedstock consumption is 3.39 t/yr, mean electrical efficiency is 59.5%

Capacity factor of the diesel generator is 19.6%, operational life is 8.66 yr, fixed generation cost is 2.60 $/hr, mean electrical output is 7.91 kW, mean electrical efficiency is 30.7%.

Nominal capacity of the battery is 365 kWh, autonomy is 57.9 hr, lifetime throughput is 139,384 kWh, energy in is 5,722 kWh/yr, energy out is 4,648 kWh/yr, losses are 994 kWh/yr.

Total carbon dioxide emission is 11,941 kg/yr, carbon monoxide is 29.5 kg/yr, unburned hydrocarbon is 3.27 kg/yr, particulate matter is 2.22kg/yr, sulphur dioxide is 24.4 kg/yr, nitrogen oxide is 263 kg/yr.

Effect of Component Cost to the System

The cost of the system components is supposed to decrease and in order to simulate those conditions for the long term analysis; a 50% decrease in component costs has been included to calculation. When only the PV cost decreases by 50%, the capacity of the components remain same in the most optimum system i.e. 15 kW PV array, 10 kW diesel generator and total 152 number of battery. Both the total net present cost and cost of energy decreases by 3% in the optimum system. The initial capital cost decreases by 5.8%.The operating cost decreases by 0.44%, but the diesel fuel consumption and the operating hour of the generator remains same.

In case of 50% decrease in biomass generator cost only, the most optimized system becomes a solar/biomass/battery energy based system consisting of 15 kW PV array, 8 kW biomass generator and total 152 number of battery. The initial capital cost and operating cost decreases by 0.46% and 30% respectively in the optimum system. Both of the net present cost and cost of energy decreases by 13.6% respectively. Renewable fraction increases to 1, consequently the emission will decrease. When only cost of diesel generator decreases by 50%, then the most optimized system include 15 kW PV array, 7 kW biomass generator and 7 kW diesel generator. One important observation is that the use of battery is eliminated in this case.

In case of 50% decrease in battery cost, the PV array capacity remains same while the diesel generator capacity decreases to 7 kW in the optimum system. The system total net present cost and energy cost also decreases about 37%. The diesel generator works more than the previous configuration. When the cost of all the components decrease by 50%, the optimal system include 15 kW PV array, 8 kW biomass generator and total 152 number of battery.

Sensitivity Analysis

We can perform a sensitivity analysis with any number of sensitivity variables. Each combination of sensitivity variable values defines a distinct sensitivity case. In this analysis five values for biomass price and seven values for diesel fuel price are considered that defines 35 distinct sensitivity cases. It performs a separate optimization process for each sensitivity case and presents the results in various tabular and graphic formats. Totally 980 systems were simulated for 35 cases. Total simulation time was 12 min.

The optimization results in graphical form are shown in Fig. 4. Here various optimal system types (OST) are displayed as function of biomass price and diesel price. This allows identification of system configuration for the typical rice mill at Tripura. Considering the present biomass price (0$/t) and diesel price (0.6$/L), a solar/diesel/battery based hybrid system is suitable for the particular location.

Now consider, the biomass price is fixed at 0$/t (as it is waste product in rice mills) and the diesel price is elevated to 0.9$/t, a solar/biomass/diesel/battery based hybrid system becomes optimal for the particular location. If we further increase the diesel price to 1.2$/L, a solar/biomass/battery based hybrid power system becomes optimal. So the system is affected by the price of diesel fuel, not by the biomass price which is evident from Fig.6.

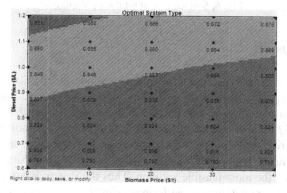

Fig. 4. Optimal system type with biomass price on x-axis and diesel price on y-axis

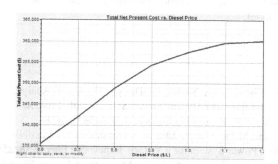

Fig. 5. Total net present cost vs. diesel price

The Fig.5 shows the variation of system total net present cost against diesel price. It can be seen from the figure that, total net present cost increases with the increase of diesel fuel price.

The Fig.6 shows the variation of total net present cost with the variation of both of the biomass price and diesel price. Here the total net present cost is plotted on the surface of the graph to see the effect of fuel price on it. The graph shows that with the increase of fuel price, the total net present cost is increasing.

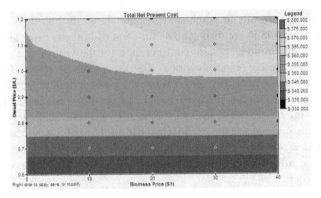

Fig. 6. Surface plot of total net present cost

4   Conclusion

A stand alone hybrid energy system model has been designed utilizing solar and biomass energy. The system is developed using meteorological and load data collected from the site. At present, a 15 kW PV, 7 kW biomass generator, 8 kW diesel generator and total 152 number of battery based hybrid system is the most feasible solution for harnessing rice husk potential for stand-alone hybrid power applications in rice mills. Above diesel fuel price 0.8$/L, a solar/biomass/diesel/battery based hybrid system would be the most feasible solution. Thus the system is affected by the price of diesel fuel. A simulation was performed with the decrease component prices in order to foresee the long term results. It has been observed that in case of 50% decrease in the gasifier cost, solar/biomass/battery based hybrid system becomes the most optimized model.

References

1. Setiawan, A.A., Chem, Y.Z., Nayar, V.: Design, Economic Analysis and Environmental Considerations of Mini-Grid Hybrid Power System with Reverse Osmosis Desalination Plant For Remote Areas. Renewable Energy 34, 374–383 (2009)
2. Rehman, S., Alam, M.M., Meyer, J.P., Al-Hadhrami, L.M.: Feasibility Study of a Wind-PV-Diesel Hybrid Power System for a Village. Renewable Energy 38, 258–268 (2012)
3. Rehmana, S., El-Amin, I.M., Ahmada, F., Shaahida, S.M., Al-Shehrib, A.M., Bakhashwainb, J.M., Shash, A.: Feasibility Study of Hybrid Retrofits to an Isolated Off-Grid Diesel Power Plant. Renewable Energy 11, 635–653 (2007)
4. Evans, A., Strezov, V., Evans, T.J.: Sustainability Considerations for Electricity Generation from Biomass. Renewable Energy 14, 1419–1427 (2010)
5. Turkay, B.E., Telli, A.Y.: Economic Analysis of Stand Alone and Gird Connected Hybrid Energy Systems. Renewable Energy 36, 1931–1943 (2011)

A Slope Compensated Current Mode Controlled Boost Converter

K.G. Remya[1], Chikku Abraham[1], and Babita R. Jose[2]

[1] Rajagiri School of Engineering and Technology, Electrical and Electronics Dept,
Rajagiri Valley, Kakkanad, India
[2] Cochin University of Science and Technology,
Electronics Dept, Cochin, Kerala, India
kgremya@gmail.com, chikkuabraham@yahoo.co.in

Abstract. This paper discusses the design and simulation of Slope compensated Current mode Controlled Boost converter. A boost converter steps up a source voltage to a higher, regulated voltage. The current mode control strategy is discussed in this paper wherein the peak inductor current is sensed and used to control the duty ratio of the switches such that the output voltage is regulated within the specified tolerance band. Simulation studies have been carried out on a 7.5 watt boost converter designed with its power circuit and IC UC 3844 is modelled to implement the current mode control of output voltage maintained at 30V irrespective of the fluctuations in input voltage. An instability is observed at duty ratio's nearing 0.5 resulting in sub harmonic oscillations. This issue is addressed by proper slope compensation. The entire circuit is designed and modelled using the power electronic software $PSIM^{©}$.

Keywords: Current mode control, Pulse width modulation, Slope compensation, Sub harmonic oscillation, Boost Converter.

1 Introduction

DC to DC converters are inevitable in almost all battery powered devices such as laptop computers, mobile phones etc. The issue of using multiple voltage sources can be dealt in a better way by using DC to DC converters[6]. Portable electronic devices with multiple batteries increases size and complexity of the device[7]. Another problem associated with usage of battery is the declining voltage of a battery as its stored power is drained, so it does not output a constant voltage level. DC to DC converters thus offer a method of generating multiple controlled voltages from a single battery voltage. The BOOST converter or the step up converter is the one which is used in regulated dc power supplies and the regenerative braking of dc motors. As the name indicates the output voltage is always greater than input voltage. In dc-dc converters, the average dc output voltage must be controlled to equal a desired level, though the input voltage and the output load may fluctuate. Switch mode dc-dc converters utilize

J. Mathew et al. (Eds.): ICECCS 2012, CCIS 305, pp. 69–76, 2012.

one or more switches to transform dc from one level to another. So the average dc output voltage may be controlled by controlling the 'on' period and 'off' period of these switches. The control strategy used may be direct duty ratio control, PWM control, voltage feed forward control or current mode control. Voltage mode control strategy is the conventional control strategy. In voltage mode control scheme a portion of output voltage is sensed and is used to regulate the output voltage[2]. There is no inherent line voltage regulation rather only load voltage regulation. Additional voltage regulation circuits are required to ensure line voltage regulation. Current mode control is the industry standard method of controlling switching power supplies where there is an inherent line voltage regulation[5]. However conventional circuits with current mode control suffer from instability phenomenon at duty ratio's nearing 0.5[12].

1.1 Boost Converter

A boost converter is a particular type of power converter with an output DC voltage greater than the input DC voltage. The specific connections are shown in Fig. 1. with an additional current sensing resistor and a MOSFET drive. The circuit can be examined for two cases switch open and switch closed. The case is referred to as 'continuous mode operation' when inductor current is continuous. Applying Kirchhoff's rules around the loops and rearranging terms yields an intuitive result [2]:

$$\frac{V_o}{V_{in}} = \frac{1}{1-D} \tag{1}$$

In some cases, the inductor current may become discontinuous and this mode is called 'discontinuous mode' of working. The focus of this paper is on continuous conduction mode.

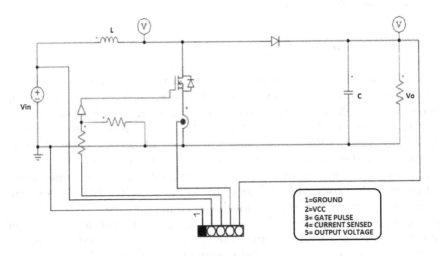

| 1=GROUND |
| 2=VCC |
| 3= GATE PULSE |
| 4= CURRENT SENSED |
| 5= OUTPUT VOLTAGE |

Fig. 1. Boost converter circuit modelled in PSIM©

2   Current Mode Control

2.1   Current Mode Control

Current mode control is one of control strategies used in boost converters to maintain the output voltage at the desired level. Current mode control as usually implemented in switching power supplies actually senses and controls peak inductor current thereby controlling the output voltage. There is an inner current loop and outer voltage loop. Fig. 2(b) is a block diagram of the concept. The net result of the two approaches voltage mode control or current mode control is the same, to regulate voltage, but this latter approach is called 'current-mode programming' since the load current is the directly controlled variable and the output voltage is controlled only indirectly. The main advantage of this method is that the output current can be limited simply by clamping the control voltage. It allows a modular design of power supplies. Another advantage is that the energy storage inductor is effectively absorbed into the current source. For higher power applications, power stages can be connected in parallel[2],[4] and [6]. A final advantage is automatic feed forward from the line voltage.

2.2   Disadvantage of Current Mode Control – Slope Compensation

The hardest part of current-mode control is measuring the current accurately and with the required bandwidth. The peak current mode control method is inherently unstable at duty ratios exceeding 0.5, resulting in sub-harmonic oscillation. Fig. 2(a) shows the sub-harmonic oscillations that may be present. A compensating ramp (with slope equal to the inductor current down slope) is usually applied to the comparator input to eliminate this instability as shown in Fig. 2(c). When the duty ratio is less than 0.5 the control is stable. While the system is stable for duty ratio less than 0.5, without a compensating ramp, even in these situations the best possible transient response is obtained only if a compensating ramp of correct slope is used.

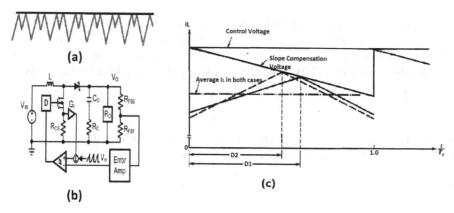

Fig. 2. (a)Subharmonic Oscillations (b)Slope compensated Boost Converter and (c)Slope compensated Inductor Current

3   Current Mode Control of Boost Converter-Design and Set Up

3.1   Circuit Design and Software Used

The simulation model was made with a boost converter having an input voltage ranging from 12V to 20V and an output that has to be stabilized at 30V and 0.25A, supporting a load of 7.5W. The current mode control strategy is implemented with current mode control IC UC 3844 as shown in Fig. 3. The UC3844, UC3845 series have good performance characteristics. They work at fixed frequency and function as current mode controllers. The inductor is designed for a ripple current of 0.5A. Assuming that inductor current rises linearly during the interval 't_{on}' and falls linearly during the interval 't_{off}' ripple current 'ΔI' is given by

$$\Delta I = \frac{(V_{in}t_{on})}{L} = \frac{(V_o - V_{in})t_{off}}{L} \tag{2}$$

And for a frequency 'f' and total time period 'T'

$$L = \frac{V_{in}(V_o - V_{in})}{f * V_o * \Delta I} \tag{3}$$

Thus equation (3) yields an inductor value L to be 200μ H. The power MOSFET has to carry about 1A and block about 20V. MOSFET drive is through the Rg and Rd. The Diode carries about 0.5A average current and blocks about 20V and is suitable for 50 KHz switching. The output capacitor C has to limit the voltage ripple to about 6 percent. Capacitor voltage ripple is given by

$$\Delta V_c = \frac{1}{c} \int_o^{t_{on}} I_o dt = \frac{I_o t_{on}}{C} \tag{4}$$

For a duty ration D

$$C = \frac{I_o D}{f \Delta V} \tag{5}$$

The controller used here is UC3844 which is a 8 pin IC with maximum supply voltage of 40V. An RC shunt is used for filtering. IC has an internal reference voltage of 5V. Oscillator section consisting of R_T and C_T is designed for a desired switching frequency of 50 KHz. For the IC UC3844 the switching frequency fs is given by[3]

$$f_s = \frac{1.72}{R_T * C_T} \tag{6}$$

Resistor R_f and Capacitor C_f form the input to the error amplifier within the IC. With an internal supply voltage of 5V and Isource=0.5mA,

$$R_f = \frac{V}{I_{source}} = 10K \tag{7}$$

Slope compensation is provided with the slope compensation circuit formed with a transistor and a resistor generating ramp waveform. There is a current sense

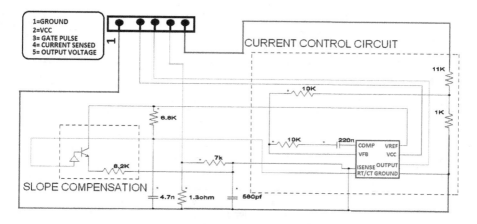

Fig. 3. Model circuit incorporating current control IC UC3844 developed in PSIM©

pin for the IC. The current sense pin senses the current from the power circuit and the slope compensation is made along with it as shown in Fig. 1 and Fig. 3. The bleeder resistance R_b ensures that the internal threshold voltage is limited to 1Volt. The calculation of bleeder resistance value follows from the equation (8).

$$R_b = \frac{1}{I_{speak}} \tag{8}$$

4 Simulation Results

The power circuit and the control circuit simulated as per the design using PSIM software gave satisfactory results. As shown in the simulation results in Fig. 4 the average output voltage remains within the tolerance limits of set value 30V even if input voltage fluctuates between 12V and 20V. There is a transient state and steady state in the actual output voltage. When input is 12V during the initial transient state the output voltage rises to 32V but the transient state exists only for a short duration of 0.01seconds and then it settles down to an average value of 30V. When input is 20V there is an initial transient state when output rises to a maximum value of 50V existing for a very short duration of 0.005seconds and then settles down on 30V. Thus even though there is an initial transient the output eventually settles on desired output voltage. The significance of slope compensation compared to that without slope compensation is the instability phenomenon that exists and this is illustrated with Fig. 6(a) and (b) respectively. As shown in Fig. 6(a) the actual output voltage without slope compensation exhibit spikes. These are significant at duty ratio near 0.6 as shown in Fig. 6(a). The spikes even though less significant can be seen in the sectional view of the output at duty ratio of 0.3 as shown in Fig. 6(b). There is a voltage ripple in the actual output of the converter. Efforts are being made to reduce this ripple.

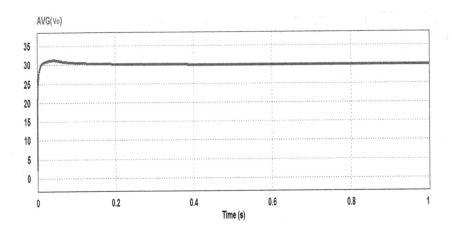

Fig. 4. Output voltage variation with time

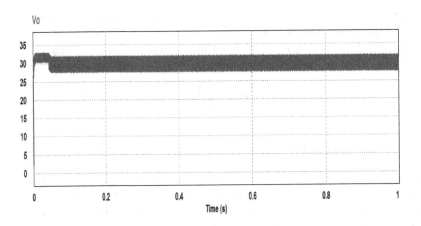

Fig. 5. Variation of actual output voltage with time with slope compensation at a duty ratio of 0.6

Comparison of the results show that subharmonic oscillations or visible disturbances in the output waveform pose a problem in the current mode controlled boost converter circuit without slope compensation. An enlarged view of the output voltage of the circuit without slope compensation is shown in Fig. 6(b). It shows that the output voltages have spikes which are the result of oscillations in the peak current. These oscillations if left unchecked will grow in amplitude with several cycles and may lead to severe problems. As can be seen in Fig. 5 these oscillations have been compensated in slope compensated system. Even in a slope compensated systems oscillations do exist but these oscillations are effectively damped out by injecting suitable slope compensation. Thus noise immunity as well as stability of the circuit is improved making the slope compensated current mode controlled boost converter an optimal choice in most applications.

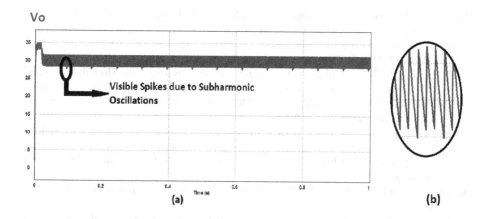

Fig. 6. (a)Variation of actual output voltage with time without slope compensation at a duty ratio of 0.6 (b)sectional view of output voltage at duty ratio of 0.3 and without slope compensation

5 Conclusion

As the simulation results make it clear the current mode control strategy used can keep the output voltage constant at 30V (.25A) inspite of the variations in input voltage. The instability phenomenon which exists naturally at duty ratio near 0.5 is compensated with the addition of compensating ramp. This method known as slope compensation of current mode controlled boost converter has been discussed in this paper. A current controlled boost converter of this type can have several applications like wind mills, solar panels and to increase the efficiency of a battery.

References

1. Raju, N., Mariappan, V., Jeya Prakash, K., Sathya, A.: Teaching Aids in SMPC Construction Project in the Curriculum of Switched Mode Power Conversion. Department of Electrical Engineering. Indian Institute of Science, Bangalore (2007)
2. Mohan, N., Undeland, T.M., Riobbins, W.P.: Power Electronics Converters Applications and Design, ch. 7. Wiley, India (2009)
3. Unitrode, UC3842/3/4/5 provides low cost current mode control. Unitrode Application Note U-100A
4. Tian, F., Kasemsan, S., Batarseh, I.: An Adaptive Slope Compensation for the Single-Stage Inverter with Peak Current-Mode Control. IEEE Trans. Ind. Electron. 26(10), 2857–2862 (2011)
5. Deisch, C.W.: Simple Switching Control Method Changes Power Converter into a Current Source. In: IEEE Power Electronics Specialists Conference, pp. 300–306. IEEE Publication 78CH1337-5AES (1978)

6. Hsu, S.-P., Brown, A., Rensink, L., Middlebrook, R.D.: Modeling and Analysis of Switching DC-to-DC in Constant-Frequency Current-Programmed Mode. In: IEEE Power Electronics Specialists Conference, pp. 284–301. IEEE Publication 79CH1461-3 AES (1979)
7. Middlebrook, R.D., Cuk, S.: A General Unified Approach to Modeling Switching Converter Power Stages. In: IEEE Power Electronics Specialists Conference, Record, pp. 18–34. IEEE Publication 76CH1084-3 AES (1976); Also International J. of Electronics 42(6), 521–550 (1977)
8. Ridley, R.: A more accurate current mode control model. Ridley Engineering Inc.
9. Richardson, C.: Taming the boost:Predicting and measuring feedback loops in current mode boost high brightness LED drivers, Part 2 of 2. National semiconductor corporation
10. On Semiconductor, UC 3844/D, Semiconductor Components Industries, LLC, USA (1999)
11. Dixon, L.: Average Current Mode Control of Switching Power Supplies, Unitrode Application Note U-140
12. Biesecker, T.E.: Current Mode Control, Venable Industries Venable Technical Paper 5

Minimizing Energy of Scalable Distributed Least Squares Localization

Diana Olivia, Ramakrishna M., and Divya S.

Department of Information and Communication Technology,
Manipal Institute of Technology, Manipal, 576104, India
{diana.olivia,ramakrishna.m,divya.sharma}@manipal.edu

Abstract. In recent years, Wireless Sensor Networks (WSN) have become a growing technology that has broad range of applications. One of the major areas of research in Sensor Networks is location estimation. Distributed Least Squares (DLS) algorithm is a good solution for fine grained localization. Here localization process is split into a complex precalculation and a simple postcalculation process. This paper presents a revised version of DLS, i.e. minimized energy Distributed Least Squares (meDLS) algorithm where each blind node collects position of neighbouring beacon nodes and directly sends it to the sink node for precalculation. The precaluated data is sent back to the blind node for postcalculation, where location of blind node is estimated. The proposed algorithm is simulated and compared with scalable DLS (sDLS) for computational and communicational cost.

Keywords: Wireless Sensor Network, Localization, Scalability.

1 Introduction

A Wireless Sensor Node (or simply sensor node) consists of sensing, computing, communication, actuation and power components. These components are integrated on a single or multiple boards, and packaged in a few cubic inches. Wireless Sensor Network (WSN) is a collection of sensor nodes that are densely deployed near the objects which needs to be sensed. A tiny sensor node has a sensing unit, data communicating unit and unit for processing of sensed data. The sensing unit senses physical values like temperature, humidity levels of environment, nitrogen content of soil, blood pressure of human beings. In earlier days, the concept of Sensor Network [1] is used in applications of military environment and disaster prospect. But advance in technology made the use of Sensor Network in commercial areas like Hospitals, traffic control systems, habitat monitoring, home automation etc.

In some applications of WSN, information of nodes' location is compulsory for a meaningful interpretation of sensed data. But due to the deployment of the sensor nodes in inaccessible terrains or disaster relief operations, the location of the sensor nodes may not be predetermined. Therefore, a localization system is required in order to provide location information to the nodes. There are

J. Mathew et al. (Eds.): ICECCS 2012, CCIS 305, pp. 77–83, 2012.

many factors which make use of localization technology for the development and operation of WSNs. These factors include the identification and correlation [2] of collected data, node addressing, management and query of nodes contained in a determined region, evaluation of nodes' density and coverage, energy map generation, geographic routing, object tracking algorithms. Due to the sensor nodes limitations like its size and energy consumption, restricted positioning within the network is preferred over utilization of familiar positioning systems like Global Positioning System (GPS). Thus, the presence of location aware sensor nodes is generally assumed, which are called beacon nodes. These sensor nodes are aware of their own position using a common positioning system. The remaining nodes, known as blind nodes, which uses communication and some kind of distance estimation or neighborhood information to know their position with the help of beacon nodes. This paper proposes a mechanism that allows blind nodes in the network to derive their approximated locations from a limited number of beacon nodes which know their position information.

The localization techniques can be divided into coarse-grained and fine-grained localization. Usually this categorization reflects the trade-off between precision and resource consumption of the corresponding techniques. Coarse-grained approaches like Centroid Localization (CL) [3] and Adaptive Weighted Centroid Localization (AWCL) [4] often fail (abstain) from exact distances, require less communications and computations and provide lower precision estimates. On the other hand, fine-grained approaches hope to have an exact localization with high precision, which is achieved by costly computations, using estimations of distances or angles. Possible precision of such approaches are commonly based on a set of linear equations.

In this paper, working of DLS and sDLS is briefed in section 2. Section 3 describes proposed work i.e. meDLS. Simulation result is presented in section 4 and section 5 concludes the paper.

2   Related Work

Distributed Least Squares (DLS) [5] is a solution for combined high precision with relatively low complexity. It splits the expensive localization estimation into precalculation and postcalculation. The complex precalculation is performed on a high performance sink, which is independent from a specific blind node. The postcalculation is less complex and performed on blind nodes.

The drawback of DLS presumes that all blind nodes in the network is capable to communicate directly with the sink and is able to estimate its distance towards each beacon node. This presumption makes the DLS infeasible for use in huge multi-hop networks. In addition, computation and communication effort on each blind node increases with the number of beacon nodes and therefore, with the applied network size. The DLS have been overcome by sDLS [6], still preserving the idea of DLS. Major changes to sDLS enabled it to be used in large WSNs. In sDLS precalculations are performed individually instead of only one precalculation for the whole network, as done in DLS. So in sDLS the computational cost becomes independent from network size and also communication effort scales better. Even though sDLS outperforms DLS in all aspects, its updating cost of the precalculation demands for improvement. This paper presents an improved DLS algorithm that reduces the energy consumption of the network by reducing the communication and computational effort [7] [8].

2.1 Algorithmic Background

The algorithm assumes that each blind node and each beacon node is able to communicate with the sink. In addition each blind node have to be able to determine its distance to all beacon nodes, which requires direct communication to all beacon nodes.

The DLS algorithm is as follows [5]:

```
Step 1 - Initialization Phase:
            All beacons send their position to the sink.
Step 2 - Precalculation Phase:
            Sink computes precalculation parameters[7].
Step 3 - Communication Phase:
            Sink sends precalculated data to all blind nodes.
Step 4 - Postcalculation Phase:
            Blind nodes determine distance to every beacon node,
            receive precalculated data and estimate their location
            by solving the postcalculation.
```

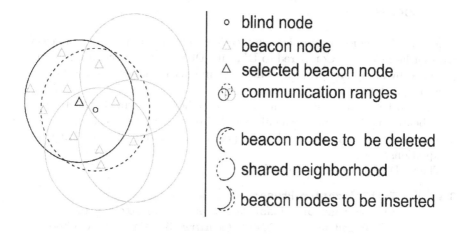

Fig. 1. Blind node selecting nearest beacon node for precalculation to minimize update operations

Fig.1 [6] shows that the number of beacon nodes that have to be added by the blind node as well as those that have to be deleted from precalculation is relatively small. Except from the beacon node which is used for linearization, all beacon nodes can be deleted from the precalculation. To ensure that the linearizing beacon node is within the communication range of the blind node, the individual precalculation of a beacon node uses itself as linearization beacon node.

The sDLS algorithm is as follows [6]:

```
Step 1-Discovery Phase:
        All beacons send a local broadcast to discover neighboring
        beacon nodes.
Step 2-Initialization Phase:
        All beacons send their position and a list of their
        neighbors to the sink.
Step 3-Precalculation Phase:
        Sink computes precalculation parameter individually for
        each beacon node.
Step 4-Distribution Phase:
        Sink sends precalculated data to beacon nodes.
Step 5-Communication Phase:
        Beacon nodes send precalculated data to blind nodes.
Step 6-Postcalculation Phase:
        Blind nodes determine distance to accessible beacon nodes,
        receive precalculated data, update precalculation and
        estimate their own position by solving the postcalculation.
```

3   Proposed Approach to Minimize Energy in Localization Process

In [6], sDLS enhances cost of communication and computation in comparison to DLS for large WSNs. Compared to DLS, in sDLS the postcalculation phase was extended by an update process by deleting and inserting beacon nodes from and to a precalculation.

Thus cost of computation consists of the following parts: 1. Delete inaccessible beacon nodes from precalculation 2. Insert accessible beacon nodes into precalculation 3. Estimate blind nodes position by solving the remaining system of equations.

The meDLS algorithm is as follows:

```
Step 1 - Initialization Phase:
           All beacons broadcast their position information.
           All blind nodes collect (minimum 3 nearest) neighbor
           beacon nodes position information and send this to the
           sink.
Step 2 - Precalculation Phase:
           Sink computes precalculation parameter individually for
           each blind node.
Step 3 - Distribution Phase:
           Sink sends precalculated data to blind nodes.
Step 4 - Postcalculation Phase:
```

```
Blind nodes determine distance to its neighbor beacon
nodes (from which it has received position information),
receive precalculated data and estimate their own
position by solving the postcalculation.
```

As in [6], the solving system of equations takes up only a small part of the computation, whereas deletion and insertion takes the most. The meDLS algorithm reduces the communication between blind node and beacon nodes and does not have additional update process like deleting and inserting beacon nodes from/to a precalculation. This leads to reduction in the cost of computation and communication. Compared to DLS, meDLS is scalable since it does not require all blind nodes to be in communication range of all the beacon nodes.

4 Experimental Result

To verify the performance of sDLS and meDLS, the NS2 [9] network simulator is used. A random deployment of n nodes within a field of 100*100 square meters was utilized. The first node is used as a sink, while the left out nodes have been randomly selected as beacon nodes (50%) and blind nodes (50%). The parameter n – the number of nodes – was varied from 10 to 100. The transmission range is set to 50 meters.

The meDLS approach is compared to sDLS in terms of localization accuracy, cost of computation and energy utilization on nodes.

4.1 Cost of Computation

To count cost of computation, the number of operations has been counted on each blind node and sink node. Three kinds of operations have been analyzed

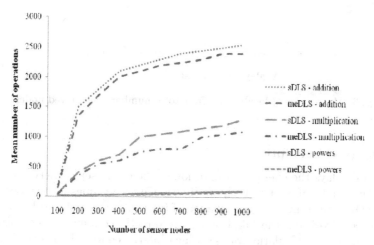

Fig. 2. Mean number of operations performed on a blind node

i.e. addition, multiplication and power operation. Additions and subtractions have been summed up as additions, multiplications and divisions are combined as multiplications, and powers include squares and square roots. The Fig. 1 depicts overall computation that included in precalculation and postcalculation of both algorithms. The proposed meDLS uses less computation operation due to reduced precalculation and postcalculation, which intern reduces energy consumption of a WSN node. The meDLS algorithm omits updating operation of postcalculation of sDLS. Similarly precalculation of meDLS uses only neighbors of blind node. Due to this the count cost of computation is reduced. Fig. 2 shows the Mean number of operations performed on a blind node.

4.2   Localization

To compare the localization precision, each blind node determines its localization error as the distance between the real and estimated positions. The results in Fig. 3 show that the averaged localization error is slightly reduced, using meDLS. But it also shows that the localization errors are larger when the sensor nodes are lesser, this is because of DLS technique which uses the neighboring node for calculation.

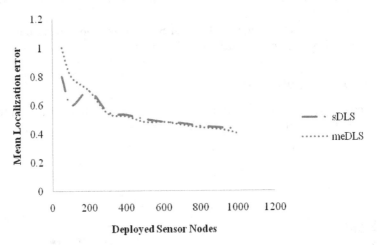

Fig. 3. Mean error of localization over total number of deployed nodes

4.3   Energy Consumption

The Fig. 4 displays the energy consumption of the entire sensor network, while using meDLS and sDLS. Here it shows that meDLS consumes less energy compared to sDLS, since in meDLS exchange of information is reduced by 20%. In the proposed algorithm, the blind nodes are directly communicating with the sink node in the precalculation phase, which intern reduces the network traffic in the Sensor Network.

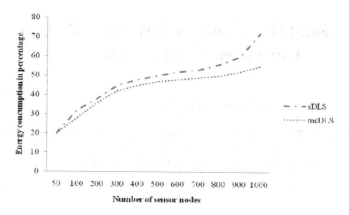

Fig. 4. Total energy consumption in sensor nodes

5 Conclusion

The proposed meDLS approach shows that energy consumption of Wireless Sensor nodes is reduced due to reduced cost of communication and computation. The simulation results prove that the energy consumption of meDLS is reduced by 25% of sDLS. While cost of computation and communication is reduced, but there is a challenge to improve localization accuracy.

References

1. Wu, G.: A Hierarchical Architecture of Heterogeneous Sensor Network (2004)
2. Bouukerche, A., Oliveira, H.A.B.F., Nakamura, E.F., Loureiro, F.A.F.: Localization Systems for Wireless Sensor Networks. IEEE Wireless Communications 14, 6–12 (2007)
3. Bulusu, N., Heidemann, J., Estrin, D.: GPS-less low. outdoor localization for very small devices. IEEE Personal Communications Magazine 7(5), 28–34 (2000)
4. Behnke, R., Timmermann, D.: AWCL: Adaptive Weighted Centroid Localization as an efficient Improvement of Coarse Grained Localization. In: 5th Workshop on WPNC 2008, pp. 243–250 (March 2008)
5. Reichenbach, F., Born, A., Timmermann, D., Bill, R.: A Distributed Linear Least Squares Method for Precise Localization with Low Complexity in Wireless Sensor Networks. In: Gibbons, P.B., Abdelzaher, T., Aspnes, J., Rao, R. (eds.) DCOSS 2006. LNCS, vol. 4026, pp. 514–528. Springer, Heidelberg (2006)
6. Behnke, R., Salzmann, J., Lieckfeldt, D., Timmermann, D.: sDLS - Distributed Least Squares Localization for Large Wireless Sensor Networks. In: International Workshop on Sensing and Acting in Ubiquitous Environments (October 2009)
7. Murphy, W.S., Hereman, W.: Determination of a position in three dimensions using trilateration and approximate distances. Tech. Rep. (1999)
8. Watkins, D.S.: Fundamentals of matrix computations, 2nd edn. Pure and Applied Mathematics. Wiley-Interscience [John Wiley & Sons], New York (2002)
9. Network Simulator- ver.2, http://www.isi.edu/nsnam/ns

Multiple Fault Diagnosis and Test Power Reduction Using Genetic Algorithms

J.P. Anita[1] and P.T. Vanathi[2]

[1] Amrita School of Engineering, Amrita Vishwa Vidyapeetham, Coimbatore, India
jp_anita@cb.amrita.edu
[2] PSG College of Technology, Coimbatore, India

Abstract. In this paper, a novel method for multiple fault diagnosis is proposed using Genetic Algorithms. Fault diagnosis plays a major role in VLSI Design and Testing. The input test vectors required for testing should be compact and optimized .Genetic Algorithm is a search technique to find approximate solutions to optimization and search problems. The proposed technique uses binary strings as a substitute for chromosomes. The chromosomes (test vectors) are initialized randomly and their fitness value is evaluated. Genetic operations selection, crossover and mutation are performed on this initialized set (initial population) to reproduce better test vectors. The test vectors thus generated are reordered by using a reordering algorithm. The total switching activity among the reordered test vectors is thus optimized and hence the reduction of test power.

Keywords: multiple faults, Genetic Algorithms (GA), test vector generation, test vector reordering, test power reduction.

1 Introduction

As the transistor density increases, the complexity of testing the chip also increases. The role of testing is to detect whether something went wrong and the role of diagnosis is to determine exactly what went wrong, and where the process needs to be altered [1]. The chip has to be tested for faults and hence fault diagnosis plays an important role in testing. The circuit is modeled and simulated for with and without faults. The value of the inputs to the circuit which determine the faults become the test vectors for those particular faults of that circuit. A single stuck at fault, is a line in a circuit where the line can be either stuck at '1' (sa1) or stuck at '0' (sa0). Earlier analyzing the circuit for single stuck at faults were adequate because the transistor density was less and hence less possibility for multiple faults to occur. But nowadays, due to increased transistor density, single stuck at fault modeling is not sufficient to completely model the circuit and hence the need for multiple fault analysis. A multiple fault represents a condition caused by the simultaneous presence of a group of single faults. An 'n' line digital circuit may have $3^n - 1$ multiple stuck at faults compared to '2n' single stuck at faults [2]. Hence diagnosis of multiple stuck at faults is very complex because of the large number of faults involved.

J. Mathew et al. (Eds.): ICECCS 2012, CCIS 305, pp. 84–92, 2012.

As the search space involved in multiple faults is very large, Genetic Algorithms (GA) can be used to provide optimal solutions. The work is organized as follows. The next section describes the background of multiple fault diagnosis. In addition, the motivation for GA based approach is described. Section 3 gives the GA approach, the flow of the approach and test vector generation using GA. Section 4 deals with the reordering algorithm and shows the amount of switching between test vectors. Section 5 gives the experimental results and Section 6 gives the conclusion.

2 Background and Motivation

Many Multiple fault diagnosis algorithms have been reported. The Single Observation – Single Location at A Time (SO-SLAT) based approach [3], starts with a list of fault candidate locations produced by any existing diagnosis method. A special kind of test called Single Observation Single Location at a Time (SO-SLAT) test is generated which detects the target fault at a single observation point and guarantees that the presence of other faults in the fault list will not mask the fault at this observation point. This is to ensure that the target fault does not activate other faults. In the 'n' detection test approach [5], the modeled fault is detected either by 'n' different test vectors or by the maximum obtainable 'm' different tests where m < n and the test sets are continuously changed. In the single and multiple fault simulation based approach [4], both single and multiple fault simulations on all suspected faults are done by repeated removal and addition of faults. In the neural network approach [7], a neural network is used to characterize the circuit. A Constraint circuit is created which consists of the faulty circuit, the fault free circuit and an interface circuit. The primary outputs of the faulty and fault free circuit are connected through the interface circuit that makes at least one primary output of the fault free circuit to differ from the corresponding faulty circuit. After the tests are generated, sample data are built to train the network and to diagnose the faults. In the Binary Decision Diagram based approach [8], decision diagrams are used to generate test vectors for those particular multiple faults. All these methods involve a very large amount of search space. Hence the proposed method makes use of Genetic Algorithms to get an optimized solution both in terms of search space and execution time. The Block Diagram of the proposed method is shown in Fig.1.

3 Test Vector Generation by Genetic Algorithms (GAs)

In Genetic Algorithm each solution is represented through a chromosome. A set of reproduction operators has to be found. Reproduction operators are applied directly on the chromosomes, and are used to perform mutations and crossover over solutions of the problem. Selection is the ability to compare each individual in the population set. It is done by using a fitness function. Each chromosome has an associated value corresponding to the fitness of the solution it represents. The fitness should correspond to an evaluation of how good the candidate solution is. The optimal solution is the one, which maximize the fitness function Then, the genetic algorithm loops over an iteration process to make the population evolve. Each iteration consists of the following steps:

- **Selection:** The first step consists in selecting individuals for reproduction. This selection is done randomly with a probability depending on the relative fitness of the individuals so that the best ones are often chosen for reproduction than poor ones.
- **Reproduction:** In the second step, offsprings are bred by the selected individuals. For generating new chromosomes, GA operations like crossover and mutation are done.
- **Evaluation:** Then the fitness of the new chromosomes is evaluated.
- **Replacement:** During the last step, individuals from the old population are killed and replaced by the new ones.
- The algorithm is stopped when the population converges towards the optimal solution.

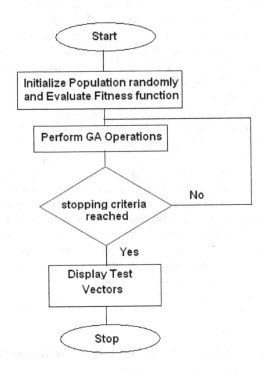

Fig. 1. Block Diagram of the proposed method

3.1   Initial Population and Chromosomes

The initial set of chromosomes is chosen randomly from the maximum possible set of chromosomes. For a circuit of 'n' inputs the maximum possible set of chromosomes is 2^n. The individual length of the chromosome, that is the number of genes in the chromosome is equal to the number of inputs given to the circuit. The order in which the input nodes are defined in the circuit file will correspond to the input value in the gene of the chromosome i.e. the second bit of the chromosome represents the second input to the circuit.For example,the ISCAS benchmark circuit C17 has five inputs and each chromosome(test vector) is a five bit pattern.The entire population set will consist of 2^5 test vectors, out of which a few are randomly taken.Associated with each chromosome is a fitness value. This value is a numerical quantification of how good is the solution given by the individual to the optimization problem. Chromosomes with better solutions have higher fitness values. During the progress of genetic algorithm, this set of chromosomes will evolve according to the fitness function. At the end of genetic algorithm, the chromosomes with maximum fitness remains and these are used as test vectors to test the circuit for the presence of the faults.

3.2   Fitness Function

The fitness function is an evaluation of the chromosome i.e. it is an estimate of how well the chromosome will perform as a solution to the given problem. Here the fitness function is defined as the number of faults or combinations of faults that a chromosome can detect for the given fault set. The fitness is calculated for the initial set of chromosomes and then for the evolving chromosomes. Thus we associate a fitness value with each chromosome and intend to choose the chromosome with the largest fitness value. The fitness value of a chromosome is given by the formula

$$\text{Fitness value} = \sum\nolimits_{All\ combination\ of\ faults} Nf * Dn$$

Where N_f is the number of faults simulated in that combination of faults and D_n is the detection number which is 0 if the chromosome cannot detect that combination of faults and 1 if the chromosome can detect that combination of faults.

All combination of the given faults is simulated and for each combination of fault the fault free and faulty response of the circuit is found to find D_n. In practice, multiple defects of large cardinality (more than four) do not happen very often [9] and hence the proposed method is simulated with a maximum of four faults.

3.3   Genetic Operators

A Simple Genetic Algorithm that yields good results in many practical problems is composed of three main evolutionary operators – Selection, Crossover and Mutation.

3.3.1   Selection

Selection is the process of replacing the parents by better performing offsprings. In this algorithm reproduction is done such that only the chromosome with maximum fitness is retained in the next iteration and others may be subjected to crossover and mutation. This makes sure that the chromosome with greater fitness among the set of chromosomes is saved for the next iteration. Thus only if a chromosome evolves such that it has a fitness value greater than the saved chromosome, it will be replaced. This makes sure that selection pressure is not more such that the algorithm will converge in local maxima.

3.3.2   Crossover

Crossover is the process in which two chromosomes (test vectors) combine their genetic material (bits) to produce a new offspring, which possesses both their characteristics. In the proposed method, one point crossover is performed where a bit is chosen randomly along the length of the chromosome and each chromosome is split into two pieces by breaking at the crossover site. The offspring is thus formed by joining the top piece of one chromosome with the tail piece of the other. The resulting offsprings can be either better or bad compared to the parents. The bad ones are eliminated by the selection operator in the next generation(iteration). An example for crossover is shown in Fig.2. Here two parents(test vectors) of binary value '10111011' and '00100010' are taken. Both the parents are of eight bits each. The crossover point is mentioned and after crossover the resulting offsprings are shown.

The amount of crossover is controlled by the crossover probability P_C, which is defined as the ratio of the number of offspring produced in each generation to the population size. A higher crossover probability allows exploration of more of the solution space and reduces the chances of settling for a false optimum.

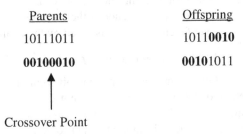

Fig. 2. One point crossover between two parents

3.3.3   Mutation

In Mutation, a string is deliberately changed to maintain diversity in the population set. It is done by flipping a bit. This is shown in Fig.3.

Fig. 3. Mutation on a chromosome

In Genetic Algorithm, mutation plays a crucial role in replacing the gene value lost from the population during the selection process so that they can be tried at new locations. The mutation probabilityP_M controls the rate at which new gene values are introduced into the population.

3.4   Number of Generations and Terminating Condition

A Generation is a cycle of selection, crossover, and mutation. These operators are repeated a number of times till the fittest chromosomes are obtained. An individual may be fitter or weaker than other population members. In every generation the weaker members are allowed to die out and members who are fit take part in the genetic operations. The net effect is the evolution of the population towards the global optimum. The terminating condition is checked after each iteration. Here the terminating condition is taken as either the maximum iterations has occurred or the maximum fitness is reached.

4   Test Vector Reordering and Test Power Reduction

The test vectors obtained are reordered. The order in which these vectors are used to test the circuit is immaterial as far as the faults are concerned but the total switching activity between the test vectors affect the test power and hence the vectors are reordered for minimum switching activity To do this the number of toggles between two consecutive test vectors is computed and the test vectors are reordered in such a way that the total switching activity for a set of test vectors is optimized. The switching activity in a circuit is given by the number of gate output transitions per unit of time. These transitions are caused by the vectors applied to the circuit inputs, and more specifically by the changing of the logic value on certain inputs when the vectors are applied, the relationship which exists between the number of changing bits in a given test pair (the Hamming distance) and the number of transitions provoked in the circuit (transition activity) is determined. The Relationship between hamming distance and the average power of a circuit plays a significant role to optimize the test power. In order to optimize the test power, the hamming distance between successive test vectors is used to arrange the test vectors in specific order so that Total Hamming Distance is minimum. Total hamming distance is defined as the Sum of the hamming distances between two consecutive test vectors in the sequence.

5   Results and Discussion

The experiment was carried out on several ISCAS benchmark circuits. For the C17 Circuit, seven test vectors were obtained for four multiple stuck at faults. The seven test vectors obtained in Fig.4 can be reordered to get the reordered vectors as in Fig.5. The initial vector may be arbitrarily chosen for hamming distance computation. Based on this initial vector the other vectors may be arranged. If the initial vector had been 0111_, then the sequence can be _1_1_, _1000, _00_0, 100_ _, 000_ _, 010_ _ because there is only one bit change between two successive vectors. The total switching activity before reordering was found to be '6' and after reordering was found to be '4' ,thereby reducing the test power.

S.No	Test vector
1	0111_
2	010_ _
3	000 _ _
4	100_ _
5	_1_1_
6	_1000
7	_00_0

S.No	Test vector
1	0111_
2	_1_1_
3	_1000
4	_00_0
5	100_ _
6	000 _ _
7	010_ _

Fig. 4. Test vectors before reordering   **Fig. 5.** Test vectors after reordering

Circuit	Number of Gates	Max Fitness for Pop Size of 3	Max Fitness for Pop Size of 30
C17	6	8	15
C432	120	12	15
C499	162	8	14
C880	320	11	15
C1355	506	11	15
C1908	603	11	14

Fig. 6. Results for various benchmark circuits with increasing chromosomes and constant No. of generations

S.No.	No. of chromosomes	No. of Generations	Max Fitness value
1	3	1	0
2	3	10	8
3	10	1	11
4	10	100	11
6	30	100	31

Fig. 7. Results for C1908 benchmark circuit with $p_c=0.8$, $p_m=0.15$

Increasing the number of chromosomes, keeping the number of generations constant also increases the fitness value. These are shown for various benchmark circuits in Fig.6.

Similarly, keeping the number of chromosomes constant and increasing the generations also increased the fitness value. These are shown for the benchmark circuit C1908 for four multiple stuck at faults in Fig.7.

References

1. Bushnell, M.L., Agarwal, V.D.: Essentials of Electronic Testing for Digital Memory and Mixed Signal VLSI Circuits. Kluwer Academic Publishers, London (2001)
2. Abramovici, M., Breuer, M.A., Friedman, A.D.: Digital Systems Testing and Testable Design. IEEE Press, New York (1990)
3. Lin, Y.-C., Cheng, K.T.: Multiple-Fault Diagnosis Based on Single fault activation and single output observation. In: Proceedings of Design, Automation and Test (2006)
4. Takahashi, H., Boateng, K.O., Saluja, K.K., Takamatsu, Y.: On diagnosing multiple stuck at faults using multiple and single fault simulation in combinational circuits. IEEE Transactions on Computer Aided Design of Integrated Circuits and Systems 21(3), 362–368 (2002)
5. Wang, Z., Tsai, K.H., Marek-Sadowska, M., Rajski, J.: Multiple fault diagnosis using n-detection tests. In: Proceedings of 21st International Conference on Computer Design (2003)
6. Wang, Z., Tsai, K.H., Marek-Sadowska, M., Rajski, J.: An efficient and effective methodology on the multiple fault diagnosis. In: International Test Conference, Charlotte, NC, pp. 329–338 (2003)
7. Pan, Z., Chen, L., Liu, S., Zhang, G.: Neural Network approach for multiple fault test of digital circuits. In: International Conference on Intelligent Systems (2006)
8. Pan, Z., Chen, L., Zhang, G.: A New Method for the Detections of Multiple Faults Using Binary Decision Diagrams. Wuhan University Journal of Natural Sciences 11(6), 1943–1946 (2006)
9. Wang, Z., Marek-Sadowska, M., Tsai, K.-H., Rajski, J.: Analysis and methodology for Multiple Fault Diagnosis. IEEE Transactions on Computer Aided Design of Integrated Circuits and Systems 25(3) (2006)
10. Takahashi, N., Ishiura, N., Yajima, S.: Fault Simulation for multiple faults by Boolean Function Manipulation. IEEE Transactions on Computer Aided Design of Integrated Circuits and Systems 13(4), 531–535 (1994)
11. Verreault, A., Aboulhamid, E.M., Karkouri, Y.: Multiple fault analysis using a fault dropping technique. In: Proceedings of 21st Fault Tolerant Computing Symposium (1991)
12. Lin, Y.-C., Lu, F., Cheng, K.T.: Multiple-Fault Diagnosis Based on Adaptive Diagnostic Test Pattern Generation. IEEE Transactions on Computer Aided Design of Integrated Circuits and Systems 26(5) (2007)
13. Huang, S.Y.: On improving the accuracy of multiple defect diagnosis. In: Proceedings of 19th IEEE VLSI Test Symposium, pp. 34–39 (2001)
14. O'Dare, M.J., Arslan, T.: Generating test patterns for VLSI circuits using a genetic Algorithm. IEEE Electronic Letters 30(10), 778–779 (1994)

15. Rudnick, E.M., Patel, J.H., Greenstein, G.S., Niermann, T.M.: A Genetic Algorithm Framework for Test Generation. IEEE Transactions on Computer Aided Design of Integrated Circuits and Systems 16(9) (1997)
16. Papa, G., Garbolino, T., Novak, F., Lawiczka, A.H.: Deterministic Test Pattern Generator Design With Genetic Algorithm Approach. Journal of Electrical Engineering 58(3), 121–127 (2007)
17. Jha, N., Gupta, S.: Testing of Digital Systems.Cambridge University Press (2003)

A Novel Design of Reduced Order Controllers Using Exponential Observers for Large-Scale Linear Discrete-Time Control Systems

Kavitha Madhavan[1] and Sundarapandian Vaidyanathan[2]

[1] Department of Mathematics, Vel Tech Dr. RR & Dr. SR Technical University, Avadi, Chennai-600 062, Tamil Nadu, India
kavi78_m@yahoo.com
[2] Research and Development Centre, Vel Tech Dr. RR & Dr. SR Technical University, Avadi, Chennai-600 062, Tamil Nadu , India
sundarvtu@gmail.com

Abstract. This paper discusses the design of observer-based reduced order controllers for the stabilization of large scale linear discrete-time control systems. This design is carried out via deriving a reduced-order model for the given linear plant using the dominant state of the linear plant. Using this reduced-order linear model, sufficient conditions are derived for the design of observer-based reduced order controllers. A separation principle has been established in this paper which demonstrates that the observer poles and controller poles can be separated and hence the pole-placement problem and observer design are independent of each other.

Keywords: Reduced order controllers, observers, linear systems, dominant state.

1 Introduction

The reduced order controller design and observer design are active research problems in the linear systems literature and there has been a significant attention paid in the literature on these two problems during the past four decades [1-10]. In the recent decades, there has been a considerable attention paid to the control problem of large scale linear systems. The observer-based reduced-order controller design is motivated by the fact that the dominant state of the linear plant may not be available for measurement and hence for implementing the pole placement law, only the reduced order exponential observer can be used in lieu of the dominant state of the given discrete-time linear system.

A recent approach for obtaining the reduced-order controllers is via the reduced-order model of a linear plant preserving the dynamic as well as static properties of the system and then devising controllers for the reduced-order model thus obtained [4-8]. In this paper, we derive a reduced-order model for any linear discrete-time control system and our approach is based on the approach of using the dominant state of the

J. Mathew et al. (Eds.): ICECCS 2012, CCIS 305, pp. 93–99, 2012.
© Springer-Verlag Berlin Heidelberg 2012

given linear discrete-time control system, *i.e.* we derive the reduced-order model for a given discrete-time linear control system keeping only the dominant state of the given discrete-time linear control system. Using the reduced-order model obtained, we characterize the existence of a reduced-order exponential observer that tracks the state of the reduced-order model, i.e. the dominant state of the original linear plant. We note that the reduced-order observer design detailed in this paper is a discrete-time analogue of the results of Aldeen and Trinh [8] for the observer design of the dominant state of continuous-time linear control systems.

2   Reduced Order Model and Observer for Discrete-Time Linear Systems

In this section, we consider a large scale linear discrete-time control system S_1 given by

$$x(k+1) = Ax(k) + Bu(k)$$
$$y(k) = Cx(k)$$

(1)

where $x \in R^n$ is the state, $u \in R^m$ is the control or input and $y \in R^p$ is the system output. We assume that A, B and C are constant matrices with real entries having dimensions $n \times n, n \times m$ $p \times n$ respectively.

First, we suppose that we have made an identification of the *dominant* (*slow*) and *non-dominant* (*fast*) states of the original linear system (1) using the modal approach as described in [9].

Without loss of generality, we may assume that $x = \begin{bmatrix} x_s \\ x_f \end{bmatrix}$, where

$x_s \in R^r$ represents the *dominant* state and $x_f \in R^{n-r}$ represents the *non-dominant* state of the system.

Then the linear system (1) becomes

$$\begin{bmatrix} x_s(k+1) \\ x_f(k+1) \end{bmatrix} = \begin{bmatrix} A_{ss} & A_{sf} \\ A_{fs} & A_{ff} \end{bmatrix} \begin{bmatrix} x_s(k) \\ x_f(k) \end{bmatrix} + \begin{bmatrix} B_s \\ B_f \end{bmatrix} u(k)$$

$$y(k) = \begin{bmatrix} C_s & C_f \end{bmatrix} \begin{bmatrix} x_s(k) \\ x_f(k) \end{bmatrix}$$

(2)

From (2), we can write the plant equations as

$$x_s(k+1) = A_{ss}x_s(k) + A_{sf}x_f(k) + B_su(k)$$
$$x_f(k+1) = A_{fs}x_s(k) + A_{ff}x_f(k) + B_fu(k)$$
$$y(k) = C_sx_s(k) + C_fx_f(k)$$

(3)

Next, we shall assume that the matrix A has a set of n linearly independent eigenvectors. In most practical situations, the matrix A has distinct eigenvalues and this condition is immediately satisfied. Thus, it follows that A is diagonalizable. Thus, we can find a non-singular (*modal*) matrix P consisting of n linearly independent eigenvectors of A such that $P^{-1}AP = \Lambda$, where Λ is a diagonal matrix consisting of the n eigenvalues of A.

Now, we introduce a new set of coordinates on the state space given by

$$\xi = P^{-1}x \tag{4}$$

In the new coordinates, the plant (1) becomes

$$\begin{bmatrix} \xi_s(k+1) \\ \xi_f(k+1) \end{bmatrix} = \begin{bmatrix} \Lambda_s & 0 \\ 0 & \Lambda_f \end{bmatrix} \begin{bmatrix} \xi_s(k) \\ \xi_f(k) \end{bmatrix} + P^{-1}Bu(k)$$

$$y(k) = CP \begin{bmatrix} \xi_s(k) \\ \xi_f(k) \end{bmatrix} \tag{5}$$

where Λ_s and Λ_f are $r \times r$ and $(n-r) \times (n-r)$ diagonal matrices respectively, consisting of the dominant and non-dominant eigenvalues of A.

Define matrices $\Gamma_s, \Gamma_f, \Psi_s$ and Ψ_f by

$$P^{-1}B = \begin{bmatrix} \Gamma_s \\ \Gamma_f \end{bmatrix} \text{ and } CP = \begin{bmatrix} \Psi_s & \Psi_f \end{bmatrix} \tag{6}$$

where $\Gamma_s, \Gamma_f, \Psi_s$ and Ψ_f are $r \times m$, $(n-r) \times m$, $p \times r$ and $p \times (n-r)$ matrices respectively.

From (5) and (6), we see that the plant (3) has the following simple form in the new coordinates (4).

$$\xi_s(k+1) = \Lambda_s \xi_s(k) + \Gamma_s u(k)$$
$$\xi_f(k+1) = \Lambda_f \xi_f(k) + \Gamma_f u(k) \tag{7}$$
$$y(k) = \Psi_s \xi_s(k) + \Psi_f \xi_f(k)$$

Next, we make the following assumptions:

(H1) As $k \to \infty$, $\xi_f(k+1) \approx \xi_f(k)$, i.e. ξ_f takes a constant value in the steady-state.

(H2) The matrix $I - \Lambda_f$ is invertible.

Then it follows from (7) that for large values of k, we have

$$\xi_f(k) \approx (I - \Lambda_f)^{-1} \Gamma_f u(k) \tag{8}$$

Substituting (8) into (7), we obtain the reduced-order model as

$$\xi_s(k+1) = \Lambda_s \xi_s(k) + \Gamma_s u(k)$$
$$y(k) = \Psi_s \xi_s(k) + \Psi_f (I - \Lambda_f)^{-1} \Gamma_f u(k) \tag{9}$$

To obtain the reduced-order model of the linear plant (1) in the x coordinates, we proceed as follows. By the linear change of coordinates (4), it follows that

$$\xi = P^{-1} x = Qx.$$

Thus, we have

$$\begin{bmatrix} \xi_s(k) \\ \xi_f(k) \end{bmatrix} = Q \begin{bmatrix} x_s(k) \\ x_f(k) \end{bmatrix} = \begin{bmatrix} Q_{ss} & Q_{sf} \\ Q_{fs} & Q_{ff} \end{bmatrix} \begin{bmatrix} x_s(k) \\ x_f(k) \end{bmatrix} \tag{10}$$

Using (9) and (10), it follows that

$$Q_{ff} x_f(k) = -Q_{fs} x_s(k) + (I - \Lambda_f)^{-1} \Gamma_f u(k) \tag{11}$$

Next, we assume the following:

(H3) The matrix Q_{ff} is invertible.

Using the assumption (H3), the equation (11) becomes

$$x_f(k) = -Q_{ff}^{-1} Q_{fs} x_s(k) + Q_{ff}^{-1} (I - \Lambda_f)^{-1} \Gamma_f u(k) \tag{12}$$

To simplify the notation, we define the matrices

$$R = -Q_{ff}^{-1} Q_{fs} \quad \text{and} \quad S = Q_{ff}^{-1}(I - \Lambda_f)^{-1} \Gamma_f \tag{13}$$

Using (13), the equation (12) can be simplified as

$$x_f(k) = Rx_s(k) + Su(k) \tag{14}$$

Substituting (14) into (3), we obtain the reduced-order model S_2 as

$$x_s(k+1) = A_s^* x_s(k) + B_s^* u(k)$$
$$y(k) = C_s^* x_s(k) + D_s^* u(k) \tag{15}$$

where the matrices A_s^*, B_s^*, C_s^* and D_s^* are defined by

$$A_s^* = A_{ss} + A_{sf} R, \ B_s^* = B_s + A_{sf} S, \ C_s^* = C_s + C_f R, \ D_s^* = C_f S \tag{16}$$

where A_s^*, B_s^*, C_s^* and D_s^* are defined as in (16).

To estimate the dominant state x_s of the system S_1, consider the candidate observer S_3 defined by

$$z_s(k+1) = A_s^* z_s(k) + B_s^* u(k) + K_s^* \left[y(k) - C_s^* z_s(k) - D_s^* u(k) \right] \qquad (17)$$

Theorem 1. If the estimation error is defined as $e = z_s - x_s$, then $e(k) \to 0$ exponentially as $k \to \infty$ if and only if the matrix K_s^* is such that $E = A_s^* - K_s^* C_s^*$ is convergent. If $\left(C_s^*, A_s^* \right)$ is observable, then we can always construct an exponential observer of the form (19) having any desired speed of convergence. ∎

3   Observer-Based Reduced Order Controller Design

In this section, we first state an important result that prescribes a simple procedure for stabilizing the dominant state of the reduced-order linear plant derived in Section 2.

Theorem 2. Suppose that the assumptions (H1)-(H3) hold. Let S_1 and S_2 be defined as in Theorem 1. For the reduced-order model S_2, the state feedback control law

$$u(k) = -F_s^* x_s(k) \qquad (18)$$

stabilizes the dominant state x_s of the reduced-order model S_2 having any desired speed of convergence. ∎

In practical applications, the dominant state x_s of the reduced-order model S_2 may not be directly available for measurement and hence we cannot implement the state feedback control law (18). To overcome this practical difficulty, we state an important theorem, usually called as the *Separation Principle*, which first establishes that the observer-based reduced-order controller indeed stabilizes the dominant state of the given linear control system S_1 and also demonstrates that the observer poles and the closed-loop controller poles can be separated. The proof follows by Lyapunov stability theory.

Theorem 3 (Separation Principle). Suppose that the assumptions (H1)-(H3) hold. Suppose that there exist matrices F_s^* and K_s^* such that $A_s^* - B_s^* F_s^*$ and $A_s^* - K_s^* C_s^*$ are both convergent matrices. By Theorem 1, we know that the system S_3 defined by (17) is an exponential observer for the dominant state x_s of the control system S_1. Then the observer poles and the closed-loop controller poles are separated and the control law

$$u(k) = -F_s^* z_s(k) \tag{19}$$

also stabilizes the dominant state x_s of the control system S_1.   ■

4   Numerical Example

In this section, we consider a fourth-order linear discrete-time control system described by

$$x(k+1) = A\, x(k) + B\, u(k)$$
$$y(k) = Cx(k) \tag{20}$$

where

$$A = \begin{bmatrix} 2.0 & 0.6 & 0.2 & 0.3 \\ 0.4 & 0.4 & 0.9 & 0.5 \\ 0.1 & 0.3 & 0.5 & 0.1 \\ 0.7 & 0.9 & 0.8 & 0.8 \end{bmatrix}, \quad B = \begin{bmatrix} 1 \\ 1 \\ 1 \\ 1 \end{bmatrix} \text{ and } C = \begin{bmatrix} 1 & 2 & 1 & 1 \end{bmatrix}. \tag{21}$$

The eigenvalues of the matrix A are

$$\lambda_1 = 2.4964, \quad \lambda_2 = 1.0994, \quad \lambda_3 = 0.3203 \text{ and } \lambda_4 = -0.2161. \tag{22}$$

From (22), we note that λ_1, λ_2 are unstable (slow) eigenvalues and λ_3, λ_4 are stable (fast) eigenvalues of the system matrix A.

For this linear system, the dominant and non-dominant states are calculated. A simple calculation using the procedure in [9] shows that the first two states $\{x_1, x_2\}$ are the dominant (slow) states, while the last two states $\{x_3, x_4\}$ are the non-dominant (fast) states for the given system (20).

Using the procedure described in Section 2, the reduced-order linear model for the given linear system (20) can be obtained as

$$x_s(k+1) = A_s^* x(k) + B_s^* u(k)$$
$$y(k) = C_s^* x(k) + D_s^* u(k) \tag{23}$$

where

$$A_s^* = \begin{bmatrix} 2.0077 & 1.1534 \\ 0.3848 & 1.5581 \end{bmatrix}, \quad B_s^* = \begin{bmatrix} 0.8352 \\ 1.1653 \end{bmatrix}$$

$$C_s^* = \begin{bmatrix} 1.0092 & 4.0011 \end{bmatrix} \text{ and } D_s^* = -0.2904$$

For this reduced order model, a reduced order observer and controller can be constructed as detailed in the Sections 2 and 3.   ■

5   Conclusions

In this paper, using the dominant state analysis of the given large-scale linear plant, we obtained the reduced-order model of the linear plant. Then we derived sufficient conditions for the design of observer-based reduced-order controllers. The observer-based reduced order controllers are constructed by combining the reduced order controllers for the original linear system which require the dominant state of the original system and reduced order observers for the original linear system which provide an exponential estimate of the dominant state of the original linear system. We also established a separation principle in this paper which shows that the pole placement problem and observer problem are independent of each other.

References

[1]   Cumming, S.D.: Design of observers of reduced dynamics. Electronics Letters 5, 213–214 (1969)

[2]   Fortman, T.E., Williamson, D.: Design of low-order observers for linear feedback control laws. IEEE Trans. Automat. Control 17, 301–308 (1972)

[3]   Litz, L., Roth, H.: State decomposition for singular perturbation order reduction – a modal approach. International J. Control 34, 937–954 (1981)

[4]   Lastman, G.J., Sinha, N.K., Rozsa, P.: On the selection of states to be retained in a reduced-order model. IEEE Proceedings – Control Theory 131, 15–24 (1984)

[5]   Anderson, B.D.O., Liu, Y.: Controller reduction: concepts and approaches. IEEE Trans. Automat. Control 34, 802–812 (1989)

[6]   Mustafa, D., Glover, K.: Controller reduction by H_∞ balanced truncation. IEEE Trans. Automat. Control 36, 668–692 (1991)

[7]   Aldeen, M.: Interaction modelling approach to distributed control with application to power systems. International J. Control 53, 1035–1044 (1991)

[8]   Aldeen, M., Trinh, H.: Observing a subset of the states of linear systems. IEE Proceedings – Control Theory 141, 137–144 (1994)

[9]   Sundarapandian, V.: Distributed control schemes for large-scale interconnected discrete-time linear systems. Mathematical and Computer Modelling 41, 313–319 (2005)

[10]   Ogata, K.: Discrete-Time Control System. Prentice Hall, New Jersey (2010)

Adiabatic Technique for Designing Energy Efficient Logic Circuits

Shari Jahan C.S. and N. Kayalvizhi

Department of Electronics and Communication Engineering, Amrita Vishwa Vidyapeetham University, Coimbatore, India
shari_jahan@yahoo.co.in, n_kayalvizhi@cb.amrita.edu

Abstract. Energy minimization is an important factor in designing digital circuits which are portable and battery operated. Irreversible logic operation causes the minimum dissipation of KT ln 2 joules of heat energy when each bit is erased. Reversible logic that employs adiabatic switching principles can be used to minimize dynamic power , which is the major contributor to total power dissipation. Reversible Energy Recovery Logic (RERL) belongs to fully adiabatic logic family and it eliminates non adiabatic energy loss by making use of reversible logic. RERL NAND/AND gate and RERL SR latch is proposed in this work using eight phase clocking scheme. This RERL circuits consume less energy compared with static CMOS logic circuits at low speed operation. The simulation result using HSPICE shows that RERL circuits consume less power compared with the static CMOS circuits.

Keywords: Adiabatic circuit, RERL, Reversible logic.

1   Introduction

By Moore's law the power dissipation will double for constant cost roughly once every two years. Also in conventional logic operation the flipping of bits cause energy dissipation. This will be accompanied by a decrease in the entropy of the system which results in the dissipation of heat. According to Landauer's research erasing of each bit in an irreversible operation causes heat dissipation of KT ln 2 joules This energy dissipation can be reduced by employing reversible logic since it is physically and logically reversible. Physically reversible means there is no heat dissipation and by logical reversibility the inputs can be retrieved from the output. The reversible operation can be performed only with the support of reversible gates. The reversible logic is implemented by adiabatic logic family by using adiabatic switching principle.

Adiabatic logic family includes fully adiabatic logic and quasi static adiabatic logic. Fully adiabatic logic includes Split Level Charge Recovery Logic (SCRL) and Reversible Energy Recovery Logic (RERL).Adiabatic circuits have adiabatic, non-adiabatic and adiabatic losses.

The adiabatic loss is inversely proportional to the transition time T of the trapezoidal clock [1, 3].The adiabatic loss in adiabatic circuits is given by

J. Mathew et al. (Eds.): ICECCS 2012, CCIS 305, pp. 100–107, 2012.
© Springer-Verlag Berlin Heidelberg 2012

$$E_{ADIABATIC} = \frac{R_{ON}C_L}{T} C_L V_{DD}^2 \tag{1}$$

where R_{ON} is the on-resistance of the switch and C_L is the load capacitance. By making the transition time very much greater than $R_{ON}C_L$, adiabatic loss can be reduced The non-adiabatic losses are due to the potential difference across the terminals of a switch when the switch is turned on. It does not depend on the frequency. At low frequency the non-adiabatic loss becomes prominent.

$$E_{NON\_ADIABATIC} = \frac{1}{2} \frac{C_1 C_2}{(C_1 + C_2)} (V_1 - V_2)^2 \tag{2}$$

where C_1 and C_2 are the capacitance across the terminals of the switch and V_1 and V_2 are their voltages .When there is a difference in voltage across the switch it causes non adiabatic losses as it is proportional to the square of the voltage difference. So zero voltage switching should be satisfied that is the transistor should not be turned on when there is a potential difference between the drain and source to eliminate non-adiabatic losses. In the case of adiabatic circuits with non-adiabatic losses, reversible logic can be used to recycle the energy. To recycle the energy, separate charge recovering path is required and the input has to be constructed from the output which is possible only with reversible logic. Even though SCRL uses reversible logic, RERL is preferred as RERL requires less clock rails compared to SCRL [2].

The rest of the paper is organized as follows: section 2 describes the switching scheme in adiabatic logic. The Reversible Energy Recovery Logic (RERL) and RERL inverter/buffer is explained in section 3.Section 4 explains the proposed RERL NAND/AND gate and RERL SR latch. Section 5 shows the simulation results using HSPICE and the conclusion is given in section 6.

2   Switching Scheme in Adiabatic Logic

The adiabatic switching scheme is shown in Fig.1.It consists of a charge up and a charge down path with a clock signal Φ_i to power the circuit. The unlabelled boxes are single transmission gate which is provided with input signals g_f and g_b to guard against the non-adiabatic loss. The labeled boxes indicate a network of transmission gates which are connected in series or parallel to implement Boolean functions [5].

The reversible operation can be achieved by ramping up and ramping down the clock signal Φ_i and by providing inputs to the labeled and unlabelled transmission gates. When Φ_i ramps up, the load capacitor will be charged to d_i via charge-up path and then the operation of the circuit is reversed by ramping down Φ_i. Thus the energy is recycled back to the power supply by using the charge down path. The logic functions are computed by making use of charge up path. The output of one stage is given as the input to the next stage in a cascading fashion.This becomes impractical when the number of cascaded stages is high and hence pipelining is used.

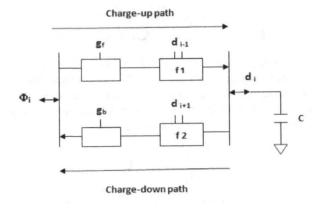

Fig. 1. A gate structure suitable for reversible logic [5]

By using pipelining, inputs of the earlier stages can be changed before the final output is ready. It can be done by computing the inverse of each function as data passes from one stage to another. The function is performed by the transmission gates and its inputs in the charge down path. The inputs to the transmission gates f2 is obtained from the output of the next stage. The energy is recycled through the charge down path when the clock signal Φ_i ramps down. The pipelining connection is shown in Fig.2.

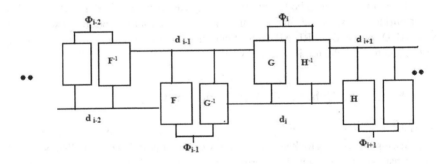

Fig. 2. Interconnection of logic gates in a pipeline [5]

3   Reversible Energy Recovery Logic

RERL is a dual rail adiabatic logic circuit that uses the concept of reversible logic to eliminate non-adiabatic energy loss by satisfying Zero-Voltage Switching (ZVS).Due to inherent micropipeling RERL computes one logic level per clock phase. So in RERL a long pipelined stage is there to implement a logic circuit.

The RERL buffer or inverter [1,7] is shown in Fig.3 and its pipeline connection is shown in Fig 4.The circuit is implemented using 18 transistors to reduce the power dissipation.Therefore the area overhead is higher in reversible logic as compared with static CMOS logic.The transmission gates are used in the circuit design to avoid non-adiabatic losses due to signal degradation [8].

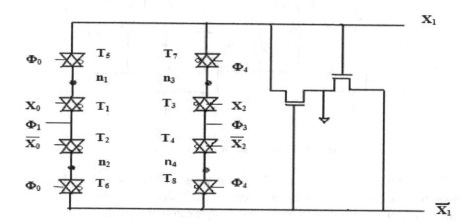

Fig. 3. RERL inverter or buffer [1]

RERL inverter or buffer uses 8 phase clocking scheme for its operation.X_0 is the input signal and X_1 is the pipelined output signal.X_2 is the output of the next stage whose input is X_1. The transmission gates T_1 and T_2 implement forward logic function and will determine the charging path of output node whereas T_3 and T_4 implement backward logic function and determine the discharging path of output node.T_5 and T_6 are the forward isolation switch which isolate the charging path and $T_7$and T_8 are backward isolation switch which isolate the discharging path.

The operation of the circuit is as follows [1]:

In RERL inverter all the internal nodes are initially grounded and they do not have any abrupt voltage change to satisfy ZVS condition. During the time interval t_0, when the input X_0 goes high and the clock signal Φ_0 ramps up, the forward isolation switch T_5 and T_6 are turned on. At t_1 the clock signal Φ_1 goes high, X_1 and internal node n_1 follows it.During this time interval n_2 remains low as the transmission gate T_2 is off. As X_1 goes high during t_1 it will become the input of the next stage and the output X_2 of the next stage goes high during t_2.During t_3 transmission gate T_3 will be turned on as X_2 is high and thereby n_3 goes high. Forward isolation switches T_5 and T_6 are turned off during the time interval t_4 as Φ_0 ramps down. But Φ_4 goes high and hence the backward isolation switches T_7 and T_8 are turned on. During t_5, Φ_1 falls and discharges n_1 and at t_6 falling Φ_2 in the previous stage discharges X_0.During t_7 falling Φ_3 discharges X_1 and n_1.Transmission gates T_7 and T_8 are turned off during t_8 to repeat the operation.

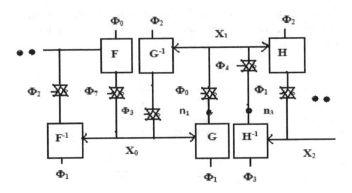

Fig. 4. Reversible pipeline connection [1]

The circuit will provide dual rail logic that is we get inverted as well as buffered output from the same circuit. It is very useful in the design of complex circuits as it limits the use of not operation.RERL reversible style can be used to design any circuit for reducing the power dissipation.

4   Proposed Work

The RERL AND/ NAND gate and RERL SR latch are proposed in Fig. 5 and Fig.6.Since NAND gate is the universal gate, all boolean functions can be implemented using it. In Fig. 5 transmission gates T_2, T_3, T_4 and T_5 implement forward logic function and transmission gates T_8, T_9, T_{10} and T_{11} implement backward logic function.Transmission gates T_1 and T_6 are forward isolation switches that determine the charging path of the output node whereas T_7 and T_{12} are backward isolation switches that determine the discharging path of the output node.The circuit uses 8 phase clocking scheme similar to that of RERL inverter or buffer. X_0 and X_{01} are the input signals and X_1 is the pipelined output signal.X_2 is the output of the next stage whose one input is X_1.The operation of the circuit is as follows:

During the time interval t_0 when the input X_0 and X_{01} goes high and the clock signal Φ_0 ramps up, the forward isolation switch T_1 and T_6 are turned on.At t_1 the clock signal Φ_1 goes high ,X_1 follows it.During this time interval the transmission gates T_4 and T_5 are off. As X_1 goes high during t_1 it will become one of the input of the next stage and the output X_2 of the next stage goes high during t_2 if the other input is high. During t_3 transmission gates T_8 and T_9 will be turned on as X_2 is high. Forward isolation switches T_1 and T_6 are turned off during the time interval t_4 as Φ_0 ramps down. But Φ_4 goes high and hence the backward isolation switches T_7 and T_{12} are turned on. During t_5, Φ_1 falls and at t_6 falling Φ_2 in the previous stage discharges X_0 and X_{01}. During t_7 falling Φ_3 discharges X_1. Transmission gates T_7 and T_{12} are turned off during t_8 to repeat the operation. This RERL AND/ NAND gate implementation reduces power compared with static CMOS logic.

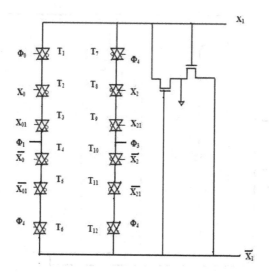

Fig. 5. RERL AND or NAND gate

RERL SR latch is implemented by the cross coupling of two NAND gates as shown in Fig.6.The area consumed by the circuit is high as it requires 24 transmission gates.But the power dissipation is reduced.

5   Simulation Results

The circuits are simulated using HSPICE with 90nm technology. The power dissipated by the static CMOS inverter, NAND gate and SR latch is compared with its corresponding RERL circuit in Table 1. The power dissipated by the RERL circuit is less compared to its static CMOS circuit. But the delay and area is high for RERL circuits.

Table 1. Simulation Results

CIRCUIT	POWER	NO OF TRANSISTORS	DELAY
Inverter(static CMOS)	180µW	2	3.001 ns
RERL inverter	5.5nW	16	26 ns
NAND Gate(static CMOS)	160 µW	4	2.4 ns
RERL NAND Gate	300nW	24	66 ns
SR Latch(static CMOS)	172 µW	8	30 ns
RERL SR Latch	20 µW	52	104 ns

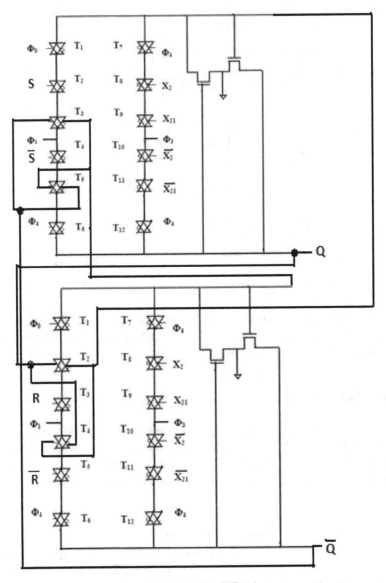

Fig. 6. RERL SR Latch

6   Conclusion

In this paper, circuits based on RERL is discussed. RERL is dual rail adiabatic logic circuit. The non-adiabatic energy loss in RERL is removed by using the concept of reversible logic. The simulation done using HSPICE proved that the energy dissipation in RERL circuits is less compared with the static CMOS circuits. In conclusion, RERL is suitable for ultra low power application where high performance is not required.

References

1. Lim, J., Kim, D.-G., Chae, S.-I.: Reversible energy recovery logic circuits and its 8 phase clocked power generator for ultra low power application. EICE Trans. Electron E82-C(4) (1999)
2. Khatir, M., Ejlali, A., Moradi, A.: Improving the energy efficiency of reversible logiccircuits by the combined use of adiabatic styles. Integration, The VLSI Journal 44 (2010)
3. Sunil Gavaskar Reddy, Y., Rajendra Prasad, V.V.G.S.: Comparison of CMOS and Adiabatic Full Adder Circuits. International Journal of Scientific & Engineering Research 2 (2011)
4. Athas, W.C., Svensson, L.J., Koller, J.G., Tzartzanis, N., Chou, Y.: Low-power digital systems based on adiabatic-switching principles. IEEE Trans. VLSI Systems 2(4), 398–406 (1994)
5. Athas, W.C., Svensson, L.J.: Reversible Logic Issues in Adiabatic CMOS. In: IEEE Conf. on Physics and Computation (1994)
6. Kim, S., Ziesler, C.H., Papaefthymiou, M.C.: Charge recovery computing on silicon. IEEE Trans. on Computers 54 (2005)
7. Lim, J., Kim, D.G., Chae, S.-I.: A 16-bit carry-lookahead adder using reversible energy recovery logic for ultra-low-energy systems. IEEE J. Solid-State Circuits 34(6), 898–903 (1999)
8. Lim, J., Kim, D.-G., Chae, S.-I.: nMOS Reversible Energy Recovery Logic for Ultra-Low-Energy Applications. IEEE J. Solid-State Circuits 35(6) (2000)

Comparative Study of Recent Compressed Sensing Methodologies in Astronomical Images

Nidhin Prabhakar T.V.[1], Hemanth V.K.[1], Sachin Kumar S.[1],
K.P. Soman[1], and Arun Soman[2]

[1] Centre for Excellence in Computational Engineering and Networking
Amrita University, Coimbatore-641112, India
[2] Department of Information Technology,
Rajagiri School of Engineering & Technology

Abstract. Compressed sensing(CS) which serves as an alternative to Nyquist sampling theory, is being used in many areas of applications. In this paper, we applied recent compressed sensing algorithm such as DALM, FISTA and Split-Bregman on astronomical images. In astronomy, physical prior information is very crucial for devising effective signal processing methods. We particularly point out that CS-based compression scheme is flexible enough to account for such information. We try to compare these algorithms using objective measures like PSNR, MSE et al. With these measures we intend to verify the image quality of reconstructed and original images.

Introduction

Compressed Sensing(CS) also known as compressive sensing, is a new field of interest based on sparsity that has emerged and rapidly attracted much attention in recent years. CS created a complete paradigm shift to the area of sampling. The theory of compressed sensing, or compressive sampling [4],[5],[6] states that signals which are sparse in some basis may be perfectly reconstructed when under-sampled at more than Nyquist rate, subject to a specific constraint. The solution i.e. the image reconstruction can be obtained by using ℓ1-norm minimization methods like DALM, FISTA, Split-Bregman reconstruction etc.

Here we are dealing with astronomical images which are of very large size and the advantage of applying CS on astronomical images are:1.One can use a single sensor to capture astronomical images(Research is still going on for developing a single-pixel camera, though RICE university has proposed a design for one.)2.Imaging time and power consumption can be reduced.3.Only less data required to recover the super-resolution photos and thus the storage space for acquired image reduces.4.The highly compressed data can be taken by CS cameras and can be easily transmitted back to earth[2].

Application of compressed sensing in astronomy Bobin et al. in [1] was very recently proposed in which on board data compression for the future Herschel space observatory was discussed. It was required by ESA to obtain a compression ratio of atleast 2.5 which can be easily achieved using CS. The versatility of

J. Mathew et al. (Eds.): ICECCS 2012, CCIS 305, pp. 108–116, 2012.

the compressed sensing framework to account for specific prior information on signals(here images) was already pointed out in that context. The potential of compressed sensing for interferometry was pointed in the signal processing community since the time when the theory emerged (Donoho in [6]; Cand'es et al. in [5].

1 Theory behind Compressed Sensing and ℓ1-Norm Minimization Based Reconstruction Methods

1.1 Compressed Sensing

Compressive sensing can be stated as : Given x of length N, only M measurements $(M < N)$ is required to fully recover x when x is K-sparse $(K < M < N)$ then

$$y = \phi x$$

and

$$x = \psi \alpha$$

Backbone of CS is sparsity and incoherence. Sparsity refers to the number of zero elements of a signal in a known transform domain[11]. Incoherence means the measurement vectors should be incoherent with basis in which the signal is sparse. Mutual coherence[1,3,11] is given by:

$$\mu(\phi, \psi) = \max_{k,j} |\langle \phi_k, \psi_j \rangle|$$

1.2 l1-Minimization Based Reconstruction Methods

The algorithm for solving the problem is mentioned in [10]. Basially three types of algorithm Split-Bregman method, Fast Iterative soft Thresholding Algorithm(FISTA) and Augmented Lagrangian Multiplier(ALM). WE describe the Augmented Lagrangian Multiplier algorithm bellow and other two algorithms are mentioned in [10].

1.**ALM**(Augmented Lagrangian Multiplier): **Algorithm for ALM**
Inputs :

$$b \in \mathbb{R}^m, A \in \mathbb{R}^{m \times n}$$

$STEP1$:**while** not converged (k=0,1,2,..) **do**
$STEP2$:

$$t_1 \leftarrow 1, z_1 \leftarrow x_k, u_1 \leftarrow x_k$$

$STEP3$: **while** not converged (l=0,1,2,..) **do**
$STEP4$:

$$u_{l+1} \leftarrow soft\left(z_1 - \frac{1}{\tau}A^T\left(Az_1 - b - \frac{1}{\mu_k}y_k\right), \frac{1}{\mu_k\tau}\right)$$

*STEP*5 :

$$t_{l+1} \leftarrow \frac{1}{2}\left(1 + \sqrt{1 + 4t_l^2}\right)$$

*STEP*6 :

$$t_{l+1} \leftarrow u_{l+1}\frac{t_l - 1}{t_{l+1}}\left(u_{l+1} - u_l\right)$$

*STEP*7 : **end while**
*STEP*8 :

$$x_{k+1} \leftarrow u_{l+1}$$

*STEP*9 :

$$y_{k+1} \leftarrow y_k + \mu_k\left(b - Ax_{k+1}\right)$$

*STEP*10 :

$$\mu_{k+1} \leftarrow \rho \cdot \mu_k$$

*STEP*11 : **end while**
Outputs :

$$x* \leftarrow x_k$$

ALM is applied to the following optimization problem

$$\max_{y} b^T y$$

$$\text{subj. to } A^T y \in B_1^\infty$$

where

$$B_1^\infty = \{x \in \mathbb{R}^n : \|x\|_\infty \leqslant 1\}$$

This method of application ALM algorithm to the dual problem is called **DALM**(**D**ual **A**ugmented **L**agrangian **M**ultiplier).

2   Image Quality Assessment

Using the following measurements we performed comparison between reconstructed image with the original image :

 – MSE (Mean Square Error)

$$MSE = \frac{1}{MN}\sum_{x=1}^{M}\sum_{y=1}^{N}\left[f(x,y) - r(x,y)\right]^2$$

where f(x,y) is the original image and r(x,y) is the reconstructed image. Lower the MSE, better the image quality.
 – PSNR (Peak-Signal-to-Noise-Ratio)

$$PSNR = 10 \log \frac{L^2}{MSE}$$

where L=maximum value of pixel in the image. Good PSNR corresponds to range of 25-35dB.

- SSIM (Structural SIMilarity Index)

$$SSIM(x,y) = \frac{(2\mu_x\mu_y + C_1)(2\sigma_{xy} + C_2)}{(\mu_x^2 + \mu_y^2 + C_1)(\sigma_x^2 + \sigma_y^2 + C_2)}$$

SSIM lies between 0 and 1.Higher the value of SSIM,better the image quality[8].

- MoD (Mean of difference): This measure is calculated as using the formula:

$$MoD(x,y) = mean(f(x,y) - r(x,y))$$

where f(x,y) is the original image and r(x,y) is the reconstructed image.

3 Results and Discussion

In this section, the experimental results are presented and provided comparisons between different $\ell1$-norm minimization algorithms.Fig.[1] (a) and (b), shows the original M49 image and original Vesta image respectively.

3.1 Results for DALM

We obtain results using 25 % measurements, 50 % measurements and 75 % measurements of M49 image and reconstructed images are as shown in Figs.[2](a),(b),(c) and (d),(e),(f)shows Obtained results using 25 % measurements, 50 % measurements and 75 % measurements of Vesta image.

The Image Quality measurements obtained are as shown in Table. 1 & Table. 2.

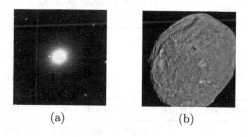

(a)   (b)

Fig. 1. Original Images

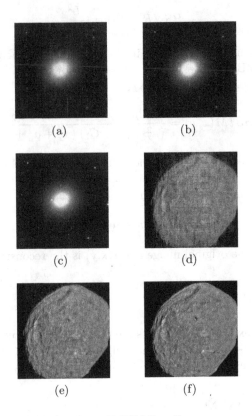

Fig. 2. DALM Results

Table 1. Objective measurements of M49 image using DALM

% of measurements	SSIM	PSNR	MSE	MoD
25	0.8092	30.8055	54.0167	2.019
50	0.8718	33.9439	26.2232	1.3546
75	0.918	37.8015	10.7878	0.9645

Table 2. Objective measurements of Vesta image using DALM

% of measurements	SSIM	PSNR	MSE	MoD
25	0.2969	19.5607	719.4733	9.1886
50	0.5645	24.4246	234.7558	4.9205
75	0.7494	28.6091	89.5716	2.8187

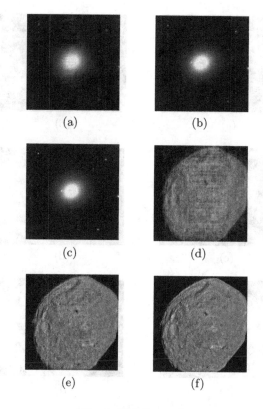

(a)   (b)

(c)   (d)

(e)   (f)

Fig. 3. FISTA results

Table 3. Objective measurements of M49 image using FISTA

% of measurements	SSIM	PSNR	MSE	MoD
25	0.7944	30.128	63.1362	0
50	0.8715	34.3148	24.0769	1.3155
75	0.9178	37.7804	10.8403	0.9314

Table 4. Objective measurements of Vesta image using FISTA

% of measurements	SSIM	PSNR	MSE	MoD
25	0.3498	20.1435	629.1178	8.6253
50	0.5635	24.5008	230.6765	4.9107
75	0.7651	28.9307	83.1784	2.6866

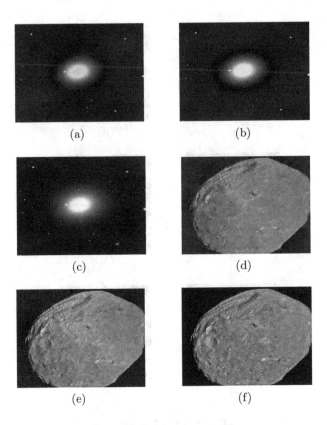

(a)

(b)

(c)

(d)

(e)

(f)

Fig. 4. Split-Bregman results

Table 5. Objective measurements of M49 image using Split-Bregman

% of measurements	SSIM	PSNR	MSE	MoD
25	0.7555	26.1538	157.6532	0
50	0.7236	25.5414	181.5256	0
75	0.785	38.106	10.0573	0

Table 6. Objective measurements of Vesta image using Split-Bregman

% of measurements	SSIM	PSNR	MSE	MoD
25	0.5026	23.0883	319.3415	0
50	0.7408	26.0929	159.8779	0
75	0.9454	30.9877	51.7978	0

3.2 Results for FISTA

We obtain results using 25 % measurements, 50 % measurements and 75 % measurements of M49 image and reconstructed images are as shown in Figs.[3](a),(b),(c) and (d),(e),(f)shows Obtained results using 25 % measurements, 50 % measurements and 75 % measurements of Vesta image.

The Image Quality measurements obtained are shown in Table. 3 & Table. 4.

3.3 Results for Split-Bregman Method

We obtain results using 25 % measurements, 50 % measurements and 75 % measurements of M49 image and reconstructed images are as shown in Figs.[4](a),(b),(c) and (d),(e),(f)shows Obtained results using 25 % measurements, 50 % measurements and 75 % measurements of Vesta image.

The Image Quality measurements obtained are shown in Table. 5 & Table. 6.

4 Conclusion

This paper presents how different ℓ1-norm minimization can be applied for astronomical images. From our experiments, out of three methods,Split-Bregman is providing better results than FISTA and DALM in terms of measurements like SSIM,PSNR,MSE and MoD for Vesta image.But DALM provides better results than FISTA and Split-Bregman for M49 image.Thus from our experiments we conclude that Split-Bregman provides better results for images which are not much sparse by itself. DALM provides better results for images which are more sparse by itself.

References

1. Bobin, J., Starck, J.-L., Ottensamer, R.: Compressed Sensing in Astronomy. IEEE Journal of Selected Topics in Signal Processing 2(5), 718–726 (2008)
2. Ma, J., Le Dimet, F.-X.: Deblurring From Highly Incomplete Measurements for Remote Sensing. IEEE Transactions on Geoscience and Remote Sensing 47(3), 792–802 (2009)
3. Scaife, A.M.M., Wiaux, Y.: The application of Compressed Sensing Techniques in Radio Astronomy. Annals of Biomedical Engineering 33(7), 937–942 (2011)
4. Candés, E., Romberg, J.: Sparsity and Incoherence in Compressive Sampling. Inverse Problems 23(3), 969–985 (2007)
5. Candés, E.J.: Compressive sampling. In: Proc. Int. Congress Math. Euro. Math. Soc., vol. 3, p. 1433 (2006)
6. Donoho, D.L.: Compressed sensing. IEEE Trans. Inform. Theory 52(4), 1289–1306 (2006)
7. Candés, E.J., Wakin, M.B.: People Hearing Without Listening. An Introduction To Compressive Sampling (2007), Preprint, http://www.dsp.ece.rice.edu/cs/
8. Wang, Z., Bovik, A.C., Sheikh, H.R., Simoncelli, E.P.: Image quality assessment: From error visibility to structural similarity. IEEE Transactions on Image Processing 13(4), 600–612 (2004)

9. Goldstein, T., Osher, S.: The Split Bregman Method For L1 Regularized Problems. Journ. Sci. Comput. 45, 272–293 (2010)
10. Yang, A.Y., Ganesh, A., Zhou, Z., Shankar Sastry, S., Ma, Y.: A Review of Fast l1-Minimization Algorithms for Robust Face Recognition. University of California at Berkeley Technical report UCB/EECS-2010-13 (2010)
11. Mourad, N.: Fundamentals of compressed sensing (CS). Ph.D Thesis

Speaker Recognition in Emotional Environment

Shashidhar G. Koolagudi[1], Kritika Sharma[1], and K. Sreenivasa Rao[2]

[1] School of Computing, Graphic Era University, Dehradun - 248002,
Uttarakhand, India
[2] School of Information Technology, Indian Institute of Technology Kharagpur,
Kharagpur - 721302, West Bengal, India
koolagudi@{ieee.org,yahoo.com}, kritikasharma.it@gmail.com,
ksrao@iitkgp.ac.in

Abstract. This paper deals with development of speaker recognition
system in emotional environments. In this paper, Mel-frequency cepstral
coefficients (MFCC) have been used to represent the speaker specific
information. A simulated speech corpus of Hindi language are used to
check the performance of speaker recognition in emotional environment.
The emotions included in this study are anger, neutral, sad, happy and
surprise. Emotion recognition models are developed using Gaussian mixture models. Performance of speaker recognition is studied in emotional
environment. The results show that emotions play vital role in speaker
recognition.

Keywords: Speaker recognition, Mel frequency cepstral coefficients,
Emotional environment, Gaussian mixture models.

1 Introduction

Speaker recognition is the process of identification of speaker on the basis of
the speaker specific characteristics of speech signal. Some of the important applications include access control of various services through voice, security control for confidential information, telephone based automated access and so on.
On the basis of spoken text, speaker recognition can be classified into :- text-
dependent (fixed text) or text-independent (free-text). Speaker identification in
text-dependent method, speaker is asked to generate the speech signal containing
the same text for both training and testing, whereas in text-independent case,
speaker recognition is performed irrespective of the training and testing speech
text. In this paper, we focus on the text-dependent case of speaker recognition.

Speaker recognition depends on some speaker specific properties. One of them
is an emotion embedded with the speech signal. Today's speaker recognition systems perform efficiently with studio recorded neutral speech. In this case speaker
has to provide the speech in neutral emotion for both training and testing purposes. This is always impossible and impractical in real world scenarios. Speaker
recognition in emotional environments is important research issue in human-
computer interaction [1]. Emotional environment can be termed as the speech
produced under the influence of different emotional states like anger, sadness

J. Mathew et al. (Eds.): ICECCS 2012, CCIS 305, pp. 117–124, 2012.

and fear. Speaker recognition in emotional situations is an acute need of today's systems as these emotions are present in our daily life and play a crucial role. The studies suggest that human speech is approximately 10% unemotional [2] and the remaining 90% of the speech communication is done through affective (emotional) speech. Hence, speaker recognition in emotional environment can be used to make most of the speech systems natural. Speaker recognition in emotional scenarios may be we find in the applications like call center conversation analysis, critical speaker recognition applications of real world.

According to available literature, speaker recognition studies and performed on the speech recorded in neutral environment in which the speech is not affected by any emotion. Yuan and Zheng used hidden Markov model for speaker recognition in neutral environment [3]. Furui stressed upon the extracting of speaker specific features from speech signal [4]. Shahin used second-order hidden Markov models to increase speaker recognition performance using features [5]. Some studies focused on speaker recognition in emotional environment. Koike explained prosodic parameters in emotional speech for speaker recognition [6]. Wu et al. studied the influence of emotion on performance of a GMM-UBM speaker recognition system [7]. Tao et al. used prosody conversion from neutral speech to emotional speech for emotion recognition [8]. Marius V. Ghiurcau et al. assessed the effect of emotional state of a speaker when text-independent speaker identification is performed using support vector machine and proposed solution for increasing the performances [9]. Shahin tested and compared three models i.e. Hidden Markov Models (HMMs), Second-Order Circular Hidden Markov Models (CHMM2s) and Suprasegmental Hidden Markov Models (SPHMMs) and concluded SPHMMs as the superior models over other models [10]. In this paper, the main aim is to check the performance of speaker recognition when testing and training speech utterances are in different emotions. For initial statics hindi emotional speech databases collected at IIT Kharaghpur (IITKGP SEHSC) has been used.

This paper is further organized as follows: Section 2 introduces the speaker database used in the study. Feature extraction is presented in Section 3. Section 4 describes the development of speaker recognition model. Finally, paper ends with conclusion and summary.

2   Database

In this study emotional databases are used as mentioned in this section. They are individually and in combination used for speaker recognition. Final system is developed with 10 speakers containing the emotional sentences of language. Following section briefly describes the database used in this work.

2.1   Hindi Emotional Speech Corpus

The proposed database is recorded using 10 (5 male and 5 female) professional artists from Gyanavani FM radio station, Varanasi, India [11]. The artists have sufficient experience in expressing the desired emotions from the neutral sentences. The male artists are in the age group of 28-48 years with varied experience of 5-20 years. Similarly female artists are from the age group of 20-30 years with 3-10 years of experience. For recording the emotions, 15 Hindi text prompts are used. All the sentences are emotionally neutral in meaning. Each of the artists has to speak the 15 sentences in 8 basic emotions in one session. The number of sessions recorded for preparing the database is 10. The total number of utterances in the database is 12000 (15 sentences X 8 emotions X 10 speakers X 10 sessions). Each emotion has 1500 utterances. The number of words and syllables in the sentences vary from 4-7 and 9-17 respectively. The total duration of the database is around 9 hours. The eight emotions considered for collecting the proposed speech corpus are: anger, disgust, fear, happy, neutral, sad, sarcastic and surprise. The speech samples are recorded using SHURE dynamic cardioids microphone C660N. The distance between the speaker and the microphone is maintained to be around 1 ft. Speech signal was sampled at 16 kHz and each sample is represented as 16 bit number. The sessions are recorded on alternate days to capture the variability in human speech production mechanism.

In each session, all the artists have given the recordings of 15 sentences in 8 emotions. The recording is done in such a way that each artist has to speak all the sentences at a stretch in a particular emotion. This provides coherence among the sentences for each emotion category. Since, all the artists are from the same organization, it ensures the coherence in the quality of the collected speech data. The entire speech database is recorded using single microphone and at the same location. The recording was done in a quiet room, without any obstacles in the recording path.

3 Feature Extraction

3.1 Mel Frequency Cepstral Coefficients (MFCCs)

A human auditory system is assumed to process a speech signal in a nonlinear fashion. It is well known that lower frequency components of a speech signal contain more phoneme specific information. Therefore a nonlinear mel scale filter bank has been the mel frequency cepstrum is a representation of the short term power spectrum on a nonlinear mel frequency scale. Conversion from normal frequency f to mel frequency m is given by the equation

$$m = 2595 \log_{10} \left(\frac{f}{700} + 1 \right)$$

The steps used for obtaining mel frquency cepstral coefficients (MFCCs) from a speech signal are as follows :

- Pre-emphasis the speech signal.
- Divide the speech signal into a sequence of frames with a frame size of 20 ms and a shift size of 5 ms. Apply the hamming window over each of the frames.

- Compute the magnitude spectrum for each windowed frame by applying DFT.
- Mel spectrum is computed by passing the DFT signal through a mel filter bank.
- DCT is applied to the log mel frequency coefficients (log mel spectrum) to derive the desired MFCCs.

4   Development of Speaker Recognition Model

In this work, Gaussian Mixture Models are used to develop speaker recognition systems. A Gaussian Mixture Model (GMM) is a parametric probability density function represented as a weighted sum of Gaussian component densities. GMMs are known to capture distribution of data points from the input feature space. Therefore, GMMs are best suited for developing speaker recognition models when one has high number of feature vectors. Fig. 1 shows the training phase of GMMs.

Fig. 1. Training Phase

Here decision regarding the speaker category of the feature vector is done based on its probability of coming from the feature vectors of the specific model. As it is among the most statistically matured methods for clustering and density estimation, the data points are classified using a multivariate Gaussian mixture density that models the probability density function. GMM find the weight of each input set usng expectation-maximization algorithm. In this paper GMM is designed with 64 centers and iterated 50 times to attain convergence of weights.

General scenario of developing speaker models is given in Fig. 1. Speech signal of Sp1 is given as an input to the feature extraction block. Extracted features are used for developing the Sp1 model.

While testing speech utterances of an unknown speaker among the set of predefined 10 speakers is given as an input to feature extraction model. The probability of each generated feature vector belonging to the speaker model is given by each of the trained speaker models. The sum of them probabilities for all feature vectors of an utterance decides the speaker that has spoken that utterance. Decision device will perform this task and outputs the speaker category. Testing of speech against trained Gaussian mixture models is shown in Fig. 2.

While testing the above models, neutral sentences of session 9 and 10 are used. This is the way today's speaker recognition models are developed (with studio recorded neutral speech). Table 1 gives the confusion matrix of the speaker recognition results. In the matrix all diagonal elements represents the correct

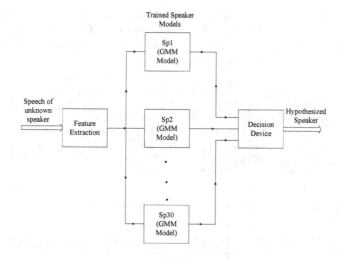

Fig. 2. Testing of speaker recognition models

Table 1. Speaker identification performance (in %),for the models trained with neutral speech and tested with neutral. Here, Sp1- Speaker1, Sp2- Speaker2, Sp3- Speaker3, Sp4- Speaker4, Sp5- Speaker5, Sp6- Speaker6, Sp7- Speaker7, Sp8- Speaker8, Sp9- Speaker9, Sp10- Speaker10.

	Sp1	Sp2	Sp3	Sp4	Sp5	Sp6	Sp7	Sp8	Sp9	Sp10
Sp1	100	0	0	0	0	0	0	0	0	0
Sp2	0	60	0	0	0	0	40	0	0	0
Sp3	0	0	97	0	3	0	0	0	0	0
Sp4	0	0	0	97	0	0	0	0	3	0
Sp5	17	0	0	0	53	3	0	0	0	27
Sp6	0	0	0	0	10	90	0	0	0	0
Sp7	0	0	0	0	0	0	100	0	0	0
Sp8	0	0	0	0	0	0	13	60	27	0
Sp9	0	0	0	0	0	0	0	0	100	0
Sp10	7	0	0	0	0	0	0	0	0	93

classification. The other members in each row indicate the miss classification pattern. For example considering the row corresponding to speaker 6 90% of utterances are correctly recognized (see column Sp6 and row Sp6 of Table 1).

The purpose of this study is to analyze speaker recognition performance during speaker's emotional conditions. This can be studied by testing the neutral GMM models using the emotional sentences spoken by the same speaker. In this experiments the models trained with neutral sentence are tested with the other emotions such as anger, happy, sad and surprise. The results are mentioned in following Table 2. It is clear from the table that, recognition of the speaker in the case of neutral test utterance is very high of around 95% (see the last column of the fourth row of Table 2).

Table 2. Speaker identification performance (in %),for the model trained with neutral speech and tested with utterances of different emotions. Here, Sp1- Speaker1, Sp2- Speaker2, Sp3- Speaker3, Sp4- Speaker4, Sp5- Speaker5, Sp6- Speaker6, Sp7- Speaker7, Sp8- Speaker8, Sp9- Speaker9, Sp10- Speaker10.

	Sp1	Sp2	Sp3	Sp4	Sp5	Sp6	Sp7	Sp8	Sp9	Sp10	Average.
Anger	37	70	100	37	7	0	63	7	74	43	43.8
Happy	47	100	100	80	20	17	100	63	100	23	65
Neutral	100	100	100	100	73	100	100	80	100	100	95.3
Sadness	43	100	100	43	23	27	100	60	100	97	69.3
Surprise	0	60	100	0	0	7	57	0	20	0	24.4

Table 3. Speaker recogntion classification performance when trained and tested with differnt emotions

Models Trained with	Models Tested with	Average Recognition
Anger	Anger	99.1
	Neutral	77
Neutral	Neutral	95.3
	Anger	43.8
	Happy	65
	Sad	69.3
	surprise	24.4
Sad	Sad	100
	Neutral	68.6
Surprise	Surprise	100
	Neutral	78.7

These findings indicate the importance of embedded emotions, present in speech utterances, in speaker recognition. The study has been extended to know the performance of speaker recognition. While speaker recognition models are trained with emotional sentences and tested with the sentences of the same emotion as well as neutral. In this experiments speaker models are developed with the emotional sentences and validation of the models is performed using neutral and emotional sentences spoken by the same speaker. Results obtained are presented in the Table 3.

Strengthening the results obtained in Table 1, the speaker recognition in the cases of training the models with emotional sentences and testing them with the sentences of the same emotion is very high. Testing the emotional models with neutral test utterances also yielded comparatively poorer speaker recognition performance. For instance it may be observed from Table 3 that, speaker recognition is 99% when training and testing utterances are of anger emotion. In the cases of happy, sad and surprise the speaker recognition performance is perfect.

5   Summary and Conclusion

In this paper, emotional speech corpus of Hindi language has been used for recognizing the speakers from speech signal. Different emotions like anger, sad, happy, neutral and surprise are used to study the influence of emotions on speaker recognition. MFCC features are used as features to represent speaker specific information. Speaker recognition models are developed using Gaussian mixture models. The result obtained shows the large influence of emotion on speaker recognition. The study can be extended to study the effect of language on speaker recognition. Transformation of features may be explored to improve the speaker recognition performance in emotional environments.

References

1. Picard, R.W.: Affective Computing, MIT Media Lab Perceptual Computing Section Tech. Rep., no. 321 (1995)
2. Portal, T.H.: Research on emotions and human machine interaction, http://emotion-research.net
3. Zheng, C., Yuan, B.Z.: Text-dependent speaker identification using circular hidden Markov models. In: Proc. IEEE Int. Conf. on Acoustics, Speech, and Signal Processing, vol. 1, pp. 580–582 (March 1988)
4. Furui, S.: Speaker-dependent-feature-extraction, recognition, and processing techniques. Speech Communication 10, 505–520 (1991)
5. Shahin, I.: Using second-order hidden Markov model to improve speaker identification recognition performance under neutral condition. In: Proc. 10th IEEE Int. Conf. on Electronics, Circuits and Systems, ICECS 2003, Sharjah, United Arab Emirates, pp. 124–127 (December 2003)
6. Koike, K., Suzuki, H., Saito, H.: Prosodic parameters in emotional speech. In: Proc. of Int. Conf. on Spoken Language Processing ICSLP 1998, November 30-December 4, pp. 679–682 (1998)
7. Wu, W., Zheng, T.F., Xu, M.X., Bao, H.J.: Study on speaker verification on emotional speech. In: Proc. of Int. Conf. on Spoken Language Processing, INTERSPEECH 2006, pp. 2102–2105 (September 2006)
8. Tao, J., Kang, Y., Li, A.: Prosody conversion from neutral speech to emotional speech. IEEE Trans. on Audio, Speech, and Language Processing 14(4), 1145–1154 (2006)
9. Ghiurcau, M.V., Rusu, C., Astola, J.: Speaker recognition in an emotional environment. In: Proceedings of SPAMEC 2011, Cluj-Napoca, Romania (2011)
10. Shahin, I.: Speaker Identification in Emotional Environments. Iranian Journal of Electrical and Computer Engineering 8(1) (Winter-Spring 2009)
11. Koolagudi, S.G., Krothapalli, R.S.: Two stage emotion recognition based on speaking rate. International Journal of Speech Technology 14 (2011)
12. Burkhardt, F., Paeschke, A., Rolfes, M., Sendlmeier, W., Weiss, B.: A database of German emotional speech. In: Proc. Interspeech 2005, pp. 1517–1520 (2005)
13. Rajeswara Rao, R., Nagesh Kamakshi Prasad, A., Ephraim Babu, K.: Text-Dependent Speaker Recognition System for Indian Languages, November 5. JNT University of Hyderabad, India (2007)

14. Zhou, Y., Wang, J., Zhang, X.: Research on Adaptive Speaker Identification Based on GMM. In: 2009 International Forum on Computer Science-Technology and Applications (2009)
15. Abushariah, A.A.M., Gunawan, T.S., Khalifa, O.O.: English Digits Speech Recognition System Based on Hidden Markov Models. In: International Conference on Computer and Communication Engineering (ICCCE 2010), Kuala Lumpur, Malaysia, May 11-13 (2010)
16. Koolagudi, S.G., Reddy, R., Sreenivasa Rao, K.: Emotion Recognition from Speech Signal using Epoch Parameters. IEEE (2010)

CBMIR: Content Based Medical Image Retrieval System Using Texture and Intensity for Dental Images

B. Ramamurthy[1], K.R. Chandran[2], V.R. Meenakshi[3], and V. Shilpa[3]

[1,2] PSG College of Technology, Coimbatore 641004
[3] Sri Ramakrishna Engineering College, Coimbatore 641 022
ramamurthy_1976@yahoo.com, chandran_k_r@yahoo.co.in

Abstract. Image retrieval systems attempt to search through a database to find images that are perceptually similar to a query image. This work aims to develop an efficient visual-Content-based technique to search, browse and retrieve relevant images from large-scale of medical image collections Features play a vital role during the image retrieval. The various features that can be extracted are texture, color, intensity, shape, resolution, global and local features etc. In this work, we concentrate on the specific medical domain. The features such as color may not prove to be a very efficient method because the medical domain largely deals with the gray scale images. The features explored in this work are intensity, texture. The first step is to extract the texture feature and the intensity feature from the given input image. Then the both features are combined to form the single feature vector of the image by using the fusion method. The resulting image is compared to the images in the database. The N top most similar images are then retrieved from the database.

Keywords: CBMIR, LBP, CBIR, Euclidean Method, Precision, Recall.

1 Introduction

In medical domain, the development of digital equipment allows to retrieve and store large number of medical data, including images. The automatic search and retrieval of images from databases poses technical challenges. Content-based Medical Image Retrieval (CBMIR) has the ability to retrieve images on the basis of content. In the medical domain CBMIR system is used to aid physicians to diagnose a patient disease, the accuracy of this method is based on retrieval of certain visual features of an image.

1.1 Content Based Image Retrieval

Content Based Image Retrieval is one of the important methods for image retrieval system. It enhances the accuracy of the image being retrieved, It is applicable for efficient query processing, automatically extract the low-level features such as texture, intensity, shape and color in order to classify the query and retrieve the similar images from the huge scale image collection of database. In CBIR, each image that is stored in the database has its features extracted and compared to the query image features.

J. Mathew et al. (Eds.): ICECCS 2012, CCIS 305, pp. 125–134, 2012.

1.2   Medical CBIR

In medical field, the usage of image database has increased. Hence there is a need to develop an efficient technique for searching and retrieving images.CBIR is one of the alternative approaches for searching and retrieval of the images. This approach is querying the images based on extracting the low-level visual features it is independent of people to find the similarity of images in the database. A one to one integrated matching is used to compare the query and collection of database image. Medical image retrieval is task specific that is for diagnostic study or organ or modality it does not transferable for other medical applications.

1.3   Objectives

The main goal of this work is to retrieve the image from the database using different visual feature extraction for comparing the similarity of query and target image at each level in terms of normalized distance measure i.e. Euclidean distance, the image retrieval system is used to provide high accurate results on the diagnostic process. The following Fig.1 shows general CBIR system architecture.

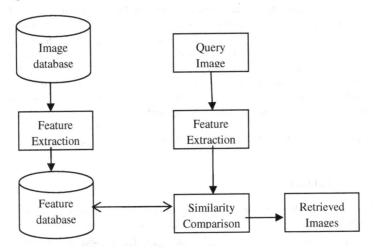

Fig. 1. General CBIR System Architecture

2   Related Work

The majority of the existing medical image retrieval systems are dedicated to specific medical domains. They have certain conveniences and inconveniences that are described below,

2.1   Unsupervised Feature Selection Applied to Content Based Retrieval of Lung Images

This method proposes the new hierarchical approach to content-based image retrieval called the "customized-Queries" Approach (CQA).In CQA search method and clustering algorithms are implemented. This method is implemented in two steps: step 1 (feature selection for classification) is used to classify the query images into one of the given disease group. In step 2 the best defined features similarity within a single disease group has been identified. This system retrieves images from only one main group at a time, it does not retrieve the images from the next most possible group to provide next best match in the group. This address the limitation of CQA [1].

2.2 Image Retrieval Based on Color and Texture Features of the Image Sub-blocks

In this method the author proposed the image retrieval system based on color and texture feature. An image is partitioned into sub blocks; color of each sub block is taking out by measuring the HSV color space into non-equal interval. The drawback of the system is the HSV color space. This method will concentrate only on the color images it does not favor for the specific medical domain. Because in specialized fields, namely in medical domain the absolute color and grey level features are very limited [2].

2.3 Using Texture-Based Symbolic Features for Medical Image Representation

In this method the author presents medical image categorization approach in the context of CISMeF Health Catalogue-capability to create queries by giving the image associated keywords to retrieve the health resources. This uses texture and high order statistical movements feature this can be improved by adding other features and classifiers to improve the results [3].

2.4 Content Based Image Retrieval Based on Pyramid Structure Wavelet

In this method the author proposed the result of image retrieval by using color, shape and texture and combination between them by using Receiver-operating characteristic curve (ROC).The hybrid technique is used with ROC technique to give best results. In hybrid technique it compares HSV query with HSV database images and it provides sorted list with sorted images and their differences. The major drawback is that it takes the longer time for calculation and comparison with other technique [4].

2.5 Integration of Color, Shape and Texture for Image Retrieval

Color will be used to differentiate objects, places and etc. colors are defined in three dimensional color spaces such as RGB (Red, Green, and Blue), HSV (Hue, Saturation, and Value) or HSB (Hue, Saturation, and Brightness). Most image formats use the RGB color space to store information. The RGB color space is defined as a unit cube with red, green, and blue axes. Thus, a vector with three co-ordinates represents the color in this space which represents black when all of them set to zeros and represents white when all three coordinates are set to 1. The histogram captures only the color distribution and it does not include any spatial correlation between individual pixels which it may cause to have limited discriminative power. [5]

3   Proposed Method

The proposed method is based on texture and intensity of images. The algorithm for the proposed system is given below:

Algorithm

STEP 1: Create a database containing various dental images.
STEP 2: Extract the Texture and Intensity feature of each image in the database.
STEP 3: Construct a combined feature vector for Texture and Intensity.
STEP 4: The new images formed are stored in another database called the
Featured Databases
STEP 5: Find the distance between feature vectors of query images and that of
Featured database images.
STEP 6: Sort the distance and Retrieve the N-top most similar images.

The Local Binary Pattern algorithm (LBP) outperforms most of the other methods for the process of texture extraction. It provides high accuracy results due to pixel-by-pixel comparison in the LBP. The retrieval process is made more efficient by extracting both texture and intensity. The extracted feature images are combined to form a single feature vector value and it is then compared using Euclidean distance method to retrieve similar images from the database. The following Fig.2 shows System Flow diagram of the proposed system.

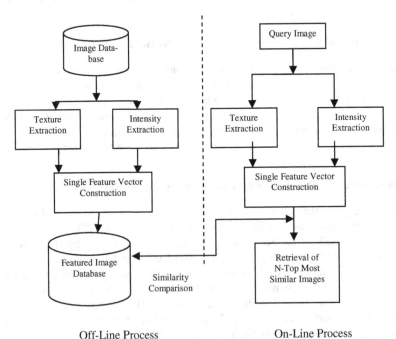

Off-Line Process   On-Line Process

Fig. 2. System Flow Diagram

3.1   Texture Feature Extraction

The texture is a manner in which the constituent parts are united. It is the Structure or repeated patterns on an image. Texture in digital images can be determined by if the neighboring pixels satisfy a specified criterion of similarity. Local binary pattern algorithm is used for texture extraction. Local Binary Pattern (LBP) is used in texture extraction of the image. The LBP operator works with eight neighbors of the pixel using the center pixel value as threshold and the LBP code for a neighborhood was produced by multiplying the threshold values with weights given to the corresponding pixels, and summing up the result. The LBP can be treated as a special case of a multi-dimensional co-occurrence statistic [6].

3.2   Intensity Feature Extraction

The Intensity is the amount of light the pixel reproduces (how bright it is).Gray scale images also known as black and white images are composed exclusively of shades of gray, varying from black at the weakest intensity to white at the strongest. The binary representations assume that 0 is black and the maximum value (255 at 8 bpp, 65,535 at 16 bpp, etc.) is white. The intensity information is extracted by determining the pixel values.

3.3   Image Retrieval System

Retrieving the most similar images from the database is can be done in two ways. Text-Based retrieval and Content-Based retrieval. In Text-Based retrieval, the description of the image cannot be interpreted easily to retrieve the images. The alternative to the Text-Based image searching is CBIR. It can enhance the accuracy of information being returned. It develops an efficient visual content based technique to search, browse and retrieve the relevant images from the large scale digital image collections. In CBIR, each image is extracted by using low level image features such as texture, intensity, shape etc.A similarity between the query image and the database images is measured using Euclidean distance vector, it provides the N most similar images. The top most similar images are displayed based on the minimum distance between the query image and the images in the database [7].

4   Implementation and Result Analysis

The implementation and experimental results below present the working of the retrieval of an image from the database.

4.1   Texture Feature Extraction Using Local Binary Pattern (LBP)

Local Binary Pattern algorithm is used for texture extraction. The LBP operator defines the texture in the image is represented by thresholding the neighborhood with the gray value of its center pixel and the results will be represented as binary code format. The pixel-to-pixel comparison in the image produces the texture and the resulting image is in the form of texture histogram [8].

Local Binary Pattern Algorithm for Texture Extraction

STEP 1: For each pixel in the cell, compare it with other 8 neighboring pixels.

STEP 2: Follow the pixel along a circle i.e. clockwise or anti-clockwise.

STEP 3: If the centre pixel's value is greater than the neighbor, write "1" otherwise Write"0".This gives an 8-bit binary number.

STEP 4: Compute the histogram over the cell, of the frequency of each "number" Occurring.

STEP 5: Optionally normalize the histogram.

STEP 6: Concatenate normalized histogram of all cell.

STEP 7: This gives the feature vector for the window. This can be used for Classification

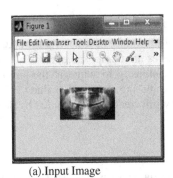   

(a).Input Image   (b).Texture Extracted Image

Fig. 3. Texture Extracted Image

4.2   Intensity Feature Extraction

Intensity information can be extracted by determining the pixel values. An intensity image is a data matrix, I, whose values represent intensities within some range. An intensity image is represented as a single matrix, with each element of the matrix corresponding to one image pixel. The matrix can be of class double, uint8, or uint16. While intensity images are rarely saved with a color map, a color map is still used to display them. In essence, handles intensity images are treated as indexed images. This figure depicts an intensity image of class double. To display an intensity image, use the imagesc ("image scale") function, which enables to set the range of intensity values. Imagesc scales the image data to use the full color map. Use the two-input form of images to display an intensity image.

4.3   Feature Vector Construction

The common single feature extraction is constructed based on the process of texture and intensity feature extraction. The one or more features are combined by using the fusion method. To generate the combined feature vector, the texture features have been fused with the intensity features. The combined single feature vector results provided to extract the images from the large collection of images based on the similarity comparison of the query image [9].

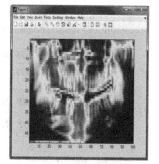

| (a). Input Image | (b). Intensity Extracted Image |

Fig. 4. Intensity Feature Extraction

 +  =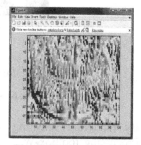

| Texture Extracted Image | Intensity Extracted Image | Single Featured Image |

Fig. 5. Single Feature Vector Image

4.4   Image Retrieval Using Euclidean Method

The Euclidean distance is calculated for every query image and the collection of images in the database to retrieve the similar images. The database images are compared with the query image. It calculates the distance between the query image and database images to implement the simplest method by using the formula,

$$d\left(A^{I}, A^{Q}\right) = \sqrt{\sum_{i=1}^{n}\left(A_{i}^{I} - A_{i}^{Q}\right)^{2}} \qquad (1)$$

Where,

AI-is the Images in the database

AQ-is the query image for retrieval

Euclidean method has an array of Euclidean distance, which is then sorted. The minimum Euclidean distance image is considered as a most similar image during retrieval. As a result of retrieval the top most images are displayed [10].

4.5   Retrieval Efficiency

For retrieval efficiency calculation, traditional measures namely precision and recall were calculated using thousand real time medical images from MATLAB workspace database

Figure 6 shows the sample snapshot of the work. Standard formulas have been used to calculate the precision and recall measures for some sample query images.

$$PRECISION = \frac{NO.OF\ RELEVENT\ IMAGES\ RETRIEVED}{TOTAL\ NO.OF\ IMAGES\ RETRIEVED}$$

$$RECALL = \frac{NO.OF\ RELEVENT\ IMAGES\ RETRIEVED}{TOTAL\ NO.OF\ RELEVENT\ IMAGES\ IN\ THE\ DATABASE}$$

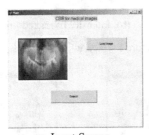 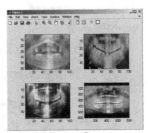

a. Input Screen   b. Output Screen

Fig. 6. Sample snapshot (GUI) of the proposed work

By randomly selecting some sample query images from the MATLAB-Image Processing toolbox-Workspace Database, the system was tested and the results are shown in the following TABLE 1 and the graphical representation for the precision and recall values are presented in Fig.7.

Table 1. Precision and Recall Values in %

Query Image	Precision	Recall
1	52.0	17.0
2	77.0	64.0
3	22.0	12.0
4	77.0	35.0
5	47.0	25.0

4.6   Result Comparisons

The proposed method gives better results than first three existing methodologies listed in the related works. CQA clusters the images into classes and returns the relevant images from only one probable class. This does not provide the user with multiple choices for the relevant results. Whereas, the proposed work returns all possible relevant images from the database as results. Image Retrieval based on the texture and color has limited efficiency for specific medical image database, due to the use of

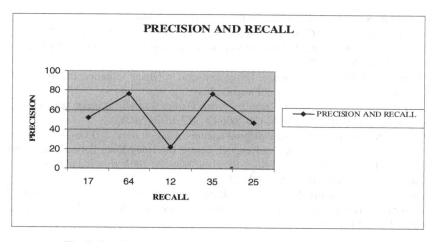

Fig. 7. Graphical representation for the precision and recall values

color feature. The performance of Texture based symbolic feature for medical image database can be improved by the use of additional features. The proposed method gives better results by using texture and intensity features.

5   Conclusion and Future Enhancement

In this work, we present effective image retrieval system based on the Texture and Intensity features. This approach uses Local Binary Pattern to extract the texture feature by comparing each pixel by pixel. This approach provides more accurate results. This texture information is then combined with Intensity information of the image to produce a single featured Image. This approach is more robust and provides accurate results. The performance of this retrieval system is compared with many others. The experiments proven a good retrieval rate and display the accurate results. In future, the system can be improved by applying new image content feature and evaluation of the technique on the larger database.

References

1. Dy, J.G., Brodley, C.E., Kak, A., Broderic, L.S., Aisen, A.M.: Unsupervised feature selection applied to context based retrieval of lung images. IEEE Transactions on Pattern Analysis and Machine Intelligence 25(3) (March 2003)
2. Kavitha, C., Prabhakara Rao, B., Govardhan, A.: Image Retrieval based on Color and Texture feature of the image sub block. International Journal of Computer Applications (0975–8887) 15(7) (February 2011)
3. Florea, F., Barbu, E., Rogozan, A., Bensrhair, A.: Using texture based symbolic features for medical image representation. In: The 18th International Conference on Pattern Recognition (ICPRO 2006). IEEE (2006)
4. Youssif, A.A.A., Darwish, A.A., Mohamed, R.A.: Content based medical image retrieval based on pyramid structure wavelet. IJCSNS International Journal of Computer Science and Network Security 10(3) (March 2010)

5. Zare, M.R., Seng, W.C.: Integration of Color, Texture and Shape for Blood Cell Image Retrieval. In: International Conference on Biomedical Engineering, vol. 21, pp. 847–850
6. Mäenpää, T., Pietikäinen, M.: Texture Analysis with Local Binary Pattern. WSPC for Review Volume (May 13, 2004)
7. Marinai, S.: A Survey of Document Image Retrieval in Digital Libraries, Dipartimento di Sistemi e Informatica University of Florence, Italy
8. Unay, D., Ekin, A., Eindhoven: Intensity versus texture for medical image search and retrieval, FP6 IRonDB Project MTK-CT-2006-047217. IEEE (2008)
9. Kong, W.K., Zhang, D., Li, W.: Palmprint feature extraction using 2-D Gabor Filters, Department of Computing,Biometrics Research Centre,The Hong Kong Polytechnic University, HungHom, Kowloon, Hong Kong Received May 15, 2002; received in revised form January 14, 2003; accepted February 14, 2003
10. Veni, S., Narayanankutty, K.A.: Image Enhancement of Medical Images using Gabor Filter Bank on Hexagonal Sampled Grids. World Academy of Science, Engineering and Technology 65 (2010)

A Ridgelet Based Symmetric Multiple Image Encryption in Wavelet Domain Using Chaotic Key Image

Arpit Jain[1], Musheer Ahmad[1], and Vipul Khare[2]

[1] Department of Computer Engineering, Faculty of Engineering and Technology,
Jamia Millia Islamia, New Delhi 110025, India
[2] Department of Computer Science & Information Technology, Jaypee Institute of Information
and Technology, Noida-201301, India

Abstract. In this paper, a novel symmetric multiple image encryption algorithm is proposed. The scheme exploits the features of discrete wavelet transform, finite ridgelet transform and chaotic maps in a way to achieve completely distorted and meaningless single encrypted image which secretly contains the information of the original images. The four original images are first transformed in wavelet domain followed by the finite ridgelet transformations of their HH bands to get two combined transformed matrices. This on pixel values rotation using chaotic key image, mixing and then their shuffling using random sequences results into a final encrypted image. The piece-wise linear chaotic map and 2D gingerbreadman chaotic map are utilized to generate the chaotic key image and shuffling sequences, respectively. The encrypted image is highly distorted and meaningless for a casual observer. The mean square error between the pairs of decrypted and original images is less than 0.004. Moreover, the experimental results and simulation analyses show that the scheme has the advantage of low correlation factor, high key sensitivity and large key space.

Keywords: Multiple image encryption, discrete wavelet transform, finite ridgelet transform, piecewise linear chaotic map, gingerbreadman chaotic map.

1   Introduction

Technological advancements in the field of networks have made information security of digital media like images, audio and video more challenging and demanding. The multimedia images are the most commonly and widely used multimedia data in various applications. The sensitive multimedia images are transmitted or shared via Internet and wired/wireless networks. To secure and prevent the multimedia images sent over the attack-prone networks against intruder's attack, the methods are employed to encrypt images before their transmission. The need of effective and standardized image encryption methods are growing continuously. To fulfill the requirement of privacy and secrecy of multimedia images having the intrinsic features of bulk data capacity and high data redundancy, the traditional encryption algorithms such as Data Encryption Standard, Advanced Encryption Standard, and International

J. Mathew et al. (Eds.): ICECCS 2012, CCIS 305, pp. 135–144, 2012.

Data Encryption Algorithm are ineffective for images [1, 2]. Alternatively, researchers came up with the design of security methods using chaos-based cryptography. They found that chaotic systems/maps have properties such as high randomness, long periods and high sensitivity to initial conditions/control parameters. These desirable features make chaotic maps based image encryption methods robust against intruder's attack. In past decade, a lot of chaos-based methods have been proposed to protect multimedia images [3-11]. Recently, multiple image encryption approach gives a new edge to these modern image encryption techniques, as it results in computation reduction and provides ample security to the multiple plain-images. Consequently, a few multiple image encryption approaches have also been suggested using mostly fractional fourier, wavelet, gyrator transforms where chaotic maps are employed to generate the phase masks, random grids etc. [12-18].

In this work, an entirely different method is proposed to achieve multiple images encryption in wavelet transform domain using finite ridgelet transform. The chaotic maps are employed in the proposed symmetric multiple image encryption process to make it more robust and key sensitive. The four LL bands of wavelet transformed images are combined together to get single transformed matrix. The finite ridgelet transforms of four HH bands of transformed images are taken, rotated and then combined into another ridgelet transformed matrix. The chaotic key image is generated using piecewise linear chaotic map to rotate the pixel values of the wavelet transformed matrix. The rotated wavelet transformed matrix is mingled with ridgelet transformed matrix to mix up the information of images. The values of combined transformed matrix are shuffled using the random sequence extracted from 2D Gingerbreadman chaotic map.

The paper is organized as follows: Section 2 discusses preliminaries of piece-wise linear chaotic map, 2D gingerbreadman chaotic map, discrete wavelet transform and finite ridgelet transform used in the scheme. Section 3 discusses the proposed encryption scheme in detail. The experimental results are discussed in Section 4 followed by conclusion.

2   Preliminaries

2.1   Piecewise Linear Chaotic Map

The 1D Piecewise linear chaotic map is composed of linear segments, in which limited numbers of breaking points are allowed. It is a dynamical system that exhibit chaotic behavior for all values of parameter $p \in (0, 1)$, defined by the following equation [19]:

$$x(n+1)=\begin{cases} \dfrac{x(n)}{p} & x(n)\in (0,p] \\ \dfrac{1-x(n)}{1-p} & x(n)\in (p,1] \end{cases} \qquad (1)$$

Where $x(0)$ is initial condition, $n \geq 0$ is the number of iterations and $x(n) \in (0,1)$ for all n. The research shows that the map has largest +ve lyapunov exponent at $p = 0.5$. The bifurcation diagram shown in Figure 1(a) depicts that for every value of the control parameter p, the trajectory visit the entire interval [0, 1]. This map is utilized to generate a chaotic key image which is used to rotate the wavelet transformed matrix.

2.2   Gingerbreadman Chaotic Map

Gingerbreadman map is a 2D chaotic map defined by the following equations [20]:

$$x_g(n+1) = 1 - y_g(n) + \left| x_g(n) \right|$$
$$y_g(n+1) = x_g(n) \tag{2}$$

Where $x_g(0)$ and $y_g(0)$ are its initial conditions. There is a stable hexagonal region forming the belly and five others forming the legs, arms and head of the gingerbreadman as shown in Figure 1(b). The map is chaotic in the filled region and stable in the six hexagonal regions. The two chaotic sequences generated by the gingerbreadman map are utilized to randomly relocate the values of combined transformed matrix.

2.3   Discrete Wavelet Transform

The DWT is applied to decompose an image at *level* =1 to get four sub-band images with different space and frequency: sub-band LL_1, horizontal low frequency and vertical high frequency sub-band LH_1, horizontal high frequency and vertical low frequency sub-band HL_1, vertical and horizontal direction of high frequency sub-band HH_1. The main energy is concentrated in the low frequency band. The remaining three sub-bands signals are known as the detail coefficients. The low frequency sub-band LL_1 can be decomposed at level 2 into four LL_2, LH_2, HL_2 and HH_2 sub-bands. Wavelet transform is used for decomposing the original images. The transformed LL and HH bands of original images are then explored to encrypt the information of images using ridgelet and chaotic maps.

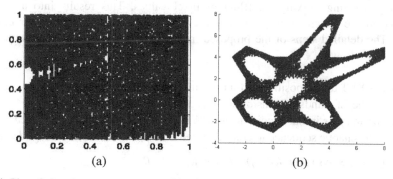

Fig. 1. Plot of chaotic sequence generated by (a) Piecewise Linear Chaotic map: $x(n)$ vs p and (b) Gingerbreadman map: $x(n)$ vs $y(n)$

2.4   Finite Ridgelet Transform

In contrast with discrete wavelet transform which deals with point singularity, finite ridgelet transform deals with line singularity in 2D space [21]. Ridgelet transform is more useful over wavelet since singularities are often joined together along edges in images. In 2D space, points and lines are related by radon transform. Wavelet and ridgelet transforms are related by radon transform. The Ridgelet transform is the result of application of 1D wavelet transform over Radon transform as shown in Figure 2.

Fig. 2. Relationship between Radon and Ridgelet transform

3   Proposed Multiple Image Encryption

An efficient encryption method modifies the original pixel gray values and breaks the correlation of adjacent pixels so as to improve the statistical properties of the image. In the proposed technique, we modify the pixel values of original images by applying wavelet transform at *level*=1 on each image and then their approximate coefficients (LL bands) are extracted to form a combined wavelet transform matrix for further processing. Two different chaotic maps are used in the encryption process. A piecewise linear chaotic map is used to generate a 2D chaotic key image to circularly rotate the pixel values in rows of wavelet transformed matrix to eliminate the correlation among the pixels. Furthermore, the finite ridgelet transform is applied to the horizontal components (HH bands) of original images to obtain one combined ridgelet transform matrix. Then both the transformed matrices (wavelet and ridgelet) are combined to lower the correlation among pixels. Now, gingerbreadman chaotic map is iterated to generate two random sequences for shuffling the pixels values to relocate them to new positions. The whole process is carried out to perform combining, rotating, mixing, shuffling of pixel values. This results into a single encrypted image which secretly contains the information of the original four plain-images. The detailed steps of the proposed multiple image encryption algorithm are described as follows:

1. Evaluate DWT decomposition of four original images to be encrypted.
2. Combine the LL bands of transformed images in one matrix $wcombined(i, j)$.
3. Take $x(0)$ and p as initial conditions for PWLCM and generate chaotic sequence.
4. Process the chaotic sequence to convert it into 2D binary chaotic key image K_C as:

   $if (x(n) > avg)$ *then* $K_C(i, j) = 1$ *else* $K_C(i, j) = 0$

5. Find the sum of each row of $K_C(i, j)$ and store it in array $Sum(i)$

6. Perform circular rotation of pixel values in each row of *wcombined(i,j)* by the value equal to *Sum(i)*.

7. Find finite ridgelet transform of all the resulting HH bands obtained in Step 1 of all images.

8. Rotate all these four components obtained above by 90° anticlockwise.

9. Combine the four components resulting from step 10 to get single ridgelet transformed matrix *rcombined(i,j)*.

10. Combine *wcombined(i,j) and rcombined(i,j)* in a fashion shown in Figure 4. Both *wcombined(i,j)* and *rcombined(i,j)* can be viewed as having four quadrants. The quadrant 1 of *wcombined(i,j)* is combined with quadrant 3 of *rcombined(i,j)*, the reason being quadrant 1 contains the energy of 1st image and quadrant 3 contains energy of 3rd image. This will mix up the transformed pixel values and hence results in lowering the correlation between adjacent pixels. The mixing of same quadrants is avoided. The mixing is carried out by inserting the values of *wcombined(i,j)* and *rcombined(i,j)* at every alternate rows and columns of *w_r_combined(i,j)* matrix.

11. Generate two random sequences S_x and S_y of each size 2N using Gingerbreadman chaotic map with initial condition as $x_g(0)$, $y_g(0)$.

12. Compare bits of S_x and S_y and save the index number to $Z(i)$ where $S_x(i) = S_y(i)$

13. Now, *w_r_combined(i,j)* matrix is divided into blocks as per the values of $Z(i)$ and $Z(i+1)$ as shown in the Figure 3. These pixel blocks are swapped as block 1 is swapped with block 2. Do it for all such alternate blocks.

14. Remaining rows of *w_r_combined(i,j)* with the index where $S_x(i)=0$ are swapped with columns of the same index.

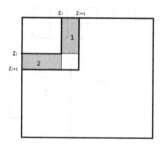

Fig. 3. Depicting the pixels Block 1 and 2 according to $Z(i)$ and $Z(i)$

This way the shuffling of values of *w_r_combined(i,j)* matrix is done and hence, getting the final encrypted image.

To illustrate the proposed approach, its block diagram is shown in Figure 4. Decryption is the inverse of encryption process at every step. While applying the inverse discrete wavelet transform as last step to complete the decryption process, the approximate coefficient of the original images are restored. Therefore, inverse DWT is taken without using detailed coefficients to obtain four decrypted images. The deviation between the original and decrypted images is very negligible and quantified by determining the MSE values.

4   Experimental Results

The proposed scheme is experimented on Matlab 2010 with four gray-scale images each of size 256×256 (*Cameraman, Clock, Man* and *Chemical plant*) which are shown in Figure 5. The initial value used for encrypting the images are: $x(0)=0.4345612$, $p=0.2234561$, $x_g(0)=0.3145265$, $y_g(0)=0.6456376$. The mother wavelet filter used is '*haar*' at *level* =1. The chaotic key image K_c of size 256×256 generated by PWLCM is shown in Figure 6. The final encrypted image obtained by applying the proposed approach is shown in Figure 7. As clear from the encrypted image, that it is highly distorted, undistinguishable and appears like a noise image. In order to evaluate the encryption and decryption performance of the proposed multiple image encryption scheme, various simulation analyses are carried out like: correlation analysis, key sensitivity analysis, mean square errors analysis and key space analysis. These analyses are done for three different set of keys $(x(0), p, x_g(0), y_g(0))$.

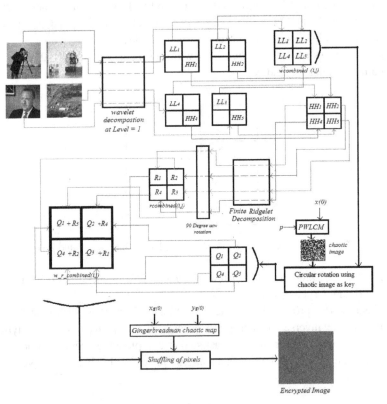

Fig. 4. Sketch of proposed multiple image encryption algorithm

(a)   (b)   (c)   (d)

Fig. 5. Four original images: (a) *Cameraman* (b) *Clock* (c) *Chemical plant* and (d) *Man*

Fig. 6. Chaotic key image K_c using PWLCM of size 256×256

Fig. 7. Result of multiple image encryption: single encrypted image of size 512×512

4.1   Correlation Analysis

An efficient encryption algorithm should be able to completely eliminate the correlation of adjacent pixels in images. The correlation coefficients obtained for encrypted and combined original images are listed Table 1. The values show that encrypted images are highly uncorrelated to each other as the values are very close to 0. Where as, the coefficients obtained between the decrypted and original images (for key used to get the encrypted image shown in Figure 7) are listed in Table 2. These values show that the two images are correlated to each other, as coefficients are close to 1. Hence, it can be said that proposed encryption algorithm satisfies the properties of an efficient encryption algorithm.

Table 1. Correlation coefficients between encrypted and combined original images

x(0)	p	xg(0)	yg(0)	corr
0.4345612	0.2234561	0.3145265	0.6456376	0.0103
0.4435654	0.2333434	0.4123512	0.5864576	0.0101
0.4543656	0.2564459	0.4654342	0.5567876	0.0090

Table 2. Correlation coefficients between decrypted and original images

Pair of images	corr
Cameraman	0.9371
Clock	0.8297
Man	0.9391
Chemical plant	0.8928

4.2   Key Sensitivity Analysis

A good encryption algorithm should be key sensitive; even a small change in the key should results into a completely different ciphertext. In our scheme slight change in the initial conditions of chaotic maps, brings a large change in encrypted content. The change of the order of $\Delta=10^{-14}$ in the key results in a percentage change of more than 96.71% in the encrypted image. Three experimental results for key sensitivity in different sets of keys are shown in Table 3.

Table 3. Percentage change in the two encrypted images

x(0)	p	Δx(0)	Δp	%change
0.4345612	0.2234561	0	10^{-14}	97.39
0.4435654	0.2333434	10^{-14}	0	96.71
0.4543656	0.2564459	10^{-14}	10^{-14}	98.19

4.3   Mean Square Error Analysis

The proposed technique has been numerically evaluated. The MSE between the original image and encrypted image is often used to check the robustness and amount of distortion takes place in the encrypted content due to the encryption algorithm. Table 4 shows the MSE values calculated for three sets of keys, the large values confirm that the encrypted image is highly deviated from the original images. Moreover, the MSE between the decrypted and original images (for key used to get the encrypted image shown in Figure 7) are listed in Table 5. The low MSE values show that the degree of deviation of decrypted images from the original images is very low, i.e. the decrypted images are almost similar to the original images. This validates the efficiency of the proposed algorithm.

Table 4. MSE between encrypted and combined original images

x(0)	p	Encrypted Vs Original
0.4345612	0.2234561	2.391×10^4
0.4435654	0.2333434	2.383×10^4
0.4543656	0.2564459	2.388×10^4

Table 5. MSE between decrypted and original images

Pair of images	Decrypted Vs Original
Cameraman	0.0038
Clock	0.0037
Man	0.0039
Chemical plant	0.0038

4.4   Key Space Analysis

Key space is the total number of different keys that can be used in a cryptographic system. In our algorithm, key consists of all initial conditions and parameters for both the PWLCM, Gingerbreadman chaotic maps i.e. $x(0)$, p, $x_g(0), y_g(0)$ and the name of mother wavelet including the level of wavelet decomposition. We have taken the initial values as *double* with 15 digit precision. Therefore, the key space of the proposed algorithm is comes out as $(10^{14})^4 \approx 2^{186}$. Thus, the key space is large enough to make our scheme robust and resistive to exhaustive intruder's attack.

5   Conclusion

In this paper, a multiple image encryption algorithm is proposed for providing security before transmitting multiple images over open networks. The proposed algorithm makes use of the features of discrete wavelet transform, finite ridgelet transform and chaotic maps. The approximate wavelet coefficients of four images are combined to encrypt the information contents of original images. The encryption process includes the chaotic-key-image based circular rotation; ridgelet transform based mixing of pixel values and their random relocation using another chaotic map. The experimental and simulation analyses show that the proposed encryption algorithm fulfills the security requirements of an efficient multiple image encryption method. As it results into low correlation coefficient, high key sensitivity, large key space and high degree of deviation from the original image(s). Moreover, the mean square error between the decrypted and original image(s) is very negligible.

References

1. Menezes, A.J., Oorschot, P.C.V., Vanstone, S.A.: Handbook of Applied Cryptography. CRC Press (1997)
2. Schneier, B.: Applied Cryptography: Protocols Algorithms and Source Code in C. Wiley (1996)

3. Fridrich, J.: Symmetric Ciphers Based on two-dimensional Chaotic Maps. International Journal of Bifurcation and Chaos 8(6), 1259–1284 (1998)
4. Zhang, L., Liao, X., Wang, X.: An Image Encryption Approach Based on Chaotic Maps. Chaos, Solitons & Fractals 24(3), 759–765 (2005)
5. Tang, Y., Wang, Z., Fang, J.: Image Encryption using Chaotic Coupled Map Lattices with Time Varying Delays. Communication in Nonlinear Science and Numerical Simulation 15(9), 2456–2468 (2009)
6. Chen, G.Y., Mao, Y.B., Chui, C.K.: A Symmetric Image Encryption Scheme Based on 3D Chaotic Cat maps. Chaos, Solitons & Fractals 21(3), 749–761 (2004)
7. Patidar, V., Pareek, N.K., Sud, K.K.: A New Substitution-Diffusion Based Image Cipher using Chaotic Standard and Logistic Maps. Communication in Nonlinear Science and Numerical Simulation 14(7), 3056–3075 (2009)
8. Behnia, S., Akhshani, A., Mahmodi, H., Akhavan, A.: A Novel Algorithm for Image Encryption Based on Mixture of Chaotic Maps. Chaos, Solitons & Fractals 35(2), 408–419 (2008)
9. Xiangdong, L., Junxing, Z., Jinhai, Z., Xiqin, H.: A New Chaotic Image Scrambling Algorithm Based on Dynamic Twice Interval-division. In: International Conference on Computer Science and Software Engineering, pp. 818–821 (2008)
10. Mao, Y., Lian, S., Chen, G.: A Novel Fast Image Encryption Scheme Based on 3D Chaotic Baker Maps. International Journal of Bifurcation and Chaos 14(10), 3616–3624 (2004)
11. Yang, Y.-L., Cai, N., Guo-Qiang, N.: Digital Image Scrambling Technology Based on the Symmetry of Arnold Transform. Journal of Beijing Institute of Technology 15(2), 216–220 (2006)
12. Singh, N., Sinha, A.: Chaos Based Multiple Image Encryption using Multiple Canonical Transforms. Optics & Laser Technology 42(5), 724–731 (2010)
13. He, M.Z., Cai, L.Z., Liu, Q., Wang, X.C., Meng, X.F.: Multiple Image Encryption and Watermarking by Random Phase Matching. Optics Communications 247(1-3), 29–37 (2005)
14. Barrera, J.F., Henao, R., Tabaldi, M., Torroba, R., Bolognini, N.: Multiple Image Encryption using an Aperture-Modulated Optical System. Optics Communications 261(1), 29–33 (2006)
15. Lin, Q.H., Yin, F.L., Mei, T.M., Liang, H.: A Blind Source Separation-Based Method for Multiple Images Encryption. Image and Vision Computing 26(6), 788–798 (2008)
16. Liu, Z., Dai, J., Sun, X., Liu, S.: Triple Image Encryption Scheme in Fractional Fourier Transform Domains. Optics Communications 282(4), 518–522 (2009)
17. Joshi, M., Shakher, C., Singh, K.: Fractional Fourier Transform Based Image Multiplexing and Encryption Technique for Four-color Images using Input Images as Keys. Optics Communications 283(12), 2496–2505 (2010)
18. Chen, T.H., Tsao, K.H., Wei, K.C.: Multiple Image Encryption by Rotating Random Grids. In: International Conference on Intelligent Systems Design and Applications, pp. 252–256 (2008)
19. Li, S., Chen, G., Mou, X.: On the Dynamical Degradation of Digital Piecewise Linear Chaotic Maps. International Journal of Bifurcation and Chaos 15(10), 3119–3151 (2005)
20. Devaney, R.L.: The Gingerbreadman. Algorithm 3, 15–16 (1992)
21. Do, M.N., Vetterli, M.: The Finite Ridgelet Transform for Image Representation. IEEE Transactions on Image Processing 12(1), 16–28 (2003)

Investigation of Quality of Metric in H.323 (VoIP) Protocol Coexisting of WLAN with WiMax Technologies of Different Network Loaded Conditions

K. Sakthisudhan, G. DeepaPrabha, A.L. Karthika, P. Thangaraj, and C. MariMuthu

Bannari Amman Institute of Technology, Erode, Tamilnadu
sakkthisudhan@gmail.com

Abstract. Intellectual mobile terminals (or users) of next generation wireless networks are expected to initiate/Establish voice over IP (VoIP) calls using session set-up protocols like H.323 or SIP (Session Initialized Protocols). To provide quality metrics of video conferences, telemedicine and other voice over broadband telephony (VoBB) applications. In this work, we analyze the performance of the H.323 call setup procedure over the wireless link. We used to call modes of operation over heterogeneous networks. The proposed model application layers in the RTP Control Protocol and Real-Time Transport Control Protocol (RTCP) protocols used in two different modes of call established. Initiate services through VoIP used for H.323 control packets. Our analytical model provides that the VoIP call set-up performance, jitter and delay in peer to peer networks. Moreover, the call setup performance can be improved significantly using the robust in application link layer such as RTP/RCTP with a comparison of heterogeneous network proposed in our paper. The analytical results were validated by our experimental measurements.

Keywords: VoIP, RTCP, SIP, multimedia, Strict Priority, Scheduler, ISAKMP.

1 Introduction

Forth coming world increasing demand by users for ubiquitous access to wireless services has led to the deployment of many wireless access technologies such as WLAN, GPRS, EDGE, 3G, and WiMAX all offer differing levels of quality, range and bandwidth. In the future there will be more multimedia devices which can access multiple radio access networks. Moreover in the future we will see greater overlap between the coverage provided by the differing access technologies. In this research work implementation of Voice over Internetworking Protocol (VoIP) and IP Multimedia Subsystem services (IMS) over the much sought after wireless standard Wi-MAX and WiFi. The multimedia transmission over wireless in the soft switching technique is compatible with WiMAX. The one of the signaling protocol is VoIP has opened a new indoor / outdoor for telephony bringing forward immense possibilities. The basic reason for the popularity of VoIP is the cost which is very low as compared to the conventional telephony services. The concept of the transmission of voice over

J. Mathew et al. (Eds.): ICECCS 2012, CCIS 305, pp. 145–152, 2012.
© Springer-Verlag Berlin Heidelberg 2012

data stream makes it possible to have VoIP transmitter and receiver using anything that uses IP - laptops, PC's, WiFi enabled handsets etc. In this VoIP uses Internet Protocol for transmission of voice as packets over IP networks. The process involves digitization of voice, the isolation of unwanted noise signals and then the compression of the voice signal using compression algorithms/codec. After the compression the voice is packetized to send over an IP network, each packet needs a destination address and sequence number and data for error checking. The signaling protocols are added at this stage to achieve these requirements along with the other call management requirements. When a voice packet arrives at the destination, the sequence number enables the packets to be placed in order and then the decompression algorithms are applied to recover the data from the packets. Here the synchronization and delay management needs to be taken care of to make sure that there is proper spacing. Jitter buffer is used to store the packets arriving out of order through different routes, to wait for the packets arriving late. H. 323 is the ITU-T standard for packet based multimedia conferencing services based on VoIP. This standard is interoperable and has both point to point and multipoint capabilities it offers specifications for call control, channel setup, codes for the transmission of Real time video and voice over the networks and this standard collaborate with RTP/RTCP layered protocol used for real time audio and video streaming.

1.1   Issues in Designing an IEEE 802.11. b and IEEE 802.11.e

Wireless Local Area Networks (WLANs) are increasingly making their way into residential, commercial, industrial and public areas. WAN issues usually result from high latency, packet loss, jitter, or temporary loss of the Internet connection and can result in everything from delayed voice to dropped calls. They can be the result of poor signal levels with your ISP's connection or unusual high delay and jitter occurring from routers and/or network congestion. The two main problems encountered when VoIP is used over WiFi are i) The system capacity for voice can be quite low for WLAN. ii) VoIP traffic and traditional data traffic such as Web traffic, emails etc.

WiMAX is a serious alternative to the wired network, such as DSL and cable modem. Besides Quality of Service (QoS) support, the IEEE 802.16 standard is currently offering a nominal data rate up to 100 Mega Bit Per Second (Mbps), and a covering area around 50 kilometers. Thus, a deployment of multimedia services such as Voice over IP (VoIP), Video on Demand (VoD) and video conferencing is now possible, WiMAX modems do exist on the market and that could be used as air interface devices, but due to their bulk and requirement of power, they become a hindrance to mobility.

The main contribution of this paper is the strict priority scheduler designed to provide the minimum guaranteed transmission rate for all active flows with the respect to their priorities and to provide a fair share of the additional bandwidth. The scheduler also rejects flows, for which the minimum rate requirements exceed the available bandwidth. The proposed solution is applicable for the WiFi wireless network, to accomplish QoS along the path. The Strict Priority Scheduling protocol is the default scheduling discipline in Qualnet. It services the highest priority queue until it is empty, and then moves to the next highest priority queue, and so on. It is possible that if there is enough high priority traffic, the lower priorities could be completely frozen out.

2   Related Works

There are many papers proposed in VOIP based video streaming technology with H.323/SIP protocols. The authors are mainly concerned with routing protocol based transmission in order to achieve high robustness and capacity of voice transmission over IP networks various methodologies have been implemented and verified the proceed.

[1] Jengfarn lee et al proposed in challenges in practical Quality of metric WLAN over VoIP networks the quality of metric values average delay against with uplink and downlink transmission over IP networks.[2] Jae-Woo So et al investigated the OFDM (Orthogonal Frequency division multiplexing) system VoIP based up/down link signaling overhead information streaming over IEEE 802.16.e networks. [3] Rahmatullah, M.M et al proposed three level multi parallelism video call transmitted over VoIP. [4] Rangel V et al examined the performance analyzed for digital video broadcasting (DVB) /Digital Audio-Visual Council (DAVIC) cable television protocol for the delivery of low rate isochronous streams for a cable population of up to 700 nodes and data rate up to 128 Kbps suitable for compressed/uncompressed voice, video delivery of supplementary services. [5] Carmona, J.V.C et al proposed analysis of VoIP over streaming video often high data rate delivery through a residential indoor power line communication (PLC) network. [6] Jianxin Liao et al proposed to provide real time voice transmission over lossy networks for introducing new SCTP transport layer protocol in wireless networks [7] Sajal K. Das et al examined call initialized setup/call setup delay and call establishment between mobile subscribers and also achieved guarantee service over H.323/SIP protocol. [8] An Chan et al proposed indoor/outdoor WLAN coexists streaming services over VoIP with different data rates.[9] Wei Wang et al investigated two major problems of low VoIP capacity in WLAN and unacceptable VoIP performance in the presence of coexisting video traffic from different user application. With each video stream requiring less than 10Kbps and IEEE 802.11.b requiring at 11Mbps support more than 500 no of VoIP sessions.[10] Nilanjan Banerjee et al examined in the undesirable delay and packet loss coexisting with heterogeneous IP based network and also achieved good quality of services in application layered SIP protocol.

3   Simulation Results

We use Qualnet simulator as our performance analysis platforms. Various evaluation parameters include the time between 1st and last packet, no of packets, Average packet size, and throughput. The simulation parameters are summarized in table1. We designed the infrastructure networks Setup containing 12 no of nodes. We are compiling the all nodes with video traffic in VoIP transmission between the source and

Table 1. Parameters for Simulation Evaluation

Parameters	IEEE 802.16.e	IEEE 802.11. b
Data rate	52Mbps	11Mbps
No. of nodes	12	12
Application	VOIP (H.323)	VOIP (H.323)
IP-queue-priority input- Queue-size	150000	150000
Routing Protocol	SIP	SIP
Traffic type	VoIP	VoIP
Running time	300s	300 S
File name	Terminal alias address file (. endpoint)	Terminal alias address file (. endpoint)
Simulation Area	900×900m²	900×900m²

destination nodes. Our work combines with demonstrated an IEEE 802.16.e and IEEE 802.11.b networks. The qualities of metrics are analyzed in the video established between two different VoIP users. Video traffic applies between source and destination. The two applications layered protocols are used in VoIP services RTP/RTCP. The parameters such as average jitter, end to end delay, RTT, VoIP initializes, establishment, and receiver parameter are taken into account.

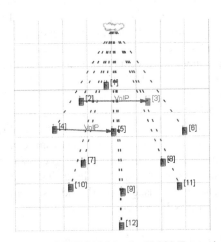

Fig. 1. Snapshot of IEEE 802.11.b/IEEE 802.16.e for H.323 Transmission via VoIP in Qualnet Simulator

The above figure 1 is the scenario model for IEEE 802.11.b and 802.16.e, which describes the video transmission over 12 numbers of mobile nodes. We are applying source and destination nodes following 2/3, 4/5 respectively. We analyzed both IEEE 802.16.e and IEEE 802.11.b network video transmission over VoIP in the RTCP protocol in application layer. In this model the setup establishment of two ray propagation. We are assign data rate up to 52 Mbps in outer and indoor environmental wireless links.

In this below figure 3 represents RRT video packets are travelling along destination node for speed test and back. IEEE 802.16.e radio link is less than IEEE 802.11.b radio networks. In order to achieve 3×10^{-6} in IEEE 802.11.b radio links.

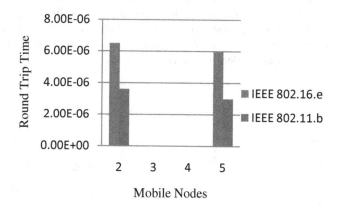

Fig. 2. Number of Mobile nodes corresponding with session average RRT over VoIP (RTCP)

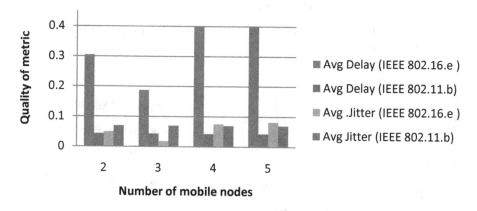

Fig. 3. Number of mobile nodes corresponding with session average jitter and average delay in RTP protocol

From this figure 3 and 4 shown as RTP session is established of each multimedia stream. A session consists of an IP address with a pair of ports for RTP and RTCP. For video streams will have a separate RTP session, enabling a receiver to deselect a particular stream. The ports which form a session are negotiated using other protocols such as RTSP (Real time streaming protocol) and Session Initiation protocol using the session description protocol in the setup method. According to the specification, an RTP port should be even and the RTCP port is the next higher odd port number. RTP and RTCP typically use unprivileged UDP ports (1024 to 65535) but may use other transport protocols most notably, SCTP (streaming control transmission protocol) and DCCP (Datagram Congestion Control Protocol) as well, as the protocol design is transport independent. In the IEEE 802.11.b protocol occupied session constant average delay is 0.43×10^{-6} and average jitter also obtained constant is 0.07 in VoIP transmission services. The below figure 5 shown the average delay and jitter with VoIP initiator video streams over IP based transmission. We are assigning the video calls in the nodes 2 and 4 and also establish same source nodes. Jitter can be described in terms of time variation in periodic signals in VoIP services at the same time qualified in all time varying signals e.g. RMS, Peak to peak displacement. Jitter can be expressed in terms of spectral density (frequency content).

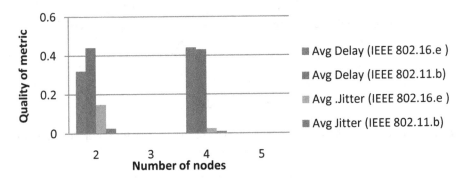

Fig. 4. Number of Mobile nodes corresponding with Average Jitter and average delay over the VoIP Initiator scheme

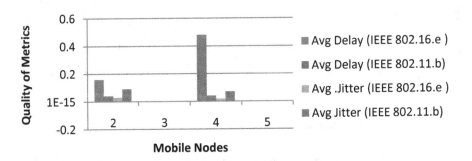

Fig. 5. Number of Mobile nodes corresponding with Average Jitter and average delay over the VoIP receiver scheme

In this figure 6 shown as average delay and jitter with a VoIP receiver over IP based transmission. We are assigning the video calls in the nodes 3 and 5 and also establishes same destination nodes.

In the below two figure 7-8 represented as the real time packet transmission over the scheduling round robin algorithm. In order to obtain the minimum guaranteed transmission rate for all active flows with the respect to their priorities and to provide a fair share of the additional bandwidth.

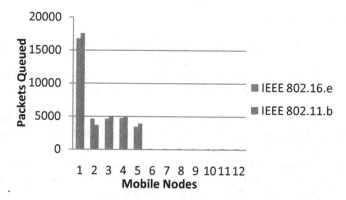

Fig. 6. Number of mobile nodes corresponding with queue packets (Strict Priority Scheduler)

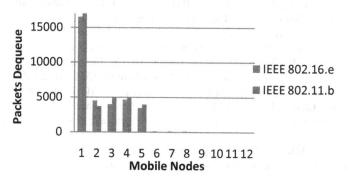

Fig. 7. Number of mobile nodes corresponding with dequeue packets (Strict Priority Scheduler)

4   Conclusion

We experimentally investigated application layered protocols to compare the quality of VoIP over peer to peer network video conservation. The RTP, RTCP, VoIP Initiator, VoIP receiver, SIP analyzed video traffic from source to destination node. RTT In this establishment of video streaming transmission WiMAX is better suited to VoIP than WiFi.

5   Future Enhancement

We will discuss security issues and challenges with radio links over VoIP particularly MANET transmission. The analyzed and comparison of application layered protocol H.323 and SIP protocols with respect to security attacks then work will extend for mobility nodes in VMANET architecture.

References

1. Lee, J., Liao, W.: A Practical QoS Solution to Voice over IP in IEEE 802.11 WLANs. IEEE Communications 47(4), 111–117 (2009)
2. So, J.-W.: Performance Analysis of VoIP Services in the IEEE 802.16e OFDMA System With In band Signaling. IEEE Transactions on Vehicular Technology 57(3), 1876–1886 (2008)
3. Rahmatullah, M.M., Khan, S.A., Jamal, H.: Carrier Class High Density VoIP Media Gateway using Hardware Software Distributed Architecture. IEEE Transaction on Consumer Electronics 53(4), 1513–1520 (2007)
4. Rangel, V., Edwards, R.M., Tzerefos, P., Schunke, K.-D.: Delivery Of Low Rate Isochronous Streams Over The Digital Video Broadcasting/Digital Audio-Visual Council Cable Television Protocol. IEEE Transaction on Broadcasting 48(4), 307–316 (2002)
5. Carmona, J.V.C., Pelaes, E.G.: Analysis and Performance of Traffic of Voice and Video in Network Indoor Plc. IEEE Transactions on Latin America 10(1), 1268–1273 (2012)
6. Liao, J., Wang, J., Zhu, X.: A Multi-Path Mechanism For Reliable VoIP Transmission Over Wireless Networks. ACM Journal Computer Networks: The International Journal of Computer and Telecommunications Networking 5(13), 2450–2460 (2008)
7. Das, S.K., Basu, K.: Performance Optimization of VoIP Calls over Wireless Links Using H.323 Protocol. IEEE Transactions on Computers 52(6), 742–752 (2003)
8. Chan, A., Liew, S.C.: Performance of VoIP over Multiple Co-Located IEEE 802.11 Wireless LANs. IEEE Transactions on Mobile Computing 8(8), 1063–1079 (2009)
9. Wang, W., Liew, S.C., Li, V.O.K.: Solutions to Performance Problems in VoIP Over a 802.11 Wireless LAN. IEEE Transactions on Vehicular Technology 54(1), 366–384 (2005)
10. Banerjee, N., Acharya, A., Das, S.K.: Seamless SIP-Based Mobility for Multimedia Applications. IEEE Transactions Networks 20(2), 6–13 (2006)

Kernel Based Automatic Traffic Sign Detection and Recognition Using SVM

Anjan Gudigar[1], B.N. Jagadale[2], Mahesh P.K.[3], and Raghavendra U.[4]

[1] Department of Electronics and Communication, M.I.T.E, Moodbidri, India
anjangudigar83@gmail.com
[2] Department of Electronics, Kuvempu University, Shimoga
basujagadale@gmail.com
[3] Department of Electronics and Communication, M.I.T.E, Moodbidri, India
mahesh24pk@gmail.com
[4] Research Scholar, M.I.T, Manipal, India
raghux109@gmail.com

Abstract. Traffic sign detection and recognition is an important issue of research recently. Road and traffic signs have been designed according to stringent regulations using special shapes and colors, very different from the natural environment, which makes them easily recognizable by drivers. The human visual perception abilities depend on the individual's physical and mental conditions. In certain conditions, these abilities can be affected by many factors such as fatigue, and observatory skills. Detection of regulatory road signs in outdoor images from moving vehicles will help the driver to take the right decision in good time, which means fewer accidents, less pollution, and better safety. In automatic traffic-sign maintenance and in a visual driver-assistance system, road-sign detection and recognition are two of the most important functions. This paper presents automatic regulatory road-sign detection with the help of distance to borders (DtBs) and distance from centers (DfCs) feature vectors. Our system is able to detect and recognize regulatory road signs. The proposed recognition system is based on the generalization properties of SVMs. The system consists of following processes: segmentation according to the color of the pixel, traffic-sign detection by shape classification using linear SVM and content recognition based on Gaussian-kernel SVM. A result shows a high success rate and a very low amount of false positives in the final recognition stage.

Keywords: Distance to Borders, Distance from Centers, Gaussian-kernel, Regulatory, Support Vector Machines (SVMs), and Traffic Sign Recognition.

1   Introduction

Recognition of highway signs has become an important object of study during the last years. Traffic signs have a dual role: First, they regulate the traffic and second, indicate the state of the road, guiding and warning drivers and pedestrians. These signs can be classified according to their color and shape, and both these

J. Mathew et al. (Eds.): ICECCS 2012, CCIS 305, pp. 153–161, 2012.
© Springer-Verlag Berlin Heidelberg 2012

characteristics constitute their content. Colors represent an important part of the information provided to the driver to ensure the objectives of road sign. So they are different from the surrounding in order to be distinguishable. The visibility of traffic signs is crucial for the driver's safety. For example, very serious accidents happen when drivers do not notice a stop sign. Of course, many other accidents are not related to traffic signs and are due to factors such as the psychological state of drivers. The causes for accidents that are related to traffic signs may be occlusion or partial occlusion of the sign, deterioration of the sign, or possible distraction of the driver.

Traffic sign detection becomes a challenging task under the circumstances such as, Color Fading, Weather Conditions, Similar colored objects, Disorientation of Signs, Motion Blur and Car Vibration, Color Information, Shadows, and Highlights. The measured color is a function of context, particularly illumination, reflectance, and orientation.

2   Related Work

There are numerous methods for the detection and recognition of traffic signs. Most of the Vision-based sign detection systems mostly suffer from adverse weather and lighting conditions. A sign detection system can be decomposed into two separate parts: detection and classification. Researchers have proposed various techniques for detection and classification. Among the commonly used techniques, we can mention fuzzy approach, Neural Networks (NNs), and Kalman filter. Kalman Filters for Traffic Sign Detection and Tracking system have additionally focused on the tracking of the signs through the image sequence. Prior to tracking phase, they have used two NNs for detecting the signs: one for color features and one for shape features. A fuzzy approach is used to create an integration map of the shape and color features, which in turn is used to detect the signs. To reduce the complexity of detection operations, the system can only detect signs of a particular size (8-pixel radius). Once the location of the sign is detected in the current frame, the size and location in the following frame is predicted by a Kalman filter. This significantly reduces the search space and increases the accuracy. Nevertheless, the detection technique proposed requires a large search space due to the complexity of the integration map [1].

Shape-based Road Sign Detection system developed to detect triangular, square and octagonal road signs. The method uses the symmetric nature of these shapes. Regular polygons are equiangular i.e., their sides are separated by a regular angular spacing. To utilize this regularity, a rotationally invariant measure was introduced. However, the algorithm has an important limitation such that, for each image frame the algorithm only seeks for predefined radii. Regarding the performance, for a 320x240 image, the algorithm was able to be run at 20Hz. The approach has strong robustness to varying illumination as it detects shapes based on edges, and will efficiently reduce the search for a road sign from the whole image to a small number of pixels. It can detect (without classification) the signs with a success rate of 95 percent [2].

Fuzzy approach for traffic sign color detection and segmentation has been developed, in which RGB images taken by a digital camera are converted into HSV and segmented by a set of fuzzy rules depending on the hue and saturation channels. The fuzzy rules are used only to segment the colors of the sign.HSV color space is used because hue is invariant to the light variations and saturation changes. Seven fuzzy (if-then) rules are applied with respect to the hue and saturation values [3]. Recognition of Traffic Signs with Two Camera System developed, in which one camera has a wide-angle lens and is directed to the moving direction of the vehicle, whereas the other camera is equipped with a telephoto lens and can change the viewing direction to focus the attention to the target sign. The detection process first identifies the candidates by color and intensity. Next, the telephoto camera is directed to the region of interest and it captures a closer view of the candidate signs. For detecting the circles they use the fact that; if an edge is a part of a circle, the centre of the circle should exist on the line which passes the edge and has the same direction as the gradient of the edge. After detecting the circles with regard to a fixed threshold value, the classification is achieved by a normalized correlation-based pattern matching technique using a traffic sign image database [4].

We present a new algorithm for detection and classification of traffic signs which is robust to wide range of weather conditions: sunny, cloudy and rainy weather. The technique is based on SVMs. Training and test is performed on vectors sets which are geometric characteristics of blobs.

3   Proposed Methodology

In this proposed method, the block level representation of the traffic sign detection and recognition system is shown in Fig.1.The detection and recognition system consists of three stages.

1) Segmentation: Candidate blobs are extracted from the input image by thresholding using HSI color space for chromatic signs.

2) Shape classification: Blobs that are obtained from segmentation are classified in this stage using linear SVMs, with the help of DtBs and DfCs .According to the color that has been used in the segmentation, only some given shapes are possible. For example, signs that are segmented using the red clues can be triangle.

3) Recognition: The recognition process is based on SVMs with Gaussian kernels. Different SVMs are used for each color and shape classification.

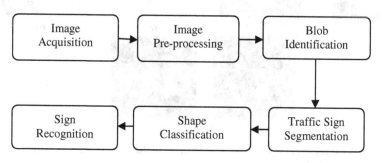

Fig. 1. Algorithm Description

3.1   Segmentation

The difficulties that we encounter in this image segmentation are related to illumination changes and possible deterioration of the signs. We believe that the hue and saturation components of the HSI space are sufficient to isolate traffic signs in a scene working with fixed thresholds. To obtain these thresholds, we have built the histograms of hue and saturation for red signs. Here, the hue and saturation components take values ranging from 0 to 255, respectively. At this point, we must consider that the response to varying wavelength and intensity of standard imaging devices is nonlinear and interdependent. Due to this reason, the images from which we analyze the hue and saturation components of traffic signs have been taken under different weather and lighting conditions in order to get well-suited thresholds. After the segmentation stage, image pixels may belong to any of the four color categories, i.e., red, blue, yellow, and/or white, and are grouped together as connected components called blobs. Then, enhancement is applied for both hue and saturation values for red, blue and yellow coloured traffic signs using Look up table (LUT)[9]. The LUTs used for enhancing hue and saturation components of red traffic signs are shown in Fig. 2.

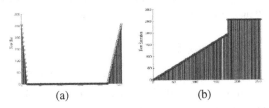

(a)   (b)

Fig. 2. a) Hue and b) Saturation LUT for RED traffic Sign

Once the segmentation process is completed, we obtain the blobs of interest (BoI) or, in other words, possible traffic signs as shown in Fig.3.

Fig. 3. a) Original Image[11] ,b) Enhanced Hue, c) Enhanced Saturation , d) Enhanced Intensity e) and f) are BoI of original image

3.2   BOI Shape Classification

The blobs that were obtained from the segmentation stage are classified in this stage according to their shape using linear SVMs [5]. Although SVMs were developed for and have been often used to solve binary classification problems, they can also be applied to regression. For shape classification, we use linear SVMs.

We have L training points, where each input x_i, has D attributes (i.e. is of dimensionality D) and is in one of two classes y_i = -1 or +1, i.e our training data is of the form $\{x_i, y_i\}$ where i=1....L, $y_i \in \{-1,1\}$, $X \in R^D$. Here we assume the data is linearly separable, meaning that we can draw a line on a graph of x_1 vs x_2 separating the two classes when D=2 and a hyperplane on graphs of x_1 , x_2 ,... ..... x_D, When D > 2.This hyperplane can be described by w . x + b = 0 where:

- w is normal to the hyperplane.
- $\frac{b}{||w||}$ is the perpendicular distance from the hyperplane to the origin.

Support Vectors are the examples closest to the separating hyperplane and the aim of Support Vector Machines (SVMs) is to orientate this hyperplane in such a way as to be as far as possible from the closest members of both classes. As shown in Fig.4.

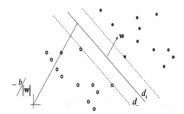

Fig. 4. Linearly separable classes

In the case of two separable classes, the training data are labelled $\{x_i, y_i\}$, where i=1....L, $y_i \in \{-1,1\}$, $X \in R^D$. In our case, the vectors x_i, are the DtB (Distance to Border)s, and the values of y_i are "1" for one class and "-1" for others, d is the dimension of the vector, and l is the number of training vectors. Once the optimization is completed, we simply determine on which side of the hyper plane a given test vector x lies. That is, to classify it to one class ("1") or to the other ("−1"), the decision function is given by the equation (1).

$$f(x) = sgn(x.w + b) \qquad (1)$$

In this approach, we present DtBs and DfCs as feature vectors for the inputs of the linear SVMs. DtBs are the distances from the external edge of the blob to its bounding box and DfCs is the distance from the centre of the blob to the external edge of the blob[10][12]. Fig.5 shows these distances for a triangular shape where D_1, D_2, D_3 and D_4, are the left, right, upper, and bottom DtBs, respectively. Four DtB vectors of 20 components are obtained, and they feed specific SVMs. SVM classifies the shape as an octagonal ("1") or not ("−1") and another four SVMs to classify the shape as a triangle ("1") or not ("−1"). Thus, four favourable votes are possible for each shape. A majority voting method has been applied in order to get the classification with a threshold; therefore, if the total number of votes is lower than this value, the analyzed blob is rejected as a noisy shape.

Area of triangle is calculated by using both minimum as well as maximum distance from centre to all points in the border. This is given as:

$$\text{Area of triangle} = \sqrt{3} \div (D_{min} + D_{max})^2 \tag{2}$$

In case of a tie, linear SVM outputs of favourable classification are used to decide which is the candidate shape. Finally, the method is invariant to scale due to the normalization of the DtB and DtC vectors to the bounding-box dimensions.

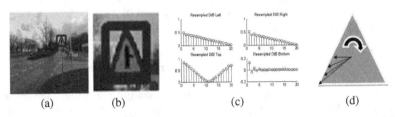

| (a) | (b) | (c) | (d) |

Fig. 5. (a) Original image, (b) BoIs of (a). (c) DtB vectors of (b) ,(d) DtC vectors of (b)

4   Recognition

Once the candidate blobs are classified into a shape class, the recognition process is initiated. Recognition is implemented by SVMs with Gaussian kernels. For the training process of SVMs, we used the library LIBSVMS [8]. However, in many cases, the data cannot be separated by linear function. A solution is to map the input data into a different space ϕ_x. Fig.6 shows the mapping of data to a higher dimensional space.

Due to the fact that the training data are used through a dot product, if there was a "kernel function" so that we satisfy $K(x_i, x_j) = (\phi(x_i), \phi(x_j))$, we can avoid computing ϕ_x explicitly and use the kernel function $K(x_i, x_j)$ .we have used a Gaussian kernel as follows:

$$K(X_i, X_j) = e^{\frac{-\|X_i - X_j\|^2}{2\sigma^2}} \tag{3}$$

And the decision function for a new input vector is

$$f(x) = sgn\left(\sum_{i=1}^{N_s} \alpha_i y_i K(s_i, x) + b\right) \tag{4}$$

Where N_s the number of support vectors , and s_i are the support vectors.

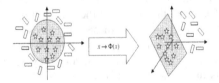

Fig. 6. Mapping of data to a higher dimensional space

This kernel requires tuning for the proper value of σ Although this approach is feasible with supervised learning, it is much more difficult to tune σ for unsupervised learning methods[6][7].

The recognition stage input, in our case, is a block of 36 × 36 pixels in gray scale image for every candidate blob; therefore, the interior of the bounding box is normalized to these dimensions. In order to reduce the feature vectors, only those pixels that must be part of the sign (pixel of interest, PoI) are used. For instance, for a triangular sign, only pixels those are inside the inscribed triangles, which belong to the normalized bounding box, are computed in the recognition module. Fig.7 shows the PoI for sign whose shape is triangular and octagonal. Different one-versus-all SVMs classifiers with a Gaussian kernel are used, so that the system can recognize every sign.

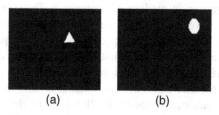

(a)   (b)

Fig. 7. PoI for a) Triangular b) Octagonal

Both the training and test are done according to the color and shape of each candidate region; thus, every candidate blob is only compared to those signs that have the same color and shape as the blob to reduce the complexity of the problem.

5   Experimental Results

This section contains the simulation results that were obtained by implementing the algorithm proposed. The algorithm has been implemented in Visual C++ 6.0. In order to classify the shape of traffic signs, we have used SVM model based on Gaussian Kernel. The main reason to select this model is its low computational complexity. The recognition and classification process has been simulated using MATLAB 7.13. To evaluate the performance of the system we have used images under various outdoor circumstances like occlusion, illumination and bad weather conditions. We have imposed certain limits on the size of objects that we are interested in, i.e., small blobs and big blobs are rejected as noise and objects of no interest. The limit for aspect ratio was empirically derived based on road images. Thus, objects with an aspect ratio greater than 1.7and less than 1are rejected.

The automated system proposed was found to be capable of detecting the traffic sign from images, segment them and recognize the sign. This system using support vector machines has yielded good results when compared to the results achieved so far. Fig.8 illustrates the results of recognition process.

Fig. 8. Experimental results

6　Conclusion

This paper describes a complete method to detect and recognize regulatory traffic signs from an image sequence, taking into consideration all the existing difficulties regarding object recognition in outdoor environments. Thus, our system can be used for application such as driver assistance system or unmanned vehicle control.

Two main modules have been developed based on the capability of SVMs as a novel technique in pattern recognition: a first module for shape classification based on linear SVMs using DtB and DfC vectors and a second module that was developed with Gaussian kernel for recognition. Experimental results indicate that the system under consideration is accurate because it allows us to detect different geometric shapes, and works correctly under difficult environmental situations. The classification results shows that better performance in shape classification is obtained while using both DtB and DfC as feature vectors instead of using DtB alone.

In addition, the algorithm can also detect signs that are partially occluded However, some improvements remain as tasks for the future, in order to recognize rectangular route-guidance signs for navigation whose pictograms can present very different icons and different conditions may occur in night.

References

[1] Fang, C.-Y., Chen, S.-W., Fuh, C.-S.: Road-sign detection and tracking. IEEE Transactions on Vehicular Technology, 52–57 (September 2003)

[2] Loy, G., Barnes, N.: Fast shape-based road sign detection for a driver assistance system. In: IEEE/RSJ International Conference on Intelligent Robots and Systems, IROS 2004, September 2- October, vol. 1, pp. 70–75 (2004)

[3] Fleyeh, H.: Road and traffic sign color detection and segmentation-a fuzzy approach (2005)

[4] Miura, J., Kanda, T., Shirai, Y.: An active vision system for real-time traffic sign recognition. In: Intelligent Transportation Systems, pp. 52–57. IEEE (2000)

[5] Burges, C.J.C.: A tutorial on support vector machines for pattern recognition. Data Mining Knowl. Discov. 2(2), 121–167 (1998)

[6] Vapnik, V.: The Nature of Statistical Learning Theory. Springer, New York (1995)

[7] Cristianini, N., Shame-Taylor, J.: Support Vector Machines and Other Kernel-Based Learning Methods. Cambridge Univ. Press, Cambridge (2000)

[8] Chang, C., Lin, C.: LIBSVM: A Library for Support Vector Machines (2001) (Online) http://www.csie.ntu.edu.tw/~cjlin/libsvm

[9] de la Escalera, A., Armingol, J.M., Pastor, J.M., Rodriguez, F.J.: Visual Sign Information Extraction and Identification by DeformableModels for Intelligent Vehicles. IEEE Transactions on Intelligent Transportation Systems 5(2), 57–68 (2004)

[10] Bascon, S.M., et al.: Road Sign Detection and Recognition Based on Support Vector Machines. IEEE Transactions on Intelligent Transportation Systems 8(2) (June 2007)

[11] Image Database, http://roadanalysis.uah.es/Documentos

[12] Lafuente Arroyo, S., Gil Jimenez, P., Maldonado Bascon, R., Lopez Ferreras, F., Maldonado Bascon, S.: Traffic Sign Shape Classification Evaluation I: SVM using Distance to Borders. In: Proceedings of IEEE Intelligent Vehicles Symposium, Las Vegas, pp. 557–562 (June 2005)

Color Segmentation of 2D Images
with Thresholding

Hepzibah A. Christinal[1,2], Daniel Díaz-Pernil[1],
Pedro Real Jurado[1], and S. Easter Selvan[3]

[1] Research Group on Computational Topology and Applied Mathematics
University of Sevilla, Avda. Reina Mercedes s/n, 41012, Sevilla, Spain
[2] Karunya University
Coimbatore, Tamilnadu, India
[3] Université catholique de Louvain
Louvain-la-Neuve, Belgium
{sbdani,real}@us.es, hepzia@yahoo.com, easterselvans@gmail.com

Abstract. Membrane Computing is a biologically inspired computational model. Its devices are called P systems and they perform computations by applying a finite set of rules in a synchronous, maximally parallel way. In this paper, we follow a new research line using tissue-like P systems to do a parallel color segmentation of images using a thresholding to look for edge pixels. We have chosen this variant of P systems because it uses a less number of computational ingredients with respect to classical variants.

1 Introduction

Natural Computing studies new computational paradigms inspired from Nature. It abstracts the way in which Nature "computes", conceiving new computing models. There are several fields in Natural Computing that are now well established as are Genetic Algorithms ([1]), Neural Networks ([2]), DNA-based molecular computing ([3]).

Membrane Computing is a theoretical model of computation inspired by the structure and functioning of cells as living organisms able to process and generate information. The computational devices in Membrane Computing are called *P systems* [4]. Roughly speaking, a P system consists of a membrane structure, in the compartments of which one places multisets of objects which evolve according to given rules. In the most extended model, the rules are applied in a synchronous non-deterministic maximally parallel manner, but some other semantics are being explored.

According to their architecture, these models can be split into two sets: cell-like P systems and tissue-like P systems. In the first type of systems, membranes are hierarchically arranged in a tree-like structure. The inspiration for such architecture is the set of vesicles inside the cell. All of them perform their biological processes in parallel and life is the consequence of the harmonious conjunction

J. Mathew et al. (Eds.): ICECCS 2012, CCIS 305, pp. 162–169, 2012.

of such processes. This paper is devoted to the second approach: tissue-like P systems.

Digital Image Processing tries to obtain a set of characteristics or parameters related to an image. Segmentation in computer vision (see [5]), refers to the process of partitioning a digital image into multiple segments (sets of pixels). The goal of segmentation is to simplify and/or change the representation of an image into something that is more meaningful and easier to analyze. Image segmentation is typically used to locate objects and boundaries (lines, curves, etc.) in images. More precisely, image segmentation is the process of assigning a label to every pixel in an image, such that pixels with the same label share certain visual characteristics.

Recently, Membrane Computing techniques have been used for solving problems from Digital Image. Different P systems models have been used for dealing with images as in [6] where cell-like P systems are used for computing the thresholding of 2D images, [7–10] where tissue-like P systems are used, or even [11], where the *symmetric dynamic programming stereo* (SDPS) algorithm [12] for stereo matching was implemented by using simple P modules with duplex channels. But in these cases, it is usual to do a pre-processing to the input images. So, they are transformed from color images to gray scale images or even binarized images.

We propose a new researching step in this paper. We obtain the first solution in Membrane Computing to the color segmentation problem. We show, from a theoretical point of view, it is possible to do a parallel segmentation of an image considering a general color alphabet. Here, the edges are not selected because a pixel has a different color from a neighbor pixel. We decide a pixel is an edge because it is enough different with a neighbor pixel.

The paper is structured as follows: in the next section we present the definition of array tissue-like P systems with input. In section 3, we design a family of systems for edge-based segmentation in 2D image ($n \times m$) using thresholding. Finally, some conclusions and future work are present in the last section.

2   Formal Framework

An *Array tissue-like P system* of degree $q \geq 1$ with input is a tuple of the form

$$\Pi = (\Gamma, V, \mathcal{E}, w_0, w_1, \ldots, w_q, A_1, \ldots, A_q, \mathcal{R}, i_\Pi, o_\Pi),$$

where

1. Γ is a finite *alphabet*, whose symbols will be called *objects*,
2. V is the *alphabet of colors* verifying $V \cap \Gamma = \emptyset$.
3. $\mathcal{E}$ is a finite subset of arrays on V.
4. $w_0, w_1, \ldots, w_q$ are strings over Γ representing the multisets of objects associated with the cells at the initial configuration,
5. $A_1, \ldots, A_n$ are arrays on V, placed on the corresponding cells at the initial configuration.

6. $\mathcal{R}$ is a finite set of communication rules of the following form: $(i, u_i W_i / u_j W_j, j)$, for $i, j \in \{0, 1, 2, \ldots, q\}, i \neq j, u_i, u_j \in \Gamma^*$ and W_i, W_j two arrays on V.
7. $i_\Pi \in \{0, 1, 2, \ldots, q\}$ is the input cell.
8. $o_\Pi \in \{0, 1, 2, \ldots, q\}$ is the output cell.

In a similar way to tissue-like P systems, an *array tissue-like P system* of degree $q \geq 1$ can be seen as a set of q cells (each one consisting of an elementary membrane) labeled by $1, 2, \ldots, q$. We will use 0 to refer to the label of the environment, i_Π and o_Π denote the input region and the output region (which can be the region inside a cell or the environment), respectively.

The strings $w_1, \ldots, w_q$ describe the multisets of objects placed in the q cells of the system. We interpret that w_0 is the set of objects placed in the environment, each one of them available in an arbitrary large amount of copies.

For each $i \in \{1, \ldots, q\}$, each A_i is an array placed in the cell i in the initial configuration and $\mathcal{E}$ is the set of arrays placed in the environment, each one of them available in an arbitrary large amount of copies. The empty array $\emptyset$ always belongs to $\mathcal{E}$. For all the non-empty copies, we will consider that the leftmost pixel of the bottom row in the array corresponds to the coordinates $(0, 0)$.

Rules are used as usual in the framework of membrane computing, that is, in a maximally parallel way (a universal clock is considered), regardless whether the environment is involved or not. In one step, each object in a membrane can only be used for one rule (non-deterministically chosen when there are several possibilities), but any object which can participate in a rule of any form must do it, i.e., in each step we apply a maximal set of rules.

3   Segmentation

We have segmented 2D digital images using array tissue-like P systems based in *edge-based segmentation* and thresholding. We consider an input 2D digital image, and an ordered color alphabet of the image. We define a family of P systems to do this task and show an overview of a computation of any system of the family. We have done complexity study for these systems. Finally, we see how one of our systems works with an example (see Fig. 1).

So, we look for pairs of adjacent pixels such that the distance between their colors is greater than a fixed threshold. We call these pixels *boundary pixels*. For each pair of border pixels, the smallest of them is called an *edge pixel*. But, we have a problem with this definition of edge pixels. The set of this type of pixel is not connected. So, we can consider a pixel adjacent to two edge pixels with near colors. If its color is near to the colors of the two edge pixels we say it is an edge pixel too.

Definition 1. The problem of segmentation of 2D digital images using thresholding (*2DEST problem*) is defined as follows:

Given a 2D digital image with pixels of (possibly) different colors establish the edge pixels (defined above).

Fig. 1. An image of size 30×30

3.1   A Family of Array Tissue-Like P Systems

Given a digital image with $n \times m$ pixels ($n, m \in \mathbb{N}$) we define an array tissue-like
P system whose input is given by the pixels of the image encoded by the objects
a_{ij}, where $1 \leq i \leq n$ and $1 \leq j \leq m$.

The key idea of how a system of this family works is the following. First, the
system marks the boundary pixels. Next, it marks the necessary pixels to connect
all the boundary pixels of the same color. Finally, the system uses a counter (z_i,
whose number of initial copies in the system is $\lceil r_1^{1/2^7} \rceil$ and $r_1 = max(n, m)$) to
send the marked objects to the environment. The output of the system is given
by the objects that appear in the output cell when it stops.

We define a family of tissue-like P systems to do the edge-based segmentation
with the thresholding of a 2D image. For each $n, m \in \mathbb{N}$, we consider the array
tissue-like P system with input of degree 2:

$$\Pi = (\Gamma, \Sigma, V, \mathcal{E}, w_0, w_1, w_2, A_1, A_2, \mathcal{R}, i_\Pi, o_\Pi),$$

defined as follows

(a) $\Gamma = \Sigma \cup \{a'_{ij} : a \in \mathcal{C}_S,\ 1 \leq i \leq n,\ 1 \leq j \leq m\} \cup \{\bar{z}_1\}$,

(b) $\Sigma = \{a_{ij} : a \in \mathcal{C}_S,\ 1 \leq i \leq n,\ 1 \leq j \leq m\}$,

(c) $V = \mathbb{N}$

(d) $\mathcal{E} = \{\, a'\, b,\, b\, a',\, \dfrac{a'\ \ b}{b\ \ a'},\, \dfrac{a\ \ b'}{c'\ \ d'},\, \dfrac{a'\ \ b}{c'\ \ d'},$
$\dfrac{a'\ b'}{c\ \ d'},\, \dfrac{a'\ b'}{c'\ \ d'},\, \dfrac{a\ \ b'}{c'\ \ d'},\, \dfrac{a'\ \ b}{c\ \ d'} : a, b, c, d \in V\}$,

(e) $w_0 = \{\bar{z}_i : 2 \le i \le 9\}$,

   $w_1 = \bar{z}_1^{\lceil r^{1/2^7} \rceil}$, where $r = max(n, m)$,

   $w_2 = \bar{z}_1^{\lceil r^{1/2^7} \rceil}$,

(f) $A_1, A_2 = \emptyset$

(g) R is the following set of communication rules:

Type 1. $(j, \bar{z}_i / \bar{z}_{i+1}^2, 0)$, for $i = 1, \ldots, 8$, $j = 1, 2$

   In this rule, we are working with a counter that it is used in the output
   of the systems.

Type 2.

   $$(1, a\,b \,/\, a'\,b\,, 0), \text{ for } a, b \in \mathcal{C}_S, \ a < b \text{ and } d(a, b) > k.$$

   $$(1, b\,a \,/\, b\,a'\,, 0), \text{ for } a, b \in \mathcal{C}_S, \ a < b \text{ and } d(a, b) > k.$$

   $$(1, \tfrac{a}{b} \,/\, \tfrac{a'}{b}\,, 0), \text{ for } a, b \in \mathcal{C}_S, \ a < b \text{ and } d(a, b) > k.$$

   $$(1, \tfrac{b}{a} \,/\, \tfrac{b}{a'}\,, 0), \text{ for } a, b \in \mathcal{C}_S, \ a < b \text{ and } d(a, b) > k.$$

These rules are used when an image has two adjacent pixels whose
associated colors have a distance greater than a threshold k (boundary
pixels). Then, the pixel with the less associated color is marked (edge
pixel).

Type 3.

   $$(1, \tfrac{a\ b'}{c'\ d} \,/\, \tfrac{a'\ b'}{c'\ d}\,, 0), \text{ for } a, b, c, d \in \mathcal{C}_S, \ d(a, b') < k \text{ and } d(a, c') < k.$$

   $$(1, \tfrac{a'\ b}{c\ d'} \,/\, \tfrac{a'\ b'}{c\ d'}\,, 0), \text{ for } a, b, c, d \in \mathcal{C}_S, \ d(b, a') < k \text{ and } d(b, d') < k.$$

   $$(1, \tfrac{a\ b'}{c'\ d} \,/\, \tfrac{a\ b'}{c'\ d'}\,, 0), \text{ for } a, b, c, d \in \mathcal{C}_S, \ d(d, b') < k \text{ and } d(d, c') < k.$$

   $$(1, \tfrac{a'\ b}{c\ d'} \,/\, \tfrac{a'\ b}{c'\ d'}\,, 0), \text{ for } a, b, c, d \in \mathcal{C}_S, \ d(c, a') < k \text{ and } d(c, d') < k.$$

The rules of this type are activated when a set of four pixels appears
in cell 1 verifying that one no-marked pixel (a, in the first set of rules)
is adjacent to two edge pixels (b', c', in the first set of rules) whose
distance between them and from the first pixel to them are less than
$k \in \mathbf{N}$ (our threshold). Finally, the fourth pixel (d, in the first set of
rules) has to be adjacent to the previous edge pixels and distance from
this last pixel to the first one is greater than k. When this pattern
appears in cell 1 the first considered pixel (a, in the first set of rules)
is marked and passed to be an edge pixel.

Type 4.

   $$(1, \bar{z}_9 a' / \bar{z}_9, 2), \text{ for } a \in \mathcal{C}_S.$$

When these rules appears, the system sends the edge pixels to the
output cell.

(h) $i_\Pi = 1$

(i) $o_\Pi = 2$.

An Overview of the Computation. A 2D image is codified by the input array that appears in the input cell and the system begins to work with them. Rules of type 1 initiate the counter $\bar{z}$. In a parallel manner, rules of type 2 identify the boundary pixels and mark the edge pixels. These rules need 4 steps to do this. From the second step, the rules of type 3 can be used with the rules of the first type at the same time. So, in the other 4 steps we can mark the rest of the (edge) pixels adjacent to two edge pixels and the other boundary pixel with a color whose distance to the colors of the others is big enough. The system can apply the types of rules 2 and 3 simultaneously in some configurations, but it always applies the same number of these two types of rules because this number is given by the edge pixels (we consider 4-adjacency). Finally, the fourth type of rules are applied in the following step, when the system finishes marking all the edge pixels in cell 1. So, with one step more the system sends the edge pixels to cell 2. Thus, we need only 9 steps to obtain an edge-based segmentation for an $n \times m$ digital image.

3.2   Complexity and Necessary Resources

Taking into account the size of the input data is $O(n \cdot m)$ and $|\mathcal{C}_S| = h$ is the number of colors of the image, the necessary resources for defining the systems of our family and the complexity of our problem is presented in the following table:

2DEST Problem	
Complexity	
Number of steps of a computation	9
Necessary Resources	
Size of the alphabet	$n \cdot m \cdot h$
Initial number of cells	2
Initial number of objects	$\lceil 2r_1^{(1/2^7)} \rceil$
Number of rules	$O(n \cdot m \cdot h^2)$
Upper bound for the length of the rules	8

3.3   Examples

In this section, we show the results obtained by the application of our method with three examples over a given image of size 30×30 (see Fig. 1). Some questions should be looked into when we begin to apply our algorithm. First, we need to choose a color space. But, this choice is hidden in the preprocessing of the systems, because we can choose a distance function.

Once we get a distance, the second question is to know which is an appropriate *threshold*. Notice that, when we work with different thresholds we obtain different

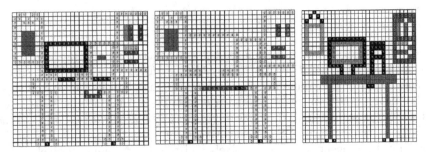

Fig. 2. (Left) Segmentation of the image of our example with thresholding $k = 3$ and considering the natural order in the colors alphabet. (Center) Segmentation of the image of our example with thresholding $k = 5$ and considering the natural order to the colors alphabet. (Right) Segmentation of the image of our example with thresholding $k = 3$ and considering the reverse order in the colors alphabet.

segmentations as Fig. 2 (Left) and Fig. 2 (Center) show. In the first image, the threshold is $k = 3$ and in the second one it is $k = 5$.

Finally, the last question is closely relate to the first one. We can consider an order in the *color alphabet*. The choice of the ordered color alphabet implies changes in the final results of our system. We take $k = 3$ and $\mathbb{N}$ as the color alphabet. We can consider two orders in this set. The natural order, $(\leq)$, or the reverse order, $(\leq^\star)$[1]. If we work with the first order we obtain the segmentation of Fig. 2 (Right), and if we work with the second order we obtain the segmentations of Fig. 2 (Left). Obviously, we can find a lot of differences between two segmentations.

4   Conclusions

We claim that Digital Imagery problems can be suitable for Natural Computing techniques in general and for Membrane Computing techniques in particular. Many of the problems in Digital Imagery share features that are very interesting for using these techniques.

We present in this paper a theoretical step to do parallel color segmentation using P systems. Until now, when we use the membrane parallel techniques of segmentation, it was essential to binarize the image first. (see [13]).

As a future work, we could try to implement this algorithm using parallel programming, as CUDA. An interesting line of research could be to apply this method to areas such as Biological and Medical Image. This algorithm could be translated to tackle higher dimensions with ease.

Acknowledgement. The first author acknowledges the support of the project "Computational Topology and Applied Mathematics" PAICYT research project

---

[1] $\forall a, b \in \mathbb{N}$, $a \leq^\star b$ if and only if $a \geq b$.

FQM-296. The third author acknowledge the support of the project MTM2006-03722 of the Ministerio español de Educación y Ciencia and the project PO6-TIC-02268 of Excellence of Junta de Andalucía.

References

1. Holland, J.H.: Adaptation in natural and artificial systems. MIT Press, Cambridge (1992)
2. McCulloch, W.S., Pitts, W.: A logical calculus of the ideas immanent in nervous activity. Bulletin of Mathematical Biophysics 5, 115–133 (1943)
3. Adleman, L.M.: Molecular computation of solutions to combinatorial problems. Science 266, 1021–1024 (1994)
4. Păun, G.: Computing with membranes. Technical Report 208, Turku Centre for Computer Science, Turku, Finland (November 1998)
5. Shapiro, L.G., Stockman, G.C.: Computer Vision. Prentice Hall PTR, Upper Saddle River (2001)
6. Christinal, H.A., Díaz-Pernil, D., Gutiérrez-Naranjo, M.A., Pérez-Jiménez, M.J.: Thresholding of 2D images with cell-like P systems. Romanian Journal of Information Science and Technology (ROMJIST) 13(2), 131–140 (2010)
7. Christinal, H.A., Díaz-Pernil, D., Real Jurado, P.: Segmentation in 2D and 3D Image Using Tissue-Like P System. In: Bayro-Corrochano, E., Eklundh, J.-O. (eds.) CIARP 2009. LNCS, vol. 5856, pp. 169–176. Springer, Heidelberg (2009)
8. Christinal, H.A., Díaz-Pernil, D., Real Jurado, P.: Region-based segmentation of 2D and 3D images with tissue-like P systems. Pattern Recognition Letters 32(16), 2206–2212 (2011); Advances in Theory and Applications of Pattern Recognition, Image Processing and Computer Vision
9. Peña-Cantillana, F., Díaz-Pernil, D., Berciano, A., Gutiérrez-Naranjo, M.A.: A Parallel Implementation of the Thresholding Problem by Using Tissue-Like P Systems. In: Real, P., Diaz-Pernil, D., Molina-Abril, H., Berciano, A., Kropatsch, W. (eds.) CAIP 2011, Part II. LNCS, vol. 6855, pp. 277–284. Springer, Heidelberg (2011)
10. Peña-Cantillana, F., Díaz-Pernil, D., Christinal, H.A., Gutiérrez-Naranjo, M.A.: Implementation on CUDA of the smoothing problem with tissue-like P systems. International Journal of Natural Computing Research 2(3), 25–34 (2011)
11. Gimel'farb, G., Nicolescu, R., Ragavan, S.: P Systems in Stereo Matching. In: Real, P., Diaz-Pernil, D., Molina-Abril, H., Berciano, A., Kropatsch, W. (eds.) CAIP 2011, Part II. LNCS, vol. 6855, pp. 285–292. Springer, Heidelberg (2011)
12. Gimel'farb, G.L.: Probabilistic regularisation and symmetry in binocular dynamic programming stereo. Pattern Recognition Letters 23(4), 431–442 (2002)
13. Sheeba, F., Thamburaj, R., Nagar, A.K., Mammen, J.J.: Segmentation of peripheral blood smear images using tissue-like p systems. In: International Conference on Bio-Inspired Computing: Theories and Applications, pp. 257–261 (2011)

Vowel Recognition from Telephonic Speech Using MFCCs and Gaussian Mixture Models

Shashidhar G. Koolagudi[1], Sujata Negi Thakur[2], Anurag Barthwal[1],
Manoj Kumar Singh[2], Ramesh Rawat[2], and K. Sreenivasa Rao[3]

[1] School of Computing, Graphic Era University,
Dehradun-248002, Uttarakhand, India
[2] Department of Computer Applications, Graphic Era University, Dehradun-248002,
Uttarakhand, India
[3] School of Information Technology, Indian Institute of Technology Kharagpur,
Kharagpur - 721302, West Bengal, India
koolagudi@yahoo.com, sujatathakur1987@gmail.com,
anubarthwal@gmail.com, manojthakur1984@rediffmail.com,
rsrawat06@gmail.com, ksrao@iitkgp.ac.in

Abstract. This paper presents vowel recognition from speech using mel frequency cepstral coefficients (MFCCs). In this work, microphone recorded speech and telephonic speech are used for conducting vowel recognition studies. The vowels considered for recognition are from Hindi alphabet namely अ(a), इ(i), उ(u), ए(e), ऐ(ai), ओ(o) and औ(au). Gaussian mixture models are used for developing vowel recognition models. Vowel recognition performance for microphone recorded speech and telephonic speech are 91.4% and 84.2% respectively.

Keywords: Gaussian mixture models, vowel recognition, mel frequency cepstral coefficients, automatic speech recognition.

1   Introduction

Vowel and consonant sound units are the basic elements of any spoken language. The difference in pronunciation of a word uttered by people of different mother tongues is mainly due to the variations in vowels and their pronounciation in respective mother tongue [1][2][3]. Hindi is normally spoken using a combination of 52 sound units containing 10 vowels, 2 modifiers and 40 consonants.

Vowel recognition is the basic step in automatic speech recognition (ASR). Spectrally well defined characteristics of vowels vitally help speech and speaker recognition systems. Speech recognition in the case of non-native speaker is also a challenging task. Attempts have been made to recognize different phonemes spoken in English by Malaysian natives [4]. Spectral features and feed forward neural network classifiers are used for classifying the phonemes.

When speech signal is observed, it is found that more than 90% of the duration is occupied by vowels. Fig. 1 shows different regions of speech signal, identified based on VOP (vowel onset point) location. From the figure it may be observed

J. Mathew et al. (Eds.): ICECCS 2012, CCIS 305, pp. 170–177, 2012.

that the syllable 'ka' contains around 300 samples. Out of which consonant /k/ comprises of only around 30 samples where as rest of the syllable is occupied by vowel /a/. The ability to recognize speech improves significantly by achieving efficient vowel recognition, both by humans as well as ASR systems [5]. Therefore vowel recognition has an important role in speech processing systems. Vowel recognition based on a speech signal is one of the intensively studied research topics in the domains of human-computer interaction and affective computing [6]. Earlier machine recognition of vowels is done to study perceptual confusions among 10 Hindi vowels. The experiments were conducted using clipped speech [7].

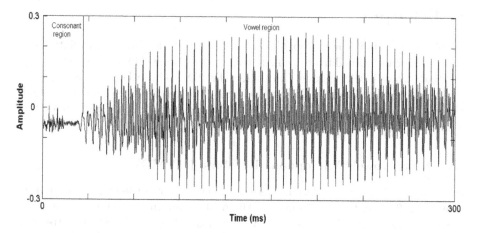

Fig. 1. Rough proportions of consonant and vowel regions in a syllable 'ka'

In today's world mobile telephonic communication has revolutionized the personal and professional communication. In this regard, in this work, the emphasis is given to the vowel recognition using telephonic speech. Recognition of vowels from telephonic speech has applications in speaker recognition for critical banking transactions, identification of criminals, access control systems and forensic applications [5]. In the field like Biometrics, goal of vowel recognition is to verify an individual's identity based on his or her voice. This is one of the trusted means of authentication as voice is one of the most natural forms of the communication. Identifying people by voice has drawn the special attention of lawyers, judges, investigators, law enforcement agencies and other practitioners of forensics [6]. Similarly list of applications may be extended to the systems such as online financial transactions, critical medical records, preventing frauds, resetting passwords, voice indexing and so on [6]. In view of above applications, accurate classification of vowels in telephonic voice expects the need for a well-trained, computationally intelligent model with an acceptable percentage of classification accuracy. In this pretext, emphasis is given to vowel recognition using telephonic speech.

In this work, a method for vowel recognition is proposed, using spectral features and Gaussian mixture models. MFCCs have been used as feature vectors. Utterances of about 30 male and 20 female speakers have been used for training and validation at the models. Various experiments are conducted to study the influence of varied number of MFCC features on telephonic/microphone based vowel recognition.

Rest of the paper is organized as follows. The creation of databases is discussed in section 2. Section 3 discusses feature extraction techniques. The proposed vowel recognition models and result are discussed in Section 4 and section 5 respectively. Paper concludes with summary in section 6.

2   Databases

In this study, Hindi vowel recognition has been performed using two databases. One is directly recorded from speakers using microphone. This database is known as Hindi Vowels Microphone Database (HVMD). The other one is recorded through telephones from different speakers. The database is known as Hindi Vowels Telephonic Database (HVTD). Following sub sections brief about two databases.

2.1   Hindi Vowels Microphone Database (HVMD)

Hindi Vowel Microphone Database used in this work is recorded using 50 speakers containing 20 females and 30 males. All speakers are in the age group of 18-32 years and speak Hindi as their native language. They are asked to utter the vowels in a neutral sense without including the emotive cues. This is done to minimize the effect of emotions during recognition task. Each of the speakers has to speak 7 vowels repeatedly 20 times in one session. The data has been collected in 4 such sessions. The seven vowels considered for collecting this speech corpus are अ(a), इ(i), उ(u), ए(e), ऐ(ai), ओ(o) and औ (au). Recording is done with Computer Headset SEHS 202 microphone. Speech signal is sampled at 16 kHz and each sample is stored as a 16 bit number using mono channel. The different sessions are recorded on alternate days to capture the variability in human speech production mechanism. The entire speech database is recorded using same microphone and at the same location.

2.2   Hindi Vowels Telephonic Database (HVTD)

Hindi Vowel Telephonic Database is recorded in Hindi language through mobile phones. Fifty(30 male + 20 female) speakers contributed to the database. Seven vowels recorded are अ(a), इ(i), उ(u), ए(e), ऐ(ai), ओ(o) and औ (au). There are multiple utterance of the same vowels and the vowels are recorded in neutral emotion. For entire recording the same handset Nokia 6303 classic is used to maintain unique recording conditions. Speaker was asked to call a specific number from the silence zone such as library. Call recording facility of a handset is used to record the calls. Call receiving handset is kept in a noise free room and the recording is done of the received call. These recorded calls are later transferred to computer for further processing. While storing these speech utterances in computer, sampling frequency of 16 kHz is used. The samples are stored as 16 bit numbers.

3   Feature Extraction

For recognition of vowels, mel frequency cepstral coefficients are used as features. Thirteen mel frequency cepstral coefficients are extracted from vowels as they are sufficient to capture acoustic information of a sound unit [8]. The speech signal is processed frame by frame with the frame size of 20 ms, and a shift of 10 ms, while extracting feature vectors, Hamming window is used while framing the speech signal. The nonlinearity of human perception mechanism was realized using mel scale of frequency bands.

The mapping of linear frequency onto Mel frequency uses a logarithmic scale as shown below:

$$m = 2595 \log_{10} \left(\frac{f}{700} + 1 \right)$$

Steps performed to obtain MFCCs from speech signal are as follows :

1. Fourier transform of a given vowel signal is obtained to get spectrum.
2. Powers of the above spectrum within the triangular overlapping windows are computed according to mel scale.
3. Logs of the power at each of the mel frequencies are calculated.
4. DCT (Discrete cosine transform) of the list of mel log powers is calculated.
5. The amplitudes of resulting spectrum give the MFCCs [9][10].

4   Development of Vowel Recognition Models

In this work, GMMs are used to develop vowel recognition systems using spectral features. GMMs capture distribution from input feature space into given number of clusters. One GMM model is developed for a vowel class. For instance all patterns of vowel अ are represented by one GMM. The number of Gausses present in the mixture model is known as the number of components. They give number of clusters in which data points are to be classified within each class. The components within each GMM capture finer level details among the feature vectors of each vowel. Depending on the number of data points, number of components may be varied in each GMM. Presence of few components in GMM and trained using large number of data points may lead to more generalized clusters, failing to capture specific details related to each class. On the other hand over fitting of the data points may happen, if too many components represent few data points. The complexity of the models increases, if they contain higher number of components [11].

Vowel recognition using pattern classifier is basically a two stage process as shown in Fig. 2. In the first stage vowel recognition models are developed by training the models using feature vectors extracted from the speech utterances of known vowels. This stage is known as supervised learning. In the next stage, testing (evaluation) of the trained models is performed by using the speech utterances of unknown vowels. Testing is performed using both microphone and telephonic data separately.

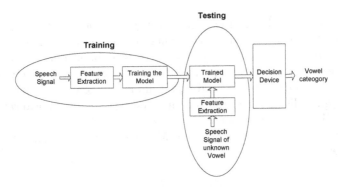

Fig. 2. Block Diagram of a Vowel Recognition System

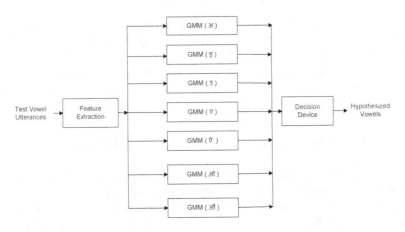

Fig. 3. Testing of vowel against trained Gaussian mixture models

5   Results

The test result obtained from the Gaussian Mixture Models(GMMs) are shown in the form of confusion matrix. Table 1 and 2 show the confusion matrices representing the test results of HVMD and HVTD respectively. Columns indicate trained models and rows indicate given test cateogory. The diagonal elements show the correct classification, whereas other elements in the row show the miss-classification among the different vowels. The average recognition rates for all seven vowels are observed to be 91.4% and 84.2% for two database in order.

Twenty utterances (10 male + 10 female) of each vowel cateogory chosen randomly from the utterances of HVMD are used for validating trained models. These utterances are not used for training. Each vowel utterance is pre processed to remove unwanted silence regions at the beginning and end of the utterance. Then the same feature vectors (as in the case of training - 13 mfcc features) are extracted. These feature vectors are given one by one to all trained GMM models. Based on the parameters of converged GMM, the likelyhood of each feature vector belonging to individual GMM is computed. Obviously likelyhood of a feature vector of matching vowel GMM is high compared to that of other GMMs. To take a decision regarding vowel cateogory, sum of log likelyhood of all feature vectors of unknown vowel utterance is computed with respect to each trained model (for all vowel cateogories). The model that yields highest log likelyhood sum is hypothesized as vowel cateogory. Decision device block, shown in the Fig. 3 performs this task.

Table 1. Vowel classification performance (in %) in the case of microphone recorded vowels)

	अ	इ	उ	ए	ऐ	ओ	औ
अ	90	0	0	0	0	0	10
इ	0	90	0	0	10	0	0
उ	0	0	85	0	0	0	15
ए	0	0	0	100	0	0	0
ऐ	0	0	0	10	90	0	0
ओ	0	0	0	0	0	100	0
औ	15	0	0	0	0	0	85

In these studies for a microphone recorded data, average vowel recognition rate obtained over 7 Hindi vowels is observed to be 91.4%. ए(e) and ओ(o) are almost cent percent recognized. Lower performance of around 85% is observed in the cases of उ(u) and औ(au). Interesting miss classification pattern may be observed in the cases of vowels अ(a), इ(i), उ(u) and diphthongs औ, ऐ. Miss classification of अ as औ, उ as औ, इ as ऐ, औ as अ are obvious as the diphthongs ae and औ inherently include the basic sounds (acoustic properties) of pure vowels अ, इ and अ, उ. Therefore the task of recognizing only pure vowels would yield 100% recognition rate. This cent percent vowel recognition performance is also observed in the cases of similar studies of other languages [3].

Similar studies have been conducted on telephonically recorded database. The average recognition rate observed is slightly lower of about 84.2% for all 7 vowels including 2 diphthongs. The miss classification pattern between pure vowels and corresponding diphthongs is also observed in the case of telephonic data.

Table 2. Vowel classification performance (in %) in the case of telephonically recorded vowels

	अ	इ	उ	ए	ऐ	ओ	औ
अ	90	0	0	0	0	10	0
इ	0	85	0	10	5	0	0
उ	0	0	85	0	0	0	15
ए	0	0	0	90	10	0	0
ऐ	0	0	0	15	85	0	0
ओ	0	0	0	0	0	80	20
औ	15	10	0	0	0	0	75

Lower performance in the case of telephonic data may be due to speech coding techniques that are adopted by different mobile vendors during transmission. It is observed from the literature that lower order frequency components (upto - 300Hz) are disturbed during speech coding [3]. However proper analysis of the reason for less performance is yet to be undertaken for our database.

6 Summary and Conclusion

In this paper, vowel recognition using 13 MFCCs as feature vectors has been studied. Results of microphone recorded vowel recognition models and telephonic recorded vowel recognition models are compared. GMMs are used for building vowel recognition systems. The average recognition in the case of telephonic data is slightly lower compared to that of microphone recorded data. Obvious overlap of classification in the case of pure vowels and related diphthongs is observed. Through analysis of telephonic data is to be carried out as a part of further studies.

References

1. Sadeghi, V.S., Yaghmaie, K.: Vowel Recognition using Neural Networks. International Journal of Computer Science and Network Security (IJCSNS) 6(12) (December 2006)
2. Rui, W., Yao, H., Gao, W.: Recognition of sequence lip images and application. In: Proc. ICSP (1998)
3. Tobely, T.E., Tsuruta, N., Amamiya, M.: On-Line Speech-Reading System for Japanese Language (2000)
4. Paulraj, M.P., Yaacob, S.B., Nazri, A., Kumar, S.: Classification of Vowel Sounds Using MFCC and Feed Forward Neural Network. In: 5th International Colloquium on Signal Processing & Its Applications (CSPA) (2009)
5. Chauhan, R., Yadav, J., Koolagudi, S.G., Sreenivasa Rao, K.: Text Independent Emotion Recognition Using Spectral Features. In: Aluru, S., Bandyopadhyay, S., Catalyurek, U.V., Dubhashi, D.P., Jones, P.H., Parashar, M., Schmidt, B. (eds.) IC3 2011. CCIS, vol. 168, pp. 359–370. Springer, Heidelberg (2011)

6. Gheidi, M., Sayadian, A.: Vowel Detection and Classification using Support Vector Machines (SVM). In: 4th International Conference: Sciences of Electronic, Technologies of Information and Telecommunications, TUNISIA, March 25-29 (2007)
7. Gupta, J.P., Agrawal, S.S., Ahmed, R.: Perception of (Hindi) Vowels in Clipped Speech. Journal of Acoustic Society of America 49(2B), 567–568 (1971)
8. Li, Y., Zhao, Y.: Recognizing emotions in speech using short-term and long-term features. In: Proc. of the International Conference on Speech and Language Processing, pp. 2255–2258 (1998)
9. Benesty, J., Sondhi, M.M., Huang, Y.: Springer handbook on speech processing. Springer (2008)
10. Sreenivasa Rao, K., Yegnanarayana, B.: Duration modification using glottal closure instants and vowel onset points. Speech Communication 51, 1263–1269 (2009), doi:10.1016/j.specom.2009.06.004
11. Koolagudi, S.G., Kumar, N., Sreenivasa Rao, K.: Speech emotion recognition using segmental level prosodic analysis. In: Proc of IEEE International Confrence on Device Communication BIT MESRA, India (February 2011)

A Robust Binary Watermarking Scheme Using BTC-PF Technique

Chinmay Maiti[1] and Bibhas Chandra Dhara[2]

[1] Department of Computer Science & Engineering,
College of Engineering & Management, Kolaghat, West Bengal
chinmay@cemk.ac.in
[2] Department of Information Technology, Jadavpur University, Kolkata
bibhas@it.jusl.ac.in

Abstract. In this work, a robust binary watermarking technique is proposed. For the embedding purpose, the host image is encoded by BTC-PF method and two pattern books (PB1 and PB2) are used to insert the watermark. Depending on the bit-value of the watermark, the current block of the host image is encoded either by using PB1 or PB2. In the present method, to enhance the security level, the original watermark is randomly permuted with a secret key K and then embedding method is executed. The experimental result shows that the proposed watermarking technique is a robust against different type of attacks like blurring, salt and pepper noise, Gaussian filter, cropping, sharpening, rotation, histogram equalization and JPEG compression.

Keywords: Binary watermarking, copyright protection, robust watermarking, watermark attacks, BTC-PF coding.

1 Introduction

Due to the rapid development of communication technology, volume of the multimedia data such as audio, video or image are increases enormously and can be duplicated, modified or even stolen easily. Therefore, it is a big challenge for the copyright owners to protect their data from various unauthorized access.

Digital watermarking techniques [1–5] provides a reliable and secure copyright protection of digital data. In the watermarking technique, a digital watermark is embedded into the host image and produce a watermarked image. Then, the watermarked image is distributed over the Internet. When a dispute over the copyright of the digital data occurs, the actual owner does extract the hidden watermark through extraction process and proves the ownership of the data.

Many researchers have come up with many watermark techniques in terms of their application areas and purposes. These techniques can be classified according to different view points. Firstly, the watermark techniques can be considered as spatial domain methods [3, 6–9] and transform domain methods [1, 2, 10, 11]. Secondly, they can be classified as non-blind methods [2, 12, 13], semi-blind methods [5, 8, 14, 15] and blind methods [3, 6, 16] depending on the requirement of host image and/or watermark image or any other side information to

J. Mathew et al. (Eds.): ICECCS 2012, CCIS 305, pp. 178–185, 2012.

extract the watermark. In [3], to embedded the watermark some pixels are selected randomly and the intensity of the selected pixels are modified. Statistical hypothesis testing is used for detection of watermark. LSB based watermark techniques have been proposed in [7]. Cox et al. [1] proposed a DCT domain based watermarking technique. Here, a watermark length of 1000 was added to the image by modifying 1000 largest coefcients of the DCT (excluding the DC term). A non-blind binary watermarking algorithm is proposed in [2]. In this work, watermark is embedded into the host image by selectively modifying the middle frequency components of DCT.

Most of the current watermarking techniques embed watermark into the original image, but for Internet based applications watermarking in compressed domain is an important issue. Moreover, the spatial domain based watermarking technique in compressed domain has lot of importance. Recently, in [17–19] VQ based watermarking technique have been proposed. The BTC based watermarking techniques are proposed in [6, 8, 9].

In this work, a robust binary watermarking technique has been proposed in the compressed domain. Here, BTC-PF method [20–22] is used to compress the host image. For the encoding purpose, two different pattern books (PB1 and PB2) are considered and depending on the bit value of the watermark the current block of the host image is encoded either using PB1 or PB2. To enhance the security level, first the watermark is permuted with a secret key and then embedded. The experimental results established the robustness of the proposed method against different types of image processing attacks [23] such as salt and pepper noise, Blurring, sharpening, Gaussian filter, cropping, rotation, histogram equalization and JPEG compression.

The rest of the paper is organized as follows. In Section 2, the BTC-PF method is described. The proposed watermarking technique (both embedding and extraction) is presented in Section 3. The experimental results are reported in Section 4. Finally, conclusions are drawn in Section 5.

2   BTC-PF Method

The BTC-PF [20–22] is a lossy spatial domain based compression method that uses a Q-level quantizer to quantize a local region of the image. This method encompasses a hybrid combination of BTC [24] and VQ [25] method. In this method, to encode an image, it is divided into blocks of size $n \times n$ $(n = 2^k)$ and each block is quantized by Q different gray values. The quantization is executed using Q-level patterns of size $n \times n$. The patterns are stored as a pattern book and let the size of the pattern book is M. The quantization levels are determined respect to the pattern. The method of selection of the best fit pattern for block B_i is as follows.

Let the pixel coordinates of the image block B_i are x_j $(j = 1, 2, \ldots, n^2)$ and corresponding pixel intensity is $f(x_j)$. Let the patterns in the pattern book are P_t $(t = 0, 1, \ldots, M - 1)$ and the value at position x_j is r, i.e., $P_t(x_j) = r$

where $0 \leq r \leq Q - 1$. To select the best fit pattern, each pattern is fit to the block B_i and error is computed for this. The error between the block B_i and the pattern P_t is define as $e_t = \sum_{r=0}^{Q-1} e_{tr}$, $e_{tr} = \sum_{P_t(x_j)=r}(f(x_j) - m_r)^2$ and $m_r = \frac{1}{|n_r|} \sum_{P_t(x_j)=r} f(x_j)$ where n_r is the number of positions with level r in the pattern P_t. The pattern with minimum error is considered as the best fit pattern and the index of the best fit pattern is $P_i = arg\{min\{e_t\}_{t=0, 1, ..., M-1}\}$. The image block B_i can be decoded using P_i and corresponding means m_{ir} ($r = 0, 1, ..., Q-1$), i.e., the tuple $t_i = (P_i, m_{i0}, m_{i1}, ..., m_{i(Q-1)})$ can be used to reconstruct B_i as B_i'.

3   Proposed Method

In this method, a binary watermark image is embedded into the host image which is encoded by BTC-PF method. The embedding process and extraction process are described in the following subsections.

3.1   Embedding Process

In the embedding process, the host image is encoded by BTC-PF method. Two pattern books PB1 and PB2 are used in the encoding step, no pattern is common in PB1 and PB2. Depending on the bit-value of the watermark pixels the current block of the host image is encoded either using PB1 or PB2. Consider the size of the host image is $m \times n$ and watermark is of size $m/4 \times n/4$. In this work, the watermark is permuted first, using a secret key, and then it is embedded. This permutation increases security level of the proposed method. The block diagram of the embedding process is shown in Fig. 1(a) and the algorithmic steps of the proposed embedding technique is given below.

Algorithm: Embedding Process

Input: Host image (I), Watermark image (W) and Secret key K
Output: Watermarked image (I')

1. I is divided into blocks of size 4×4.
2. W is randomly permuted with K which gives W^p.
3. Each Block B_{ij} of I is encoded by BTC-PF with following the logic
   if ($w_{ij}^p = 0$) encode B_{ij} using PB1
   else encode B_{ij} using PB2
4. Compressed watermarked image I' is obtained.

3.2   Extraction Process

An important aspect of any watermarking scheme is its robustness against different types of attack. It is obvious that the unauthorized owners try to remove/destroy the watermark from the watermarked image so that the actual owner fails to prove the ownership of the image. In the extraction process, only the watermarked image or attacked watermark image needed (i.e., the proposed

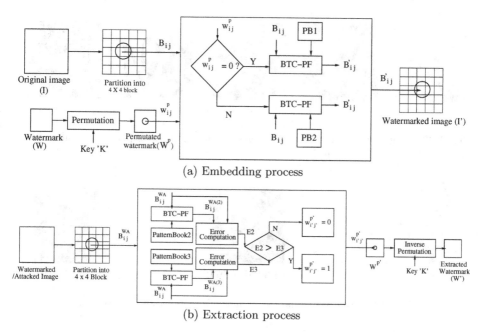

(a) Embedding process

(b) Extraction process

Fig. 1. Block diagram of the proposed watermarking technique

method is a blind watermarking technique). In this sub-section, the symbol I^{WA} is used to represent watermarked image or attacked watermarked image. The block diagram of the extraction method is shown in Fig. 1(b) and steps of the extraction process is as follows.

Algorithm: Extraction Process

Input: Watermarked image (I^{WA}) and Secret key K
Output: Extracted watermark (W')

1. Divide I^{WA} into blocks of size 4×4
2. For each block B_{ij}^{WA} do
   (a) Encode B_{ij}^{WA} by BTC-PF with PB1 which gives $B_{ij}^{WA(1)}$
   (b) Encode B_{ij}^{WA} by BTC-PF with PB2 which gives $B_{ij}^{WA(2)}$
   (c) Compute $E1 = error\ (B_{ij}^{WA}, B_{ij}^{WA(1)})$ and $E2 = error\ (B_{ij}^{WA}, B_{ij}^{WA(2)})$
   (d) if $(E1 > E2)\ w_{ij}^{p'} = 1$ else $w_{ij}^{p'} = 0$
3. Apply inverse permutation on $W^{p'}$ with K which gives W'

4   Experimental Result

In this experiment, three gray scale images 'Lena', 'Man' and 'Zelda' are used as the host image of size 512×512 and a binary image of size 128×128 as the watermark. The original images are shown in Fig. 2. In this work, two different

(a)   (b)   (c)   (d)   (e)

Fig. 2. Original images: (a)- (c) host images {Lena, Man, Zelda}, (d) original watermark, (e) permuted watermark

(a) 31.87, 0.9817   (b) 30.01, 0.9712   (c) 35.44, 0.9856

Fig. 3. Attacked watermarked images with PSNR and SSIM value

pattern books are used: one is with two level patterns and other is with three level patterns. To measure the quality of images, PSNR and SSIM [26] are used in the experiment. In Fig. 3, the watermarked images with PSNR and SSIM value are shown. As PSNR is greater than 30dB and SSIM is also above 0.97, we can say that watermarked images are good enough for general applications [21]. The similarity (or quality) of the extracted watermark is measured using SSIM and normalized coefficient (NC). The NC between watermark (W) and extracted watermark (W') is define as

$$NC = \frac{\sum_{i=1}^{m} \sum_{j=1}^{n} \overline{W(i,j) \oplus W'(i,j)}}{m \times n} \times 100\% \tag{1}$$

where $m \times n$ is the size of the watermark. The robustness of the proposed method is evaluated under different attacks like salt and pepper noise, blurred, Gaussian filter, cropping, sharpening, rotation, histogram equalization, and JPEG compression with varying image quality. The similarity value of the extracted watermarks under different attacks are given in Table 1. Also, different attacks on 'Lena' and corresponding extracted watermark are shown in Fig. 4. The present work is implemented in MATLAB 7.8.0.347. The parameters of different attacks are specified in Table 1. For JPEG compression, Adobe Photoshop (version 8.0) is used, where maximum quality factor is 12 and higher value indicates better quality image. Experimental result shows that proposed scheme is robust under different attacks.

(a) Salt/Pepper noise: 96.73, 0.7153

(b) Cropping: 71.01, 0.3677

(c) Blurred: 99.95, 0.9867

(d) Gaussian filter: 95.70, 0.6367

(e) Histogram equalization: 97.76, 0.7602

(f) Sharpen: 85.70, 0.3148

(g) Rotation: 94.34, 0.4830

(h) JPEG(QF=4): 61.48, 0.0608

(i) JPEG(QF=6): 68.31, 0.1333

(j) JPEG(QF=8): 71.47, 0.1661

(k) JPEG(QF=10): 86.06, 0.3578

(l) JPEG(QF=12): 98.61, 0.8034

Fig. 4. Different attacks on 'watermarked Lena' and corresponding extracted watermark with NC and SSIM value

Table 1. Similarity value of the extracted watermark against different attacks

Watermark Attack	Lena		Man		Zelda	
	NC	SSIM	NC	SSIM	NC	SSIM
Salt/pepper (noise density 0.005)	96.73	0.7153	96.95	0.7058	96.81	0.7202
Cropping (upper 50% set to 255)	71.01	0.3677	70.97	0.3662	71.03	0.3681
Blurring (disk radius 0.6)	99.95	0.9867	99.69	0.9305	99.97	0.9930
Gaussian filter (3×3, $\sigma = 0.5$)	95.70	0.6367	96.15	0.6285	96.25	0.6817
Histogram Equalization	97.76	0.7602	99.27	0.8684	98.93	0.8391
Sharpening (3×3, $\alpha = 0.8$)	85.70	0.3148	86.92	0.3350	83.50	0.2724
Rotation (5^0, clock-wise)	94.34	0.4830	94.57	0.4959	94.56	0.4935
JPEG (QF=4)	61.48	0.0608	64.43	0.0821	59.38	0.0491
JPEG (QF=6)	68.31	0.1333	73.70	0.1795	64.74	0.1079
JPEG (QF=8)	71.47	0.1661	77.49	0.2173	68.17	0.1358
JPEG (QF=10)	86.06	0.3578	89.65	0.4138	85.61	0.3370
JPEG (QF=12)	98.61	0.8034	98.64	0.8008	99.70	0.9348

5   Conclusions

In this work, a binary watermarking scheme is presented. The current method is based on BTC-PF method and this method embeds watermark in compressed domain, which is useful in Internet based applications. The proposed method is blind and robust under different attacks. Here, security level is enhanced by considering the random permutation of the watermark. The future scope is to extend this work for gray scale watermarking.

References

1. Cox, I.J., Kilian, J., Leighton, T., Shamoon, T.: Secure spread spectrum water-marking for multimedia. IEEE Trans. on Image Processing 6, 1673–1687 (1997)
2. Hsu, C., Wu, J.: Hiden digital watermarks in images. IEEE Trans. on Image Processing 8(1), 58–67 (1999)
3. Nikolaidis, N., Pitas, I.: Robust image watermarking in the spatial domain. Signal Processing 66, 385–403 (1998)
4. Potdar, V., Han, S., Chang, E.: A survey of digital image watermarking techniques. In: Proc. of IEEE International Conference on Industrial Informatics, pp. 709–716 (2005)
5. Shieh, J., Lou, D.C., Chang, M.: A semi-blind digital watermarking scheme based on singular value decomposition. Computer Standards & Interfaces 28(4), 428–440 (2006)
6. Lin, M.H., Chang, C.C.: A novel information hiding scheme based on btc. In: Proc. of International Conference on Computer and Information Technology, pp. 66–71 (2004)
7. Schyndel, R., Tirkel, A., Osborne, C.: A digital watermark. In: Proc. of IEEE International Conferences on Image Processing, pp. 86–90 (1994)

8. Tu, S., Hsu, C.S.: A btc-based watermarking scheme for digital images. International Journal on Information & Security 15(2), 216–228 (2004)
9. Yang, C.N., Lu, Z.M.: A blind image watermarking scheme utilizing btc bitplanes. Journal of Digital Crime and Forensics 3(4), 42–53 (2011)
10. Liang, T., Zhi-jun, F.: An adaptive middle frequency embedded digital watermark algorithm based on the dct domain. Pattern Recognition 40, 2408–2417 (2008)
11. Reddy, A., Chatterji, B.: A new wavelet based logo-watermarking scheme. Pattern Recognition Letters 26(7), 1019–1027 (2005)
12. Hsieh, M., Tseng, D., Huang, Y.: Hiding digital watermarks using multiresolution wavelet transform. IEEE Trans. on Industrial Electronics 48(5), 875–882 (2001)
13. Kundur, D., Hatzinakos, D.: A robust digital image watermarking method using wavelet based fusion. Optics Express 3(12), 485–490 (1998)
14. Dugad, R., Ratakonda, K., Ahuja, N.: A new wavelet-based scheme for watermarking images. In: Proc. IEEE Int. Conf. Image Processing (ICIP 1998), vol. 2, pp. 419–423 (1998)
15. Lu, C., Huang, S., Sze, C., Liao, H.: A new watermarking technique for multimedia protection, multimedia image and video processing, ch. 18, pp. 507–530. CRC Press, Boca Raton (2001)
16. Lin, S., Chin, C.: A robust dct-based watermarking for copyright protection. IEEE Trans. Consumer Electronics 46, 415–421 (2000)
17. Chang, C.W., Chang, C.C.: A novel digital image watermarking scheme based on the vector quantization technique. Journal of Computer & Security 24, 1460–1471 (2005)
18. Lu, M., Pan, S., Sun, H.: Vq-based digital image watermarking method. IEE Electronic Letters 36(14), 1201–1202 (2001)
19. Makur, A., selvi, S.: Variable dimension vector quantization based image watermarking. Signal Processing 81, 889–893 (2001)
20. Dhara, B.C., Chanda, B.: Block truncation coding using pattern fitting. Pattern Recognition 37(11), 2131–2139 (2004)
21. Dhara, B.C., Chanda, B.: Color image compression based on block truncation coding using pattern fitting principle. Pattern Recognition 40, 2408–2417 (2007)
22. Dhara, B.C., Chanda, B.: A fast progressive image transmission scheme using block truncation coding by pattern fitting. Journal of Visual Communication and Image Representation 23(2), 313–322 (2012)
23. Dukhi, R.G.: Watermarking: A copyright protection tool. In: Proc. of 3rd International Conference on Electronics Computer Technology, pp. 36–41 (2011)
24. Delp, E.J., Mitchell, O.R.: Image compression using block truncation coding. IEEE Trans. on Communication 27, 1335–1342 (1979)
25. Nasrabadi, M.N., King, R.B.: Image coding using vector quantization: a review. IEEE Trans. on Communication 36, 957–971 (1988)
26. Wang, Z., Bovik, A.C., Sheikh, H., Simoncelli, E.P.: Image quality assessment: From error visibility to structural similarity. IEEE Trans.on Image Processing 13(4), 600–612 (2004)

Uniform Based Approach for Image Fusion

Radhika Vadhi[1], Veeraswamy Kilari[2],
and Srinivaskumar Samayamantula[1]

[1] Department of ECE, University College of Engineering,
JNTUK, Kakinada, A.P, India
radhikav139@gmail.com, samay_ssk2@yahoo.com
[2] Department of ECE, QIS College of Engineering & Technology,
Ongole, A.P, India
kilarivs@yahoo.com

Abstract. This paper presents uniform based image fusion algorithm. Image fusion is a process of combining the source images to acquire the relevant information which is nearer to the original image. Source images are divided into sub blocks. Smoothness of the each block is calculated using variance of the block. In general most of the images are affected by Gaussian noise [1]. Hence, in this work a new image is generated based on blocks which have more smoothness. By considering smoothed blocks alone in both the images, as a result, most of the Gaussian noise is eliminated. Further, different pixel based algorithms (average, max-abs, and min-abs) are tested with the uniform based algorithm. Performance of different fused algorithms is assessed by using Peak Signal to Noise Ratio (PSNR), Mutual Information (MI), Edge Strength and Orientation Preservation (ESOP), Normalized Cross Correlation (NCC), and Feature Similarity (FSIM).

Keywords: Smoothness, Image fusion, and uniform.

1   Introduction

Image Fusion is a process of combining the desired data from different source images into a single image which contains more data compared with the individual source image. The Image fusion has wide variety of applications like medicine, remote sensing, automatic change detection, machine vision, biometrics, robotics, tracking and microscopic vision, etc. Several pixel based image fusion algorithms are available in the literature [2,3,4,5,6,7]. The pixel based image fusion algorithm has to fuse entire salient information contained in the input source images, not to introduce any blocking artifacts, and it must be robust, tolerant to noise, removes blur or mis registrations. In general fusion algorithms are categorized into spatial domain and transform domain algorithms. Spatial domain fusion algorithms use the local spatial features such as pixel based methods, gradient, and weighted integrated

J. Mathew et al. (Eds.): ICECCS 2012, CCIS 305, pp. 186–194, 2012.

methods. These methods are designed to reduce the uncertainty, extended range of operation, extended spatial and temporal coverage. Different kinds of multi-resolution transforms are designed to represent the sharpness and edges of an image. The image can be reconstructed using the transformed coefficients which provide the information of an image. Image fusion can be performed to satisfy the requirements, without discarding any salient information from the source images. This should not introduce any artifacts or inconsistencies which may cause the false diagnosis by the human observer or machine in image processing. Image fusion can be performed at different levels, pixel level, feature level, symbol level, and decision level. In the pixel level, fusion can be performed according to a set of pixels from the source images using fusion rules. The simple fusion rules in image fusion are simple averaging, selection of max-abs and min-abs. The performance of fusion algorithm depends on registered input images. Any mis-registration results in poor performance of image fusion.

In the uniform based approach, a new image is generated from source images based on smoothness of the blocks. Smallest variance blocks are more uniform (smoother) than the highest variance blocks. Hence, this approach is named as uniform based approach. Gaussian noise is eliminated by considering uniform blocks. The use of Uniform based algorithm is explored due to its advantages of have less computations, thereby less hardware complexity when compared to transform based [8] techniques. Further, uniform based approach eliminates the Gaussian noise.

This paper is organized as follows. Simple average, max-abs, and min-abs methods are reviewed in section 2. The details of proposed method are presented in section 3. Experimental results are discussed in section 4. Concluding remarks are given in section 5.

2   Pixel Based Fusion Techniques

The objectives of Pixel based fusion schemes are extracting all of the useful information from the source images without introducing artifacts or inconsistencies that will distract human observers. Those objectives could be achieved though creating combined images using most common fusion rules, that are more suitable for human perception and machine image processing such as simple average, max-abs, and min-abs [9,10,11,12].The following steps are helpful in performing the fusion rules:

— Compute the salience measures of individual source images.
— Fuse them according to the corresponding fusion rules.

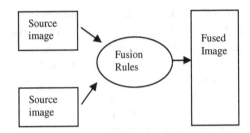

Fig. 1. General fusion process

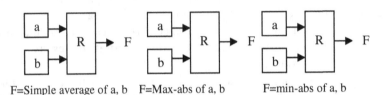

F=Simple average of a, b   F=Max-abs of a, b   F=min-abs of a, b

Fig. 2. Fusion Rules

In the fusion process, two source images are fused using fusion rules and finally results a single, more information fused image. In the Fig. 2, where a and b are the source image and F is the fused image. Formulae's for average rule, Max-abs rule, and min-abs rule are given below:

$$F_{avg}(x,y) = \frac{\sum_{x=1}^{N} \sum_{y=1}^{N} a(x,y) + b(x,y)}{2} \qquad (1)$$

$$F_{max}(x,y) = \max|a(x,y), b(x,y)| \qquad (2)$$

$$F_{min}(x,y) = \min|a(x,y), b(x,y)| \qquad (3)$$

In the fusion process, pixel based algorithms are easy to implement. Processing involves simple manipulations than transform based image fusion algorithms. Pixel based algorithms have less computations, thereby less hardware complexity when compared to transform based techniques.

3   Proposed Algorithm

The steps of the Uniform based fusion algorithm are as follows.

1. Two multi focused, registered images are considered with the same size {N, N}.
2. Divide each image into equal number of sub blocks with the size {n, n}.
3. Compute smoothness of each block using the following steps[13]

- Compute the length of the block(L) where length is n
- Compute the variance of the block
- Normalize the variance

$$Var_{nor} = \frac{Var}{(L-1)^2}$$

- Compute smoothness of the block using:

$$S = 1 - \frac{1}{1 + Var_{nor}}$$

4. Consider one of the images as reference image.
5. More smoothened blocks are identified by using the value of S. The blocks which possess more value of S are placed in reference image, resultantly processed image is generated.
6. Processed image and one of the source images are considered for fusion by any pixel based technique (average, max-abs, and min-abs)

Block diagram for the proposed method is shown in Fig. 3.

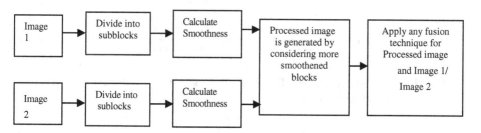

Fig. 3. Proposed Uniform based Fusion Block Diagram

4   Experimental Results

The size of each input source images is 128 x 128. The objective performance evaluation is done by Peak Signal-to-Noise Ratio (PSNR)[13], and Normalized Cross Correlation (NCC) as given in equations 4 to 6. Mean square error (MSE) is

$$MSE = \frac{1}{(N-1)(N-1)} \sum_{x=0}^{N-1} \sum_{y=0}^{N-1} [s(x,y) - F(x,y)]^2 \qquad (4)$$

$s(x,y)$ and $F(x,y)$ are the pixel values in the ground truth and fused image.

$$PSNR = 10\log_{10}\left(\frac{255^2}{MSE}\right) \qquad (5)$$

where, the usable gray level values range from 0 to 255.

Normalized Cross Correlation (NCC) is given as

$$NCC = \frac{\sum_i \sum_j (S(x,y) - \bar{S}(x,y))(F(x,y) - \bar{F}(x,y))}{\sqrt{\left[\sum_i \sum_j (S(x,y) - \bar{S}(x,y))^2\right]\left[\sum_i \sum_j F(x,y) - \bar{F}(x,y))^2\right]}} \tag{6}$$

where, $\bar{s}(x,y)$ indicates the mean of the original image and $\bar{f}(x,y)$ indicates the mean of the fused image. The Mutual Information (MI) [14] as a measure of image fusion performance between two images. Mutual information is given as:

$$M_F^{S_1 S_2} = I_{FS_1}(f, s_1) + I_{FS_2}(f, s_2) \tag{7}$$

where, $I_{FS_1}(f, s_1), I_{FS_2}(f, s_2)$ are the Information that contains about s1,s2 and are given as:

$$I_{FS_1}(f, s_1) = \sum_{f, s_1} P_{FS_1}(f, s_1) \log \frac{P_{FS_1}(f, s_1)}{P_F(f) P_{S_1}(s_1)} \tag{8}$$

$$I_{FS_2}(f, s_2) = \sum_{f, s_2} P_{FS_2}(f, s_2) \log \frac{P_{FS_1}(f, s_2)}{P_F(f) P_{S_2}(s_2)} \tag{9}$$

where, s1,s2 are source images.The Edge Strength and Orientation Preservation values [15] can be calculated as

$$Q_g^{SF}(x,y) = \frac{\Gamma_g}{1 + e^{K_g(G^{SF}(x,y) - \sigma_g)}} \tag{10}$$

$$Q_\alpha^{SF}(x,y) = \frac{\Gamma_g}{1 + e^{K_\alpha(S^{SF}(x,y) - \sigma_\alpha)}} \tag{11}$$

$Q_g^{SF}(x,y)$ and $Q_\alpha^{SF}(x,y)$ model perceptual loss of information in F, in terms of how well the strength and orientation values of a pixel $p(x,y)$ in S are represented in the fused image. The constants Γ_g, K_g, σ_g and $\Gamma_\alpha, K_\alpha, \sigma_\alpha$ determine the exact shape of the sigmoid functions used to form the edge strength and orientation preservation values.

$$S^{SF}(x,y) = \frac{\left| |\alpha_S(x,y) - \alpha_F(x,y)| - \pi/2 \right|}{\pi/2} \tag{12}$$

$$G_{SF}(x,y) = \begin{cases} \dfrac{g_F(x,y)}{g_S(x,y)} & \text{if } g_S(x,y) > g_F(x,y) \\ \dfrac{g_S(x,y)}{g_F(x,y)} & \text{otherwise} \end{cases} \tag{13}$$

where $g(x,y)$ is the edge strength and $\alpha(x,y)$ is the orientation [15]. The Feature Similarity (FSIM) [16] index can be measured as

$$FSIM = \frac{\sum_{x \in \Omega} S_L(x).PC_m(x)}{\sum_{x \in \Omega} .PC_m(x)} \tag{14}$$

where Ω means the whole image spatial domain, PC_m means Phase congruent structure for f_m.

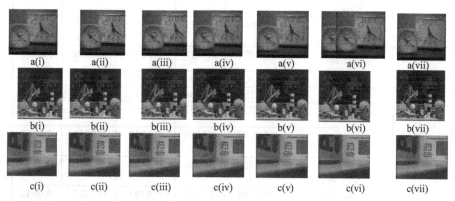

Fig. 4. Comparisons between traditional fusion rules and proposed Uniform based fusion algorithm

a(i),b(i),c(i) are results of Simple average rule, a(ii),b(ii),c(ii) are Max-abs rule, a(iii),b(iii),c(iii) are min-abs rule, a(iv),b(iv),c(iv) are Uniform based algorithm: a(v),b(v),c(v) are Uniform based average algorithm: a(vi),b(vi),c(vi) are Uniform based Max-abs rule: a(vii),b(vii),c(vii) are Uniform based min-abs rule.

Experiments are performed with all categories of images. More emphasis is given to multifocus images than medical and panchromatic images. Experimental results are shown in Fig.4. Numerical results are presented in Table 1. Uniform based methods are giving better results than other methods for the multifocus images. Medical images and panchromatic images are also tested with different algorithms. Experimental results indicate that in medical images case, max abs is giving better results. In Panchromatic images, uniform based max abs is giving better results.

ESOP and NCC values are better in uniform based average method for all multifocused images. Pure uniform based algorithm has given better FSIM value than other methods for the clock images. Uniform based min abs has given better MI value than other methods for the clock images.

Table 1. Experimental Results

Source image	Fusion rule	PSNR	MI	ESOP	NCC	FSIM
Clock	Average	30.2282	2.9992	0.7253	0.9815	0.9368
	Max-abs	29.3145	3.1006	0.6945	0.9794	0.9455
	min-abs	28.7124	3.2643	0.7573	0.9762	0.9508
	uniform based	34.5720	4.1499	0.7606	0.9929	**0.9791**
	Uniform+ Average	**36.2596**	3.8262	**0.7680**	**0.9952**	0.9774
	Uniform+ Max-abs	33.8952	3.4647	0.7475	0.9919	0.9756
	Uniform+ Min-abs	34.7546	**3.9371**	0.7643	0.9933	0.9756
Toy	Average	24.5782	2.9412	0.7271	0.9719	0.8551
	Max-abs	34.3204	2.9127	0.8721	0.9950	0.9816
	Min-abs	31.3153	**3.1634**	0.8704	0.9955	0.9832
	uniform based	32.9281	2.7841	0.8637	0.9929	0.9725
	Uniform+ Average	**34.8973**	3.1619	**0.8839**	**0.9965**	**0.9867**
	Uniform+ Max-abs	34.3324	2.9056	0.8740	0.9950	0.9816
	Uniform+ Min-abs	31.3243	3.1378	0.8707	0.9955	0.9833
Pepsi	Average	23.7552	2.4543	0.5561	0.9641	0.9079
	Max-abs	29.0222	3.1970	0.6959	0.9829	0.9642
	Min-abs	30.1386	3.1873	0.7395	0.9875	0.9582
	uniform based	33.7244	3.1738	0.7779	0.9933	0.9559
	Uniform+ Average	**34.8616**	**3.2551**	**0.7877**	**0.9948**	**0.9592**
	Uniform+ Max-abs	33.9490	3.2343	0.7745	0.9937	0.9569
	Uniform+ Min-abs	34.5438	3.2505	0.7876	0.9945	0.9574

Pepsi images are shown in Fig.5 to understand the visual quality. Fused image quality of uniform based average method is better than the average method. The same is valid for all the multifocus images.

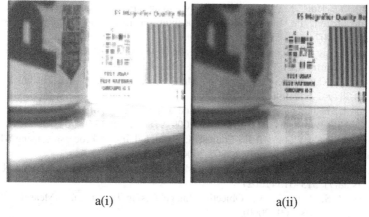

a(i)   a(ii)

Fig. 5. a(i) Result of simple average rule (ii) Result of Uniform based average rule

5   Conclusions

In this paper, uniform based algorithm is presented. Considerable improvement is observed with the uniform based approach with all existing fusion methods like average, max-abs, and min-abs fusion   algorithms. The Uniform based approach can be extended to any image fusion technique to improve the results further. The uniform based approach with average rule gives better results than other methods. The uniform based approach offer a significant advantage in terms of processing (simpler manipulation) compared to transform based fusion algorithms.

References

1. Stathaki, T.: Image Fusion Algorithms and Applications. Academic Press (2008)
2. Paella, G.: A General Frame Work for Multiresolution Image Fusion from Pixels to Regions. Information Fusion 4, 259–280 (2003)
3. Abidi, A.M., Gonzalez, R.C.: Data Fusion in Robotics and Machine Intelligence. Academic Press (1992)
4. Smith, M.I., Heather, J.P.: Review of Image Fusion Technology. In: Proc. SPIE, pp. 29–45 (2005)
5. Zeng, J., et al.: Review of Image Fusion Algorithms for Unconstrained Outdoor Scenes. In: International Conference on Signal Processing, pp. 16–20 (2006)
6. Deepali, D.: Wavelet based Image Fusion using Pixel based Maximum Selection Rule. International Journal of Engineering Science and Technology (IJEST) 3(7) (2011)
7. Mitianoudis, N., Stathaki, T.: Pixel- based and Region-based Image Fusion Schemes using ICA bases. Information Fusion 8, 131–142 (2007)
8. Chiorean, L., Vaida, M.: Medical Image Fusion based on Discrete Wavelet Transform using Java Technology. In: International Conference on Information Technology Interfaces, pp. 22–25 (2009)

9. Zheng, Y., Hou, X., Bian, T.: Effective Image Fusion Rules of Multi-scale Image Decomposition. In: International Symposium on Image and Signal Processing and Analysis (2007)
10. Zhang, X., Liu, X.: Pixel Level Image Fusion Scheme based on Accumulated Gradient and PCA Transform. IEEE Press (2008) 978-1-4244-3291-2/08
11. Luo, R.C., Kay, M.G.: A Tutorial on Multi Sensor Integration and Fusion. IEEE Press (1990) 087942-600-4/90/1100-0707
12. Kumar, U., Mukhopadhyay, C., Ramachandra, T.V.: Fusion of Multi Sensor Data: Review and Comparative Analysis. IEEE Press (2009) 978-0-7695-3571-5/09
13. Gonzalez, R.C., Woods, R.E., Eddins, S.L.: Digital Image Processing using MATLAB. Prentice-Hall (2004), Copyright 2002-2004
14. Qu, G.H., Zhang, D.L.: Information Measure for Performance of Image. Electronic Letters 38(7), 313–315 (2002)
15. Xydeas, C.S., Petrovic, V.: Objective Image Fusion Performance Measure. Electronic Letters 36(4), 308–309 (2000)
16. Zhang, L., Mou, X.: FSIM: A Feature Similarity Index for Image Quality Assessment. IEEE Transactions on Image Processing 20(4), 2378–2386 (2011)

A Fast Image Reconstruction Algorithm Using Adaptive R-Tree Segmentation and B-Splines

Ravikant Verma and Rajesh Siddavatam

Department of Computer Science & IT,
Jaypee University of Information Technology,
Waknaghat, Himachal Pradesh, 173215, India
ravikant.verma@juit.ac.in, srajesh@ieee.org

Abstract. The image reconstruction using adaptive R tree based segmentation and linear B- splines is addressed in this paper. We used our own significant pixel selection method to use a combination of canny and sobel edge detection techniques and then store the edges in an adaptive R tree to enhance and improve image reconstruction. The image set can be encapsulated in a bounding box which contains the connected parts of the edges found using edge-detection techniques. Image reconstruction is done based on the approximation of image regarded as a function, by B-spline over adapted Delaunay triangulation. The proposed method is compared with some of the existing image reconstruction spline models.

Keywords: Image Segmentation, Delaunay triangulation, B-splines, Image Reconstruction.

1   Introduction

Image reconstruction using regular and irregular samples have been developed by many researchers recently. Rajesh Siddavatam et. al. [1,2,3,4] has developed a fast progressive image sampling using Non-uniform B-splines. Eldar et. al [5] has developed image sampling of significant samples using the farthest point strategy. Arigovindan [6] developed Variational image reconstruction from arbitrarily spaced samples giving a fast multiresolution spline solution. Carlos Vazquez et al, [7] has proposed interactive algorithm to reconstruct an image from non-uniform samples obtained as a result of geometric transformation using filters. Cohen and Matei [8] developed edge adapted multiscale transform method to represent the images. Aldroubi and Grochenig, [12] have developed nonuniform sampling and reconstruction in shift invariant spaces. Delaunay triangulation [9] has been extensively used for generation of image from irregul ar data points. The image is reconstructed by either by linear or cubic splines over Delaunay Triangulations of adaptively chosen set of significant points. This paper concerns with reconstruction using B- splines from adaptive R-tree segmentation. The reconstruction algorithm deals with generating Delaunay triangulations of scattered image points, obtained by detection of edges using Sobel and Canny edge detection algorithms.

J. Mathew et al. (Eds.): ICECCS 2012, CCIS 305, pp. 195–203, 2012.

Section 2 describes the significant pixel selection method. and in section 3 the modeling of the 2D images using the B- splines is elaborated. Section 4 deals with Adaptive R-tree based Segmentation and the proposed novel reconstruction algorithm is discussed in section5. Significant Performance Measures are presented in section 6. The experimental results and conclusion by using the proposed method are discussed in section 7.

2   Significant Sample Point Selection

We use our own Significant Sample Point algorithm proposed by Rajesh Siddavatam et.al [1]   which   involves following steps :

Let M be a m X n matrix representing a grayscale image
The algorithm involves following steps:-
1) Initialization: initialization of variables
2) Edge Detection: Edge detection using sobel and canny filters.
3 Significant Points for Strong edges.
4) Significant Points for Weak edges
5) Overview of Delaunay Triangulation.
6) Delaunay Triangulation (First pass)
7) Re-Triangulation Algorithm

3   Image Reconstruction Using B- Splines

A 2D spline model is used to represent images due to their interpolation and approximation properties[12]. Splines are piecewise polynomial functions where in a spline function of degree n, each piece is a polynomial of degree n , connected to its neighbors in such a way that the whole function and its derivatives upto the order n-1 are continuous.

$$f(x) = \sum_{k \in z} c[k] \, \beta^n (x-k).$$
(1)

Where f is a spline function and β^n is B-spline function of degree n $[c(k)]$ are the spline coefficients.

For modeling images, we need to construct 2D B-spline functions. They can be constructed very easily using a tensor product of 1D B-spline functions in the directions of x_1 and x_2 respectively.

$$\varphi^{(n)}(x_1, x_2) = \beta^{(n)}(x_1) \, \beta^{(n)}(x_2)$$
(2)

A 2D spline function is easily constructed as a linear combination of translated versions of this 2D B-spline function to the positions of the samples such that

$$f(x) = \sum c[n] \, \varphi^{(n)}(x - n), \forall \; x \in R^2$$
(3)

4   Adaptive R- Tree Based Segmentation

R-trees are tree data structures used for indexing multi-dimensional information such as geographical coordinates, rectangles or polygons. It is data structure to store data objects in group and represent them with minimum bounding boxes (MBB). Using Canny- Sobel mix edge detection technique we have found the edges of the image. The edges so found are not fully connected owing to the various kinds of masks applied. The connectivity of the edges changes according to the mask applied. Thus each connected edge is encapsulated in a bounding box of the least possible size. Hence the 2D image is spatially segmented into a set of bounding boxes each with varying dimensions up to the size of the image as shown in Fig.10 based on usual R-tree segmentation.

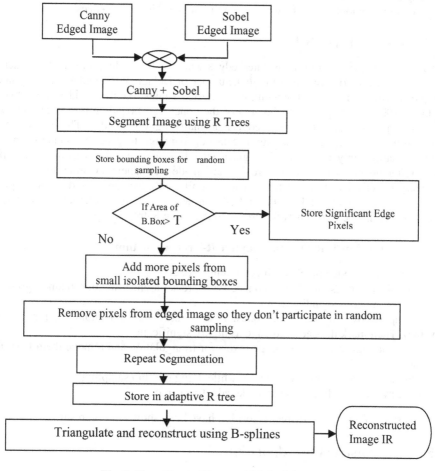

Fig. 1. Flow Chart of Proposed Methodology

The usual R-tree approach has a major fault. It gives us much more bounding boxes than that are required. As we are going to follow random sampling to find vertices for Delaunay triangulation apart from the significant pixels from significant pixel selection algorithm of section 2.The R-Tree approach gives us two types of random pixels. One type which is part of the high density edges and others which are located in isolated edges depending on the test image. In case of normal random sampling we will get approximately less no of pixels (or vertices) for triangulation in the isolated regions resulting in haziness near the isolated edges.

To avoid the same, we can take two types of pixels for efficient reconstruction:

4.1   Non-uniform Pixels

These pixels are derived by randomly selecting a fixed number of pixels from the image edges (mixture of canny and sobel), the same are responsible for uniform reconstruction throughout the image due to presence of many vertices in the high edge density region. Some of the random samples are also from the isolated significant edges.

4.2   Isolated Edge Pixels

These are the pixels from the isolated edges which will now be permanent in order to get better reconstruction due to higher number of Delaunay triangulations in isolated regions. The area of each bounding box is tested against the threshold value between 100 to 1000 square pixels. If the area of the concerned bounding box is greater than the threshold value, then it is treated as a normally significant edge. The segmentation algorithm is run again on the significant edged image to give the bounding boxes encapsulating only the normally significant edges and the pixels from the smaller bounding boxes in the isolated region are made permanent in order to give high density for efficient triangulation. These significant edges are stored in an adaptive R-tree as shown in the Fig. 11 of Lena on which the reconstruction algorithm is implemented. The Flow chart for proposed Adaptive R-tree is depicted in Fig. 1.

4.3   Algorithm 3: R Tree to Adaptive R-Tree Algorithm

1. Obtain set of edge pixels of image(canny+sobel)
2. For all the pixels, Wherever connectivity breaks a encapsulating bounding box is drawn for the corresponding edge.
3. Compute area of bounding boxes; if area more than threshold(a set minimum value). Then mark the enclosed edge as highly significant.
4. Store the highly significant edge pixels for triangulation and remove them from the image.
5. This will remove pixel overlapping while doing random sampling.
6. Redraw the bounding boxes to make adaptive r tree.

The results of adaptive R-tree Algorithm have been shown in Figures 6,7,8 and 9.

5   Reconstruction Algorithm

The following steps are used to reconstruct the original image from set of pixels from Significant pixel selection algorithm of section 2 defined as significant (Sig) and isolated significant pixels defined as (Iso-sig) from adaptive R-tree Algorithm of section 4.

5.1 Input
1. Let S_N = data set
2. z_O: luminance
3. S_O: set of regular data for initial triangulation

Step1. Use Significant pixel selection algorithm to find a set of new significant pixels (SP)
Step2: Add adaptive R-tree pixels set to the above set.
Step3. Use Delaunay triangulation and B-Splines to produce unique set of triangles and image.
Step4. Get SIG = sig + Iso-sig
Step5. Repeat steps 1 to 3 to get the image IR (y)
Step6. Return SIG and IR (y)

5.2 Output:
SIG and Reconstructed Image   IR (y)

6   Algorithm Complexity

In general, the complexity of the non-symmetric filter is proportional to the dimension of the filter n2, where n * n is the size of the convolution kernel. In canny edge detection, the filter is Gaussian which is symmetric and separable. For such cases the complexity is given by n+1 [20]. All gradient based algorithms like Sobel do have complexity of O(n). The complexity of well known Delaunay algorithm in worst case is O(n^ceil(d/2)) and for well distributed point set is ~ O(n). N is number of points and d is the dimension. So in 2D, Delaunay complexity is O(N) is any case.

Step 1: Sobel Edge Detector: O(n)
Step 2: Canny Edge Detector: O(n)
Step 3   Algorithm1: O(2n-1)=O(n)
Step 4: Algorithm2: O(3n-2)=O(n)
Step 5: Algorithm3: O(n)
Step 6: Image Reconstruction: O(n)

Hence the total complexity of the proposed algorithm is O(n) which is quite fast and optimal.

7   Significance Measures for Reconstructed Image

7.1   Peak Signal to Noise Ratio

A well-known quality measure for the evaluation of image reconstruction schemes is the Peak Signal to Noise Ratio (PSNR),

$$PSNR = 20*\log 10(b/RMS)$$

(4)

where **b** is the largest possible value of the signal and RMS is the root mean square difference between the original and reconstructed images. PSNR is an equivalent measure to the reciprocal of the mean square error. .

8   Results and Discussions

In this paper, a novel algorithm based on Adaptive R-tree based significant pixel selection is applied for image reconstruction. Experimental results on the popular images of Lena and Peppers are presented to show the efficiency of the method. Set of regular points are selected using Canny and Sobel edge detection and Delaunay triangulation method is applied to create triangulated network. The set of significant sample pixels are obtained and added in the preceding set of significant pixels samples at every iteration. The gray level of each sample point is interpolated from the luminance values of neighbor significant sample point. The original image and its reconstruction results along with the error image are shown for LENA and PEPPERS images. The reason for a better PSNR = 30.78 dB for our proposed method as shown in Table 1 is due to the fact that the missing/isolated edges in Fig 11 are due to adaptive R-tree algorithm of section 4,   and these pixel sets will now participate in greater majority in reconstruction than the normal random pixels from pixel selection algorithm of section 2.

Fig. 2. Original Lena Image

Fig. 3. Adaptive FPS[2] 4096 samples (18.08 dB)

Fig. 4. Reconstructed Lena Image 4096 samples (PSNR=29.22dB)

Fig. 5. Reconstructed Lena Image 4096 Non uniform samples (PSNR = 30.78 dB)

Fig. 6. R-tree Segmentation of Peppers Image

Fig. 7. Adaptive R-tree of Peppers

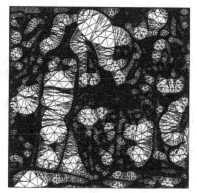

Fig. 8. Triangulation of Peppers

Fig. 9. Reconstructed Peppers (PSNR=29.89 dB)

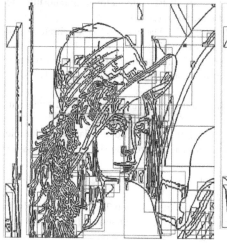

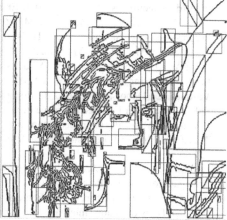

Fig. 10. R-tree segmentation of Lena

Fig. 11. Adaptive R-tree of Lena

Table 1. Comparative Evaluation of proposed Method

Test Case	Method	PSNR (dB)
Lena **512x512**	Proposed Adaptive R-tree	**30.78**
	Significant pixel selection [1]	29.22
	Progressive Image Sampling [4]	21.45
	Farthest Point Sampling(FPS)[5]	18.08
Peppers **512x512**	Proposed Adaptive R-tree	**29.89**
	Significant pixel selection [1]	29.01
	Progressive Image Sampling [4]	22.06
	Farthest Point Sampling(FPS)[5]	18.18

References

1. Siddavatam, R., Verma, R., Srivastava, G.K., Mahrishi, R.: A Fast Image Reconstruction Algorithm Using Significant Sample Point Selection and Linear Bivariate Splines. In: IEEE TENCON, pp. 1–6. IEEE Press, Singapore (2009)
2. Siddavatam, R., Verma, R., Srivastava, G.K., Mahrishi, R.: A Novel Wavelet Edge Detection Algorithm For Noisy Images. In: IEEE International Conference on Ultra Modern Technologies, pp. 1–8. IEEE Press, St. Petersburg (2009)
3. Siddavatam, R., Verma, R., Srivastava, G.K., Mahrishi, R.: A Novel Image Reconstruction Using Second Generation Wavelets. In: IEEE International Conference on Advances in Recent Technologies in Communication and Computing, pp. 509–513. IEEE Press, Kerala (2009)
4. Siddavatam, R., Sandeep, K., Mittal, R.K.: A Fast Progressive Image Sampling Using Lifting Scheme And Non-Uniform B-Splines. In: IEEE International Symposium on Industrial Electronics, pp. 1645–1650. IEEE Press, Spain (2007)
5. Eldar, Y., Lindenbaum, M., Porat, M., Zeevi, Y.Y.: The Farthest Point Strategy For Progressive Image Sampling. IEEE Trans. Image Processing 6(9), 1305–1315 (1997)
6. Arigovindan, M., Suhling, M., Hunziker, P., Unser, M.: Variational Image Reconstruction From Arbitrarily Spaced Samples: A Fast Multiresolution Spline Solution. IEEE Trans. on Image Processing 14(4), 450–460 (2005)
7. Vazquez, C., Dubois, E., Konrad, J.: Reconstruction of Nonuniformly Sampled Images in Spline Spaces. IEEE Trans. on Image Processing 14(6), 713–724 (2005)
8. Cohen, A., Mate, B.: Compact Representation Of Images By Edge Adapted Multiscale Transforms. In: IEEE International Conference on Image Processing, Tessaloniki, pp. 8–11 (2001)
9. Laurent, D., Nira, D., Armin, I.: Image Compression by Linear Splines over Adaptive Triangulations. Signal Processing 86(4), 1604–1616 (2006)
10. Tzu-Chuen, L., Chin-Chen, C.: A Progressive Image Transmission Technique Using Haar Wavelet Transformation. International Journal of Innovative Computing, Information and Control 3, 6(A), 1449–1461 (2007)

11. Eldar, Y., Oppenheim, A.: Filter Bank Reconstruction of Bandlimited Signals from Non-Uniform and Generalized Samples. IEEE Trans. Signal Processing 48(10), 2864–2875 (2000)
12. Aldroubi, A., Grochenig, K.: Nonuniform Sampling and Reconstruction in Shift Invariant Spaces. SIAM Rev. 43, 585–620 (2001)
13. Wu, J., Amaratunga, K.: Wavelet Triangulated Irregular Networks. Int. J. Geographical Information Science 17(3), 273–289 (2003)
14. Barber, C.B., Dobkin, D.P., Huhdanpaa, H.T.: The Quickhull Algorithm for Convex Hulls. ACM Transactions on Mathematical Software 22(4), 469–483 (1996)
15. Preparata, F.P., Shamos, M.I.: Computational Geometry. Springer, New York (1988)

A Novel Algorithm for Hub Protein Identification in Prokaryotic Proteome Using Di-Peptide Composition and Hydrophobicity Ratio

Aswathi B.L., Baharak Goli, Renganayaki Govindarajan,
and Achuthsankar S. Nair

Department of Computational Biology and Bioinformatics,
University of Kerala,
Trivandrum 695581, India
aswathi.bl@gmail.com

Abstract. It is widely hypothesized that the information for determining protein hubness is found in their amino acid sequence patterns and features. This has moved us to relook at this problem. In this study, we propose a novel algorithm for identifying hub proteins which relies on the use of dipeptide compositional information and hydrophobicity ratio. In order to discern the most potential and protuberant features, two feature selection techniques, CFS (Correlation-based Feature Selection) and ReliefF algorithms were applied, which are widely used in data preprocessing for machine learning problems. Overall accuracy and time taken for processing the models were compared using a neural network classifier RBF Network and an ensemble classifier Bagging. Our proposed models led to successful prediction of hub proteins from amino acid sequence information with 92.94% and 92.10 % accuracy for RBF network and bagging respectively in case of CFS algorithm and 94.15 % and 90.89 % accuracy for RBF network and bagging respectively in case of ReliefF algorithm.

Keywords: Hub proteins, Protein- protein interaction networks, machine learning, Feature vectors.

1 Introduction

The dawn of the genomics era is deepening our focus from keying the molecular components of life in terms of their functions to a more systems level outlook, where the prominence is on the interactions between these cellular components. For better realization of cellular level processes, the study of the complex networks they define is critical. Majority of cellular mechanisms are mediated by Protein-protein interactions. They build complex networks that have been shown to have a scale-free topology [1]. This topological property is characterized by the number of connections of each node which is termed as connectivity. This connectivity is predominantly significant in biological systems, as it imparts heftiness [1]. Since many of the scale -free networks follows power law, most of the nodes are sparsely connected while a few nodes are highly interactive. Such proteins

J. Mathew et al. (Eds.): ICECCS 2012, CCIS 305, pp. 204–211, 2012.

with high degree of connectivity are termed as 'Hubs'. The network remains fully connected even after the sparingly connected nodes are removed. Instead the removal of the most central or highly connected hub nodes may causes the collapse of the system [1]. In fact, it has been established that connectivity and essentiality in the yeast protein interaction network are positively correlated [1]. Thus, hub proteins which literally 'hold the protein interaction networks together' are more likely to be essential [3]. Hub proteins are known to have high density of binding sites [4], which enable them to have multiple interactions.

Hub protein analysis presumes vital importance, since the possibility of their involvements in multiple pathways are higher [4]. Also this can lead to better realization of cellular functions as well as discovering novel drug targets and predicting the side effects in drug discovery by understanding the pathways, topologies and dynamics of them. Many of the well-known and widely examined proteins including p53, associated with many diseases, are hubs. Probing these hub proteins can provide useful information for predicting the possible side effects in drug discovery [2, 5, 6].

There have been various attempts to predict hub proteins in protein-protein interaction networks using various data such as gene ontology [7], gene proximity [8, 9], gene fusion events [10, 11] and gene co-expression data [12-13]. One of the major limiting factors for using the above mentioned data is the lack of availability of them for the entire protein interaction data of an organism. Application of existing methods which use structural information is also severely limited as PDB structures are not available for many of the proteins [2].

In this study we have developed a statistics-based approach to classify hub proteins using soft computational algorithms. We experimented with amino acid sequence information alone to overcome the imitations of availability of structural and functional data which are slow in advent.

2   Materials and Methods

2.1   The Dataset

For this study, we selected Escherichia coli k12 as the model organism, which has well annotated and have modest protein interaction information. The protein interaction data was generated from IntAct [14] database release 2012-February 10 and corresponding amino acid sequences of varying lengths were compiled from Uniprot [15]. We used CD-HIT-2D [16] webserver to remove sequences that have similarity greater than or equal to 50%. To avoid any bias, protein sequence having amino acid sequence length less than 60 were also eliminated. The curated dataset included 1,915 proteins and total number of protein interactions was 15,222 with an average degree of interaction 7.982. Training and testing sets were constructed in such a way that no pair of proteins from either sets had significant sequence similarity.

2.2   Identification of Hub-Threshold

Degree of connectivity of proteins in our E.coli PPI dataset ranged from 1 to 139. In order to identify a protein as hub we had to choose a degree threshold. According to the previous studies the connectivity cut-off of hub proteins are species specific [6] and so far there is no consensus on the exact connectivity thresholds for these proteins [6].We followed fold change definition for connectivity threshold by taking the ratio of connectivity and average connectivity to determine appropriate cutoff to identify hubs in E.coli PPI network [2]. A node with connectivity fold change greater than or equal to 2 was the criterion applied for considering a protein as hub (cutoff, P-value < 0.003, using distribution of standard normalized fold change values in E.coli). The final number of highly connected protein was 210 and sparsely connected protein was 1,705.

2.3   Feature Vectors

Since many of the machine learning techniques require property vectors as input for classification, each amino acid sequence should be replaced by a set of numeric values representing its properties. Capturing meaningful biological information from the amino acid sequences of varying length is an important and most crucial step in classification problems. In this work, di-peptide composition and hydrophobicity ration were used as the feature vectors for hub classification.

Di-Peptide Composition

Di-Peptides are molecules comprising two amino acids joined by a single peptide bond. Since there are 20 amino acids, 400 different di-peptide combinations are possible. Di- peptide composition captures information about amino acid sequence local order effects as well as amino acid distributions [17]. Here we computed the di-peptide composition in a sequential way. Consider a sample sequence C C C P A C Q C C A C. Di-Peptide composition of the CC pair, DP_{CC} is calculated as 3.

Fig. 1. Di-Peptide sequential count

If a protein sequence P is given with N amino acids, we can map the sequence with the feature vector as, FP = { F_1, F_2, F_3,...F_{400} }. Since there are 20 amino acids, we obtained 400 different binary residue combinations.

Hydrophobicity Ratio

From the previous studies it is assumed that, hydrophobic effect has a prominent role in driving protein-protein interactions [18]. We computed the ratios of strong and weak hydrophobic residues using Chaos Game Representation (CGR), which is one of the graphical representation methods for biological sequences [19]. ProtScale tool of Expasy [20] was used to acquire kyte-doolittle hydrophobicity scale which is one of the most commonly using hydrophobicity index. We divided the 20 amino acids into 4 groups as, least Hydrophobic (Arginine, Lysine, Asparagine, Glutamine, Glutamic Acid, Histidine, Aspartic Acid), Weak Hydrophobic (Proline, Tyrisine, Tryptophan, Threonine, Glycine, Serine), Medium Hydrophobic (Cysteine, Alanine, Phenylalanine, Methionine) and Strong Hydrophobic (Isoleucine, Leucine, Valine) based on the hydrophobicity values and represent each group at each corner of the CGR Plot.

After getting the CGR graph, it is divided by a hyper plane and hence the total amino acid distribution is divided into two groups- Least Hydrophobic and strong Hydrophobic. Linear sum of each group is calculated and the ratio has taken. Using this feature we extracted the hydrophobicity ratio for each amino acid sequence. Fig. 2 illustrates the Hydrophobicity- ratio computation using CGR plots. Where, WH= Weak Hydrophobic, LH= Least Hydrophobic, MH= Medium Hydrophobic and SH= Strong Hydrophobic regions.

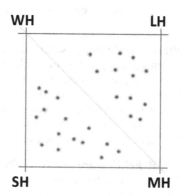

Fig. 2. Hydrophobicity- ratio plot using CGR for any amino acid sequence

We compared CGR plots of highly connected hub nodes and sparsely connected non hub nodes having same lengths. The CGR plots obtained for a highly connected protein with length 1536 and a sparsely connected protein with having 1536 are shown below.

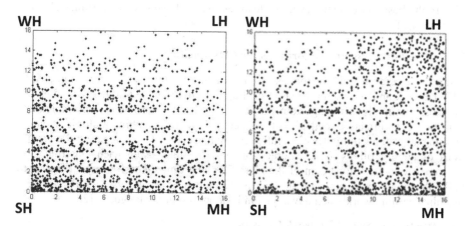

Fig. 3. CGR plot for a highly connected sequence of length 1536 in *E.Coli*

Fig. 4. CGR plot for a sparsely connected sequence of length 1536 in *E.Coli*

2.4   Feature Pruning

Generally, the performance of any classifier relies on the consistency of the features chosen and the size of the training set [21]. Feature selection is one of the significant techniques in data preprocessing for machine learning and data mining problems, which trashes out irrelevant and redundant features and speeds up the data mining algorithm and improves classification accuracy [22, 23, 24]. For this we adopted CFS (correlation-based feature selection) [25] and ReliefF feature selection algorithm [26] which are two well-known feature selection techniques. In this study 401 features generated from the transformation step explained above, 400 from di-peptide composition and 1 feature from hydrophobicity ratio were taken. After feature pruning a total of 189 features remained. Of these 188 features represents di-peptide composition. Hydrophobicity ratio also yielded well. We briefly describe these feature selection algorithms below.

Feature Trimming Algorithm: Correlation-Based Feature Selection (CFS)
This is one of the powerful techniques in purging uncorrelated and redundant features. Using a best first-search heuristic approach it estimates the prominence of subsets of features [25]. This heuristic algorithm contemplates the importance of individual features for predicting the class as well as the level of correlation among them. The basic logic in CFS is that good feature subsets include those features that are highly correlated with the target class and uncorrelated with each other.

Feature Selection Algorithm: Relief Feature Selection (ReliefF)
This is an extension of Relief algorithm developed to use in classification problems [24, 26]. It evaluates the relevance of these features by considering the strong correlation between the features. Here an instance i is selected randomly from the dataset. The weight for each feature is rationalized based on the distance, 'd' to its nearest neighbors from the same class and nearest neighbors from each of the different classes at each step of an iterative process[27]. This process is iterated 'n' times, where 'n' is a predefined parameter and is equal to the number of samples in dataset. Finally the best subset includes those features, which have relevance values above a chosen cut-off.

2.5   Machine Learning Algorithms for Classification

With the selected feature vectors as the backbone, we have modeled two classifiers, RBF network and an ensemble classifier, Bagging. Artificial neural network is one of the most commonly using algorithms to solve classification problems. Ensemble classification is an effective method that has been adopted to combine multiple machine learning algorithms to improve overall prediction accuracy by aggregating the predictions of all algorithms. The accuracy of classifications using each classifier was measured. For the comparison of the classifiers, the time taken by each classifier to build the model was also noted. For the implementation we used, weka suite, a machine learning workbench developed in java programming language [28].

Construction of Neural Network Classifier
RBF networks are supervised neural networks which are popular substitute to multilayer perceptions which employ reasonably lesser number of locally tuned units and are adaptive in nature [29, 30, 31]. They are widely used for classification and pattern recognition problems. In this study, the training set consisting of 250 hubs and 850 non-hub elements were given to the each network in the 10-fold cross-validation scheme.

Ensemble Classifier

This classifier helps to incorporate multiple classification models, each having different input feature vectors. The aim of using ensemble classifier is to obtain more accurate classification as well as better generalization. We experimented with one of the most popular ensemble methods, bagging [32]. The name bagging is derived from bootstrap aggregation. This samples each input data repeatedly with replacement from the training dataset. Each bootstrap sample has the same size as that of the training data set. Performance of bagging depends on the stability of the base base classifiers [33].

3 Results

The performance of our proposed classification models were estimated using standard 10-fold cross-validation in which the whole dataset is randomly partitioned into ten evenly-sized subsets. Performance is measured for each test set, and the mean is reported as overall accuracy. Several measures were used to evaluate the performance of the classifiers (True positive (TP), True negative (TN), False positive (FP) and False negative (FN), respectively).These measures include, Specificity=TN/ (TN+FP)*100, Sensitivity=TP/ (TP+FN)*100, Accuracy= TP+TN/ (TP+TN+FP+TN) and time taken to build the models in seconds. Table 1 summarizes the performance of different classifiers in 10-fold cross-validation.

Table 1. Performance of Hub classification algorithms in 10-fold cross-validation

Classification method	Sensitivity (%)	Specificity (%)	Accuracy (%)	Time Taken in Sec.
RBF Network + cfs	93.71	92.17	92.94	2.67
Bagging+ cfs	92.73	91.48	92.10	3.28
RBF Network + relief-f	94.76	93.54	94.15	1.98
Bagging +relief-f	87.05	93.41	90.89	2.89

To evaluate the classification model, Self-consistency test and independent test were also done. The results are shown in Table 2. Self-consistency test checks the consistency of the developed model. In independent dataset the training set was composed two equal halves of hub and non-hub proteins. The remaining sequences were used as the testing set.

Table 2. Accuracy of each classifier for self-consistency and independent data test

Classification Method	Self-consistency (%)	Independent Test (%)
RBF Network + cfs	94.21	90.71
Bagging+ cfs	95.73	91.80
RBF Network + cfs	95.32	93.25
Bagging+ cfs	96.81	92.91

4 Discussion

In this study, we propose a novel hub classification algorithm which relies only on amino acid sequence information. Analysis of structural and functional phenomena from sequence information is not a novel approach. It has been widely used with the advent of bioinformatics approaches in genomics and proteomics studies. There have been many computational Biology works which applies this approach to various problems including gene finding [34], protein subcellular localization [35] and protein allostery prediction [36].

Our results show that the chosen amino acid features, di-peptide composition and hydrophobicity ratio have strong correlation in identifying hubs from non- hub proteins. With Correlation based feature selection and the Relief-F algorithm followed by two classification algorithms, RBF Networks and bagging, we could effectively trace out useful amino acid features which are significant in the hub protein identification. The biological importance of the chosen amino acid properties in this work are yet to be explained. It would be remarkable to investigate the significance of these properties in the formation of PPINs.

References

1. Albert, R., Jeong, H., Barabási, A.-L.: Error and attack tolerance of complex networks. Nature 406, 378–382 (2000)
2. Latha, A.B., Nair, A.S., Sivasankaran, A., Dhar, P.K.: Identification of hub proteins from sequence. Bioinformation 7 (2011)
3. Tun, K., Rao, R.K., Samavedham, L., Tanaka, H., Dhar, P.K.: Rich can get poor: conversion of hub to non-hub proteins. Systems and Synthetic Biology 2, 75–82 (2009)
4. He, X., Zhang, J.: Why do hubs tend to be essential in protein networks? PLoS Genetics 2, e88 (2006)
5. Patil, A., Kinoshita, K., Nakamura, H.: Hub promiscuity in protein-protein interaction networks. International Journal of Molecular Sciences 11, 1930–1943 (2006)
6. Hsing, M., Byler, K.G., Cherkasov, A.: P The use of Gene Ontology terms for predicting highly-connected "hub" nodes in protein-protein interaction networks. BMC Systems Biology 2, 80 (2006)
7. Srihari, S.: Detecting hubs and quasi cliques in scale-free networks. In: 2008 19th International Conference on Pattern Recognition, pp. 1–4 (2008)
8. Dandekar, T., Snel, B., Huynen, M., Bork, P.: Conservation of gene order: a fingerprint of proteins that physically interact. Trends Biochem. Sci. 23, 324–328 (1998)
9. Overbeek, R., Fonstein, M., D'Souza, M., Pusch, G.D., Maltsev, N.: The use of gene clusters to infer functional coupling. Proc. Natl. Acad. Sci. USA 96, 2896–2901 (1999)
10. Marcotte, E.M., Pellegrini, M., Ng, H.L., Rice, D.W., Yeates, T.O., Eisenberg, D.: Detecting protein function and protein-protein interactions from genome sequences. Science 285, 751–753 (1999)
11. Enright, J., Iliopoulos, I., Kyrpides, N.C.,, C.: Protein interaction maps for complete genomes based on gene fusion events. Nature 402, 86–90 (1999)
12. Ge, H., Liu, Z., Church, G.M., Vidal, M.: Correlation between transcriptome and interactome mapping data from Saccharomyces cerevisiae. Nat. Genet. 29, 482–486 (2001)
13. Pellegrini, M., Marcotte, E.M., Thompson, M.J., Eisenberg, D., Yeates, T.O.: Assigning protein functions by comparative genome analysis: protein phylogenetic profiles. Proc. Natl. Acad. Sci. USA 96, 4285–4288 (1999)
14. Kerrien, S., Alam-Faruque, Y., Aranda, B., Bancarz, I., Bridge, A., Derow, C., et al.: IntAct–open source resource for molecular interaction data. Nucleic Acids Research 35, D561-D565 (2007), http://www.ebi.ac.uk/intact/main.xhtml

15. Apweiler, R., Bairoch, A., Wu, C.H., Barker, W.C., Boeckmann, B., Ferro, S., et al.: Uni-Prot: the Universal Protein knowledgebase. Nucleic Acids Research 32, D115–D119 (2004), http://www.uniprot.org

16. Li, W., Godzik, A.: Cd-hit: a fast program for clustering and comparing large sets of protein or nucleotide sequences. Bioinformatics 22(13), 1658–1659 (2006)

17. Garg, A., Gupta, D.: VirulentPred: a SVM based prediction method for virulent proteins in bacterial pathogens. BMC Bioinformatics 9, 62 (2008)

18. Young, L., Jernigan, B.L., Covell, D.G.: A role for surface hydrophobicity in protein-protein recognition. Protein Sci. 3, 717–729 (1994)

19. Jeffrey, H.J.: Chaos game representation of gene structure. Nucleic Acids Res. 18, 2163–2170 (1990)

20. http://web.expasy.org/protscale/pscale/Hphob.Doolittle.html

21. Goli, B., Aswathi, B.L., Nair, A.S.: A Novel Algorithm for Prediction of Protein Coding DNA from Non-coding DNA in Microbial Genomes Using Genomic Composition and Dinucleotide Compositional Skew. In: Meghanathan, N., Chaki, N., Nagamalai, D. (eds.) CCSIT 2012, Part II. LNICST, vol. 85, pp. 535–542. Springer, Heidelberg (2012)

22. Hall, M., Holmes, G.: Benchmarking Attribute Selection Techniques for Discrete Class Data Mining. IEEE Trans. Knowl. Data Eng. 15, 1–16 (2003)

23. Wang, C., Ding, C., Meraz, R.F., Holbrook, S.R.: PSoL.: A positive sample only learning algorithm for finding non-coding RNA genes. Bioinformatics 22, 2590–2596 (2006)

24. Liu, H., Yu, L.: Towards integrating feature selection algorithms for classification and clustering. IEEE Transactions on Knowledge and Data Engineering 17(3), 1–12 (2005)

25. Hall, M.A.: Correlation based feature selection for machine learning. Doctoral dissertation, The University of Waikato, Dept. of Comp. Sci. (1999)

26. Marko, R.S., Igor, K.: Theoretical and empirical analysis of relief and rreliefF. Machine Learning Journal 53, 23–69 (2003)

27. Kira, K., Rendell, L.A.: A practical approach to feature selection. In: Proceedings of the Ninth International Workshop on Machine Learning, pp. 249–256. Morgan Kaufmann Publishers Inc. (1992)

28. Hall, M., Frank, E., Holmes, G., Pfahringer, B., Reutemann, P., Witten, I.H.: The WEKA Data Mining Software: An Update. SIGKDD Explorations 11(1) (2009)

29. Werbos, P.J.: Beyond Regression: New Tools for Prediction and Analysis in the Behavioral Sciences. PhD thesis, Harvard University (1974)

30. Parker, D.B.: Learning-logic. Technical report, TR-47, Sloan School of Management. MIT, Cambridge, Mass (1985)

31. Rumelhart, D.E., Hinton, G.E., Williams, R.J.: Learning internal representations by error-propagation in Parallel distributed processing: Explorations in the Microstructure of Cognition, vol. I. Bradford Books, Cambridge (1986)

32. Bauer, E., Kohavi, R.: An empirical comparison of voting classification algorithms:Bagging, boosting, and variants. Machine Learning 36(1/2), 105–139 (1999)

33. Breiman, L.: Bagging predictors. Machine learning 24(2), 123–140 (1996a)

34. Achuthsankar, S.N., Sreenadhan, S.P.: An improved digital _ltering technique using nucleotide frequency indicators for locating exons. Journal of the Computer Society of India 36, 60–66 (2006)

35. Cherian, B.S., Nair, A.S.: Protein location prediction using atomic composition and global features of the amino acid sequence. Biochemical and Biophysical Research Communications 391, 1670–1674 (2010)

36. Namboodiri, S., Verma, C., Dhar, P.K., Giuliani, A., Nair, A.-S.S.: Sequence signatures of allosteric proteins towards rational design. Systems and Synthetic Biology 4, 271–280 (2011)

Synchronization of Hyperchaotic Liu System via Backstepping Control with Recursive Feedback

Suresh Rasappan and Sundarapandian Vaidyanathan

Department of Mathematics, Vel Tech Dr. RR & Dr. SR Technical University
Avadi- Alamathi Road, Avadi, Chennai-600062, India
sundarvtu@gmail.com
http://www.veltech.org

Abstract. This paper investigates the backstepping control design with recursive feedback input approach for achieving global chaos synchronization of identical hyperchaotic Liu systems(2001). Our theorem on global chaos synchronization for hyperchaotic Liu systems is established using Lyapunov stability theory. Since the Lyapunov exponents are not required for these calculations, the backstepping control method is effective and convenient to synchronize the hyperchaotic Liu systems. Numerical simulations are also given to illustrate the synchronization results derived in this paper.

Keywords: Hyperchaos, synchronization, backstepping control, hyperchaotic Liu system.

1 Introduction

Chaos is a bounded unstable dynamic behavior that exhibits sensitive dependence on initial conditions and includes infinite unstable periodic motions. The most important feature of chaotic systems is the sensitive dependence on initial conditions- two nearby points in state space will separate rapidly as they evolve in time.

A hyperchaotic system is typically defined as chaotic system with at least two positive Lyapunov exponents. Combined with one null exponent along the flow and one negative exponent to ensure the boundness of the solution, the minimal dimension for a hyperchaotic system is four. In fact, the presence of more than one positive Lyapunov exponent clearly improves security by generating more complex dynamics. However, is applied to design secure communications systems. In particular, the idea is to combine conventional cryptographic methods and synchronization of chaotic systems to design hyperchaos-based cryptosystems.

In recent years, the term synchronization of chaos is used mostly to denote the area of studies lying at the interfaces between the control theory and the theory of dynamic systems studying the methods of control of deterministic systems with nonregular, chaotic behavior.

J. Mathew et al. (Eds.): ICECCS 2012, CCIS 305, pp. 212–221, 2012.

The synchronization of chaotic system was first researched by Yamada and Fujisaka [1] with subsequent work by Pecora and Carroll [2,3]. The synchronization of chaos is one way of explaining sensitive dependence on initial conditions[4]. It has been established that the synchronization of two chaotic systems, that identify the tendency of two or more systems are coupled together to undergo closely related motions. The problem of chaos synchronization is to design a coupling between the two systems such that the chaotic time evaluation becomes ideal. The output of the response system asymptotically follows the output of the drive system, *i.e.* the output of the master system controls the slave system.

A variety of schemes for ensuring the control and synchronization of such systems have been demonstrated based on their potential applications in various fields including chaos generator design, secure communication [5], physical systems [6], and chemical reaction [7], ecological systems [8], information science [9], etc. So far a variety of impressive approaches have been proposed for the synchronization of the chaotic systems such as the OGY method[10] , sampled feedback synchronization method[11], time delay feedback method, adaptive design method [13], sliding mode control method [14], active control method [15], etc.

In recent years, a backstepping method has been developed for designing controllers to control the chaotic systems [12]. The backstepping method is based on the mathematical model of the examined system, introducing new variables into it in a form depending on the state variables, controlling parameters, and stabilizing functions.

In this paper, backstepping control design with recursive feedback input approach is proposed. This approach is a systematic design approach and guarantees global stability of the hyperchaotic Liu system ([16],2001). Based on the Lyapunov function, the backstepping control is determined to tune the controller gain based on the precalculated feedback control inputs. We organize this paper as follows. In Section 2, we present the methodology of hyperchaotic synchronization by backstepping control method. In Section 3, we demonstrate the hyperchaotic synchronization of identical Liu systems. In Section 4, we summarize the results obtained in this paper.

2   The Problem Statement and Our Methodology

In general, the two dynamic systems in synchronization are called the master and slave system respectively. A well designed controller will make the trajectory of the slave system track the trajectory of the master system.

The master system is described by the dynamics

$$
\begin{aligned}
\dot{x}_1 &= f_1(x_1, x_2, ..., x_n) \\
&\vdots \quad \vdots \qquad \vdots \\
\dot{x}_n &= f_n(x_1, x_2, ..., x_n)
\end{aligned}
\tag{1}
$$

where $x(t) \in R^n$ is a state vector of the system (1).

The slave system with the controller u is defined by

$$\dot{y}_1 = g_1(y_1, y_2, ..., y_n) + u_1(t)$$
$$\vdots \quad \vdots \qquad\qquad \vdots \qquad\qquad\qquad (2)$$
$$\dot{y}_n = f_n(y_1, y_2, ..., y_n) + u_n(t)$$

where $y(t) \in \mathbb{R}^n$ is a state vectors of the system (2).

$f_i, g_i (i = 1, 2, 3, ...n)$ are linear and nonlinear functions with inputs from systems (1) and (2). If f equals g, then the systems are called *identical*. Otherwise, they are called as *non-identical* chaotic systems.

Let us define the error variables between the response system (2) that is to be controlled and the controlling drive system(1) as

$$e = y - x \qquad\qquad (3)$$

Subtracting (2) from (1) and using the notation above notation yields

$$\dot{e}_1 = g_1(y_1, y_2, ..., y_n) - f_1(x_1, x_2, ..., x_n) + u_1(t)$$
$$\vdots \quad \vdots \qquad\qquad\qquad \vdots \qquad\qquad\qquad (4)$$
$$\dot{e}_n = f_n(y_1, y_2, ..., y_n) - f_n(x_1, x_2, ..., x_n) + u_n(t)$$

The synchronization error system controls a controlled chaotic system with control input $u_i, i = 1, 2, 3, ..., n$ as a function of the error states $e_1, e_2, e_3, ..., e_n$. That means the systematic feedbacks so as to stabilize the error dynamics (3), $e_1, e_2, e_3, ..., e_n$ converge to zero as time t tends to infinity.

This implies that the controllers $u_i, i = 1, 2, 3, ..., n$ should be designed so that the two chaotic systems can be synchronized. In mathematically

$$\lim_{t \to \infty} \|e(t)\| = 0$$

Backstepping design is recursive and guarantees global stabilities performance of strict-feedback nonlinear systems.

By using the backstepping design, at the ith step, the ith order subsystem is stabilized with respect to a Lyapunov function V_i, by the design of virtual control α_i and a control input function u_i.

We consider the stability of the system

$$\dot{e}_1 = g_1(y_1, y_2, ..., y_n) - f_1(x_1, x_2, ..., x_n) + u_1(t) \qquad (5)$$

where u_1 is control input, which is the function of the error state vectors e_i, and the state variables $x(t) \in \mathbb{R}^n$, $y(t) \in \mathbb{R}^n$.

As long as this feedback stabilize the system (5)converge to zero as $t \to \infty$, where

$$e_2 = \alpha_1(e_1)$$

is regarded as an virtual controller.

For the design of $\alpha_1(e_1)$ to stabilize the subsystem (4), we consider the Lyapunov function defined by

$$V_1(e_1) = e_1^T P_1 e_1 \tag{6}$$

The derivative of V_1 is

$$\dot{V}_1 = -e_1^T Q_1 e_1 \tag{7}$$

where Q_1 is a positive definite matrix, then $\dot{V}_1$ is a negative definite function on $\mathbb{R}^n$.

Thus by Lyapunov stability theory [17], the error dynamics (4) is asymptotically stable.

The virtual control $e_2 = \alpha_1(e_1)$ and the state feedback input u_1 makes the system(4) asymptotically stable. The function $\alpha_1(e_1)$ is estimative when e_2 is considered as controller.

The error between e_2 and $\alpha_1(e_1)$ is

$$w_2 = e_2 - \alpha_1(e_1) \tag{8}$$

Study the (e_1, w_2) system

$$\begin{aligned}
\dot{e}_1 &= g_1(y_1, y_2, ..., y_n) - f_1(x_1, x_2, ..., x_n) \\
\dot{w}_2 &= g_2(y_1, y_2, ..., y_n) - f_2(x_1, x_2, ..., x_n) - \dot{\alpha}_1(e_1) + u_2
\end{aligned} \tag{9}$$

Consider e_3 as a virtual controller in system (8), assume when it is equal to $\alpha_2(e_1, w_2)$ and it makes system (8) asymptotically stable.

Consider the Lyapunov function defined by

$$V_2(e_1, w_2) = V_1(e_1) + w_2^T P_2 w_2 \tag{10}$$

The derivative of V_1 is

$$\dot{V}_2 = -e_1^T Q_1 e_1 - w_2^T Q_2 w_2 < 0 \tag{11}$$

where Q_1, Q_2 are positive definite matrices, then $\dot{V}_2$ is a negative definite function on $\mathbb{R}^n$.

Thus by Lyapunov stability theory [17], the error dynamics (8) is asymptotically stable. The virtual control $e_2 = \alpha_2(e_1, w_2)$ and the state feedback input u_2 makes the system(8) asymptotically stable.

For the n_{th} state of the error dynamics, define the error variable w_n as

$$w_n = e_n - \alpha_{n-1}(e_1, w_2, w_2, ..., w_n) \tag{12}$$

Study the $(e_1, w_2, w_2, ..., w_n)$ system

$$\begin{aligned}
\dot{e}_1 &= g_1(y_1, y_2, ..., y_n) - f_1(x_1, x_2, ..., x_n) \\
&\vdots \qquad\qquad\qquad\qquad \vdots \\
\dot{w}_n &= g_n(y_1, y_2, ..., y_n) - f_n(x_1, x_2, ..., x_n) - \dot{\alpha}_{n-1}(e_1, w_2, w_2, ..., w_n) + u_n
\end{aligned} \tag{13}$$

Consider the Lyapunov function defined by

$$V_n(e_1, w_2, w_2, ..., w_n) = V_{n-1}(e_1, w_2, w_2, ..., w_{n-1}) + w_n^T P_n w_n \qquad (14)$$

The derivative of V_n is

$$\dot{V}_n = -e_1^T Q_1 e_1 - w_2^T Q_2 w_2 - .... - w_n^T Q_n w_n < 0 \qquad (15)$$

where $Q_1, Q_2, Q_3, ..., Q_n$ are positive definite matrices, then $\dot{V}_n$ is a negative definite function on $\mathbb{R}^n$.

Thus by Lyapunov stability theory [17], the error dynamics (12) is asymptotically stable.

The virtual control $e_{n-1} = \alpha_2(e_1, w_2, w_2, ..., w_{n-1})$ and the state feedback input u_n makes the system(10) asymptotically stable.

Thus by Lyapunov stability theory [17], the error dynamics (3) is globally exponentially stable for all initial conditions $e(0) \in \mathbb{R}^n$. Hence, the states of the master and slave systems are globally and exponentially synchronized.

3   The Synchronization of Identical Hyperchoatic Liu Systems via Backstepping Control Design Design with Recursive Feedback

In this sectio,n we apply the backstepping method with recursive feedback function for the hybrid synchronization of identical hyperchaotic Liu ([16],2001) system.

The equation for the hyperchaotic Liu system is

$$\dot{x}_1 = a(x_2 - x_1) \qquad (16)$$
$$\dot{x}_2 = bx_1 + x_1 x_3 - x_4$$
$$\dot{x}_3 = -x_1 x_2 - cx_3 + x_4$$
$$\dot{x}_4 = dx_1 + x_2$$

where $x(t) \in \mathbb{R}^4$ is a state vector of the system.

The slave system also described by hyperchaotic Liu system as

$$\dot{y}_1 = a(y_2 - y_1) + u_1 \qquad (17)$$
$$\dot{y}_2 = by_1 + y_1 y_3 - y_4 + u_2$$
$$\dot{y}_3 = -y_1 y_2 - cy_3 + y_4 + u_3$$
$$\dot{y}_4 = dy_1 + y_2 + u_4$$

where $y(t) \in \mathbb{R}^4$ is a state vector of the system.

Let us define the error variables between the slave system (16) that is to be controlled and the controlling master system(15) as

$$e_i = y_i - x_i, i = 1, 2, 3, ..., n$$

Subtracting (16) from (15) and using the above notation yields

$$\dot{e}_1 = ae_2 - ae_1 + u_1 \tag{18}$$
$$\dot{e}_2 = (b + x_3)e_1 + (e_1 + x_1)e_3 - e_4 + u_2$$
$$\dot{e}_3 = -y_1 y_2 + x_1 x_2 - ce_3 + e_4 + u_3$$
$$\dot{e}_4 = de_1 + e_2 + u_4$$

We introduce the backstepping procedure to design the controller $u_i, i = 1, 2, 3, 4$. Where $u_i, i = 1, 2, 3, 4$ are control feedbacks, as long as these feedbacks stabilize system (17) converging to zero as the time t goes to infinity.

First, we consider the stability of the system

$$\dot{e}_1 = ae_2 - ae_1 + u_1 \tag{19}$$

where e_2 is regarded as virtual controller.

We consider the Lyapunov function defined by

$$V_1(e_1) = \frac{1}{2} e_1^T e_1 \tag{20}$$

the derivative of V_1 is as following

$$\dot{V}_1 = e_1(ae_2 - ae_1) \tag{21}$$

Assume the controller $e_2 = \alpha_1(e_1)$.

If $\alpha_1(e_1) = 0$ and $u_1 = 0$ then $\dot{V}_1 = -ae_1^2 < 0$ makes the system (18) asymptotically stable.

Function $\alpha_1(e_1)$ is an estimative function when e_2 is considered as a controller. The error between e_2 and $\alpha_1(e_1)$ is

$$w_2 = e_2 - \alpha_1(e_1) \tag{22}$$

Study (e_1, w_2) system (22)

$$\dot{e}_1 = aw_2 - ae_1 \tag{23}$$
$$\dot{w}_2 = (b + x_3)e_1 + (e_1 + x_1)e_3 - e_4 + u_2$$

We consider e_3 as a virtual controller in system (22). We assume when it is equal to $\alpha_2(e_1, w_2)$ and it makes system (22) asymptotically stable.

Consider the Lyapunov function defined by

$$V_2(e_3, w_2) = V_1(e_1) + \frac{1}{2} w_2^T w_2 \tag{24}$$

The derivative of $V_2(e_3, w_2)$ is

$$\dot{V}_2 = -e_1^2 + w_2^2((a + b + x_3)e_1 + (e_1 + x_1)e_3 - e_4 + u_2) \tag{25}$$

If $\alpha_2(e_1, w_2) = 0$ and

$$u_2 = e_4 - (a + b + x_3)e_1 - w_2$$

then $\dot{V}_2 = -e_1^2 - w_2^2 < 0$ makes the system (22) asymptotically stable.

Define the error variable w_3 as

$$w_3 = e_3 - \alpha_2(e_1, w_2) \tag{26}$$

Study (e_1, w_2, w_3) system (26)

$$\dot{e}_1 = aw_2 - ae_1 \tag{27}$$
$$\dot{w}_2 = (e_1 + x_1)e_3 - ae_1 - w_2$$
$$\dot{w}_3 = -y_1 y_2 + x_1 x_2 - ce_3 + e_4 + u_3$$

Consider the Lyapunov function defined by

$$V_3(e_1, w_2, w_3) = V_2(e_1, w_2) + \frac{1}{2}w_3^T w_3 \tag{28}$$

The derivative of $V_3(e_3, w_2, w_3)$ is

$$\dot{V}_3 = -ae_1^2 - w_2^2 + w_3((e_1 + x_1)w_2 + x_1 x_2 - y_1 y_2 - cw_3 + e_4 + u_3) \tag{29}$$

If $\alpha_3(e_1, w_2, w_3) = 0$ and $u_3 = y_1 y_2 - x_1 x_2 - (e_1 + x_1)w_2$ then

$$\dot{V}_3 = -e_1^2 - w_2^2 - w_3^2 < 0$$

makes the system (26) asymptotically stable.

Define the error variable w_4 as

$$w_4 = e_4 - \alpha_3(e_1, w_2, w_3) \tag{30}$$

Study (e_1, w_2, w_3, w_4) system (30)

$$\dot{e}_1 = aw_2 - ae_1 \tag{31}$$
$$\dot{w}_2 = (e_1 + x_1)e_3 - ae_1 - w_2$$
$$\dot{w}_3 = e_4 - ce_3 - (e_1 + x_1)x_2$$
$$\dot{w}_4 = de_1 + e_2 + u_4$$

Consider e_4 as a virtual controller in system (30), assume when it is equal to $\alpha_3(e_1, w_2, w_3)$ and it makes system (30) asymptotically stable.

Consider the Lyapunov function defined by

$$V_4(e_1, w_2, w_3, w_4) = V_3(e_1, w_2, w_3) + \frac{1}{2}w_4^T w_4 \tag{32}$$

The derivative of $V_4(e_3, w_2, w_3, w_4)$ is

$$\dot{V}_4 = -ae_1^2 - w_2^2) - cw_3^2 + w_4(w_3 + de_1 + e_2 + u_4) \tag{33}$$

If $u_4 = -de_1 - e_2 - w_3 - w_4$ then

$$\dot{V}_4 = -e_1^2 - w_2^2 - cw_3^2 - w_4^2 < 0$$

makes the system (30) asymptotically stable.

Thus, by Lyapunov stability theory [17], the error dynamics (17)is globally exponentially stable.

Hence, we obtain the following result.

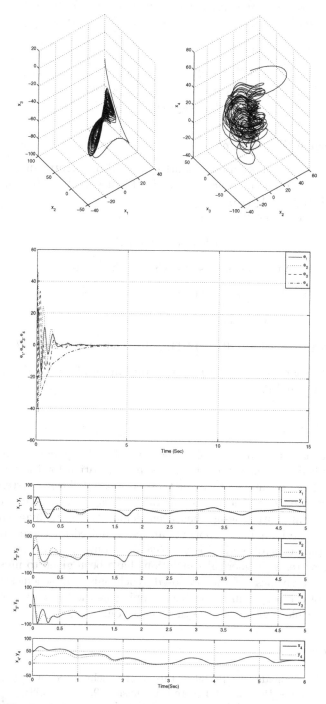

Fig. 1. (a). Phase orbit of hyperchaotic Liu system, (b). Error state of hyperchaotic Liu system, (c). Synchronization state of hyperchaotic Liu system

Theorem 1. *The identical hyper chaotic Liu systems (15) and (16) are globally and exponentially synchronized with the backstepping controls*

$$u_1 = 0$$
$$u_2 = e_4 - (a + b + x_3)e_1 - w_2$$
$$u_3 = y_1 y_2 - x_1 x_2 - (e_1 + x_1)w_2$$
$$u_4 = -de_1 - e_2 - w_3 - w_4$$

3.1   Numerical Simulation

For the numerical simulations, the fourth order Runge-Kutta method is used to solve the differential equations (15) and (16)with the backstepping controls u_1, u_2, u_3 and u_4 given by(18).

The parameters of the systems (15) and (16) are

$$a = 10, b = 35, c = 1.4, d = 5$$

and initial values of the master system (15) are chosen as

$$x_1(0) = 10.65, x_2(0) = 26.25, x_3(0) = 11.87, x_4(0) = 45.56$$

and slave system(16)are chosen as

$$y_1(0) = 21.54, y_2(0) = 42.23, y_3(0) = 63.10, y_3(0) = 9.32$$

Fig. 1(a) and (b)and (c) depicts the synchronization of identical hyperchaotic Liu systems (15) and (16).

4   Conclusion

In this paper, backstepping control method based on Lyapunov stability theory has been applied to achieve global chaos synchronization for the hyperchaotic Liu systems. The advantage of this method is that it follows a systematic procedure for synchronizing chaotic system and there is no derivative in controller. The backstepping control design has been demonstrated to hyperchaotic Liu systems. Since the Lyapunov exponents are not required for these calculations, the backstepping control method is effective and convenient to synchronize the hyperchaotic systems. Numerical simulations have been given to illustrate and validate the effectiveness of the proposed synchronization schemes for identical hyperchaotic Liu systems. The backstepping method is very effective and convenient to achieve global chaos synchronization.

References

1. Fujisaka, H., Yamada, T.: Stability theory of synchronized motion in coupled-oscillator systems. Progress of Theoretical Physics 69, 32–47 (1983)
2. Pecora, L.M., Carroll, T.L.: Synchronization in chaotic systems. Phys. Rev. Lett. 64, 821–824 (1990)

3. Pecora, L.M., Carroll, T.L.: Synchronizing chaotic circuits. IEEE Trans. Circ. Sys. 38, 453–456 (1991)
4. Alligood, K.T., Sauer, T., Yorke, J.A.: Chaos: An Introduction to Dynamical Systems. Springer, Berlin (1997)
5. Murali, K., Lakshmanan, M.: Secure communication using a compound signal using sampled-data feedback. Applied Mathematics and Mechanics 11, 1309–1315 (2003)
6. Lakshmanan, M., Murali, K.: Chaos in Nonlinear Oscillators: Controlling and Synchronization. World Scientific, Singapore (1996)
7. Han, S.K., Kerrer, C., Kuramoto, Y.: D-phasing and bursting in coupled neural oscillators. Phys. Rev. Lett. 75, 3190–3193 (1995)
8. Blasius, B., Huppert, A., Stone, L.: Complex dynamics and phase synchronization in spatially extended ecological system. Nature 399, 354–359 (1999)
9. Kocarev, L., Parlitz, U.: General approach for chaotic synchronization with applications to communications. Phys. Rev. Lett. 74, 5028–5030 (1995)
10. Ott, E., Grebogi, C., Yorke, J.A.: Controlling chaos. Phys. Rev. Lett. 64, 1196–1199 (1990)
11. Murali, K., Lakshmanan, M.: Secure communication using a compound signal using sampled-data feedback. Applied Mathematics and Mechanics 11, 1309–1315 (2003)
12. Park, J.H., Kwon, O.M.: A novel criterion for delayed feedback control of time-delay chaotic systems. Chaos, Solitons and Fractals 17, 709–716 (2003)
13. Lu, J., Wu, X., Han, X., Lu, J.: Adaptive feedback synchronization of a unified chaotic system. Phys. Lett. A 329, 327–333 (2004)
14. Yau, H.T.: Design of adaptive sliding mode controller for chaos synchronization with uncertainties. Chaos, Solitons and Fractals 22, 341–347 (2004)
15. Sundarapandian, V., Suresh, R.: Global chaos synchronization for Rossler and Arneodo chaotic systems by nonlinear control. Far East Journal of Applied Mathematics 44, 137–148 (2010)
16. Liu, L., Liu, C., Zhang, Y.: Analysis of a novel four- dimentional hyperchaotic system. Chinese Journal of Physics 46, 1369–1372 (2001)
17. Hahn, W.: The Stability of Motion. Springer, Berlin (1967)

New Feature Vector for Recognition
of Short Microbial Genes

Baharak Goli[1], Aswathi B.L.[2], Chinu Joy[3], and Achuthsankar S. Nair[2]

[1,2] Department of Computational Biology and Bioinformatics,
University of Kerala, Trivandrum 695581, India
[3] Sree Chitra Thirunal College of Engineering, Pappanamcode, Trivandrum-18, India
baharak_goli@yahoo.com

Abstract. The effectiveness of a classifier is highly dependent on the discriminative power of the feature vectors extracted from the dataset. In this study a novel feature vector is presented that aims at better classification of short protein coding DNA. For this feature vector a straightforward ensemble method, Adaboost.M1 in conjunction with Multilayer Perceptron (MLP) as the base classifier was employed. The proposed model shows 97.36% accuracy, 97.76% sensitivity and 96.82% specificity. The results demonstrate that the proposed feature vector is promising, and help in increasing the prediction accuracy.

Keywords: Short gene, Adaboos.M1, Multilayer perceptron, Physico-chemical properties, Classification.

1   Introduction

Short proteins have proven to play important roles in various biological processes including the spore development [1], stabilizing factors for larger protein complexes [2] and regulation of amino acid metabolism [3]. They are mediator in other functions such as regulation of innate immunity, cell communication and homeostasis signal transduction.

The performance of gene finding algorithms is highly dependent on the coding measures that are adopted to annotate the sequence. Coding measure describes the likelihood that a DNA sequence is coding for a protein. The currently available gene prediction algorithms in prokaryotic genomes often fail to identify short protein coding DNA due to limited prominent coding measures in short nucleotide sequences, resulting in incomplete or wrong annotations [4]. Hence, study of coding measures in short DNA sequences are of major importance in gene prediction and genome annotation. During the past two decades, a large number of gene finding tools have been developed and several review papers have been published [5-8].

Some of the remarkable of such effective coding measures for longer sequences are hydrophobicity [9], symmetry in the codon positions with respect to purine/pyrimidine, amino/keto and strong/weak hydrogen bonding nature of nucleotides represented by Z

J. Mathew et al. (Eds.): ICECCS 2012, CCIS 305, pp. 222–229, 2012.
© Springer-Verlag Berlin Heidelberg 2012

curve [10], global features obtained from multifractal analysis of DNA sequence [11], hexamer usage [12], Entropy Distance Profile [13] and GC bias in the 1st, 2nd, and 3rd positions of each codon [14].

We defined the short genes as the fragments ranging from 60 nt to 400 nt since prediction of significant fraction of short genes lying in this range are beyond the detection ability of many of the top most gene prediction tools.

Any classification problem deals with selection of feature vectors as its major initial concern. In this study discriminative power of 30 physico-chemical features for classification of short protein coding DNA from non-coding DNA sequences is scrutinized. The proposed feature vector is extracted from six prokaryotic organisms. A straightforward ensemble method, Adaboost.M1 in conjunction with multilayer perceptron as the base classifier was employed. The performance of proposed prediction model was compared with 5 base classifiers including SVM with radial basis function kernel, RBF network, Random Forest, Naïve Bayes and ADTree.

2   Materials and Methods

2.1   Dataset Construction

In this study dataset was compiled obtaining coding and non-coding sequences of two unique Ecoli strain: Escherichia coli K12-MG1655 and Escherichia coli UTI89 (UPEC) and 4 Enterobacteriaceae: Buchnera aphidicola 5A, Enterobacter 638, Klebsiella nteric a 342 and Yersinia pestis KIM 10 from Integrated Microbial Genome (IMG) database [15].

Sequences whose length is not a multiple of three were excluded from the dataset. Short coding and non-coding sequences falling within the range of [60-400 nt] were extracted. The training set was generated by taking two-thirds of coding and non-coding sequences from each organisms and the remaining one–third was allotted to the test set. Training dataset comprised of 3270 coding and 2489 non-coding sequences. Testing dataset comprised of 1637 coding and 1247 non-coding sequences.

2.2   Feature Vector Construction

We divided the DNA sequence into three subregions to enhance biological feature extractions. The length of first part, middle region and last part are 10%, 80% and 10% of the length of the DNA sequence respectively assuming that transcription starts at first and stops at last part. The Physico-chemical properties of the DNA sequences are of great relevance in gene prediction problem.10 Physico-chemical properties obtained from Hyperchem Pro 8.0 software of HyperCubeInc, USA. These properties are given in Table1.

Each base in each subregion of the DNA sequence was replaced with corresponding Physico-chemical value, summed over all bases and divided by the total number of bases. Here it is instanced with an example. Let a random DNA sequence having a length of 10 bp: ATTGCGTGTA and we take the property, 'EIIP'.

Nucleotide indices for EIIP are as shown in Table 1. The numerical representation of this DNA sequence is obtained as the sum (DNA index of index of each nucleotide) /Total length of the sequence = (0.1260+0.1335+0.1335+0.0806+0.1340+0.0806 +0.1335+0.0806+0.1335+0.1260) /10, which computes to 0.1161. The feature vector has a length of 30 for each DNA sequence.

Table 1. Physico-chemical properties of nucleotide

Physico-Chemical Properties	A	T	G	C
Hydration Energy	-13.69	-9.09	-16.61	13.67
Dipole Moments	0.4629	1.052	3.943	6.488
EIIP	0.1260	0.1335	0.0806	0.1340
Polarizability	25.54	22.47	24.82	21.48
Refractivity	71.19	64.60	71.74	62.43
Molecular Weight	13.15	126.13	111.12	151.15
Log P	4.25	4.0	3.57	4.06
Mass	312.20	303.19	328.20	288.18
Surface Area	470.47	446.42	481.25	384.14
Volume	766.40	741.70	787.34	728.99

Before training the classifier all features were normalized to lie in the range [1, 1] using z-scores.

2.3   Ensemble Classifier

Computational cost and accuracy of the learning techniques have always been of major importance for machine learning systems. Recently there has been much effort to design new learning algorithms with higher predictive accuracy by ensembling multiple classifiers. Ensemble learning [16] deals with techniques which uses multiple classifiers to solve a problem for which the final classification relies upon the combined outputs of individual classifiers. Ensemble methods are very effective since generalization ability of an ensemble is usually considerably better than that of an individual learner. Several intensive and extensive surveys have been conducted to explore and assess the ensemble methods [17-18].

Two such commonly used techniques bootstrap aggregating or bagging [19] and boosting [20] for generating and using multiple classifiers are proper to a wide variety of learning methods. Both these techniques manipulate the training data for generation various classifiers. Bagging an acronym of bootstrap aggregating, produces new training sets by sampling with substitution from the training data while boosting uses adaptive sampling adopting all instances at each iteration. Each instance is weighted in the training data. Weights for each instance are optimized causing the classifier to concentrate on different instances which consequently leads to multiple classifiers. In both methods the multiple classifiers are then combined using simple voting system and a meta classifier is made. In bagging, each classifier has the same vote whereas boosting assigns different voting strengths to classifiers based on their performance. There are large numbers of boosting algorithms. AdaBoost, very popular technique has been proved to be one of the best meta classifier since it has built upon a solid theoretical foundation, very accurate prediction, great clearness and simplicity.

2.4   AdaBoost.M1

In this study we made use of AdaBoost.M1 algorithm proposed by [21] which is the most straightforward extension of Adaboost algorithm. We begin with the description of AdaBoost.M_1 algorithm here. The training data $((x_1,y_1),...,(x_n,y_n))$, where $x_i \in X$ is an observation and $y_i \in Y$ of is the class label (in this work the set of possible labels Y is of $\{1,2\}$ since we are dealing with binary classification) is given as input to boosting algorithm. The algorithm then initializes a nonnegative weight W_1 to each observation X_i. This value is updated after each step. Learning algorithm, noted as C, is invoked iteratively for prespecified T times. In each iteration, C is provided with the distribution W_t and calculates a hypothesis h_t with respect to the distribution. Then the error ϵ_t is computed related to the distribution W_t and the hypothesis h_t. A function of the error referred to as α_t, is used for weights updating in the distribution at every iteration t. Wrongly classified examples will have a higher weights whereas the weight of correctly classified examples is decreased, making the single learner constructed in the next iteration to concentrate on the hardest examples. Normalization factor is called as N_t. The final weight is computed using α_t value as log ($1/\alpha_t$) for each hypothesis h_t. The final hypothesis $h_{fin}(x)$ is a weighted sum of its vote determined by the weights of the hypothesis.

3   Results

We carried out experiments with Adaboost.M1algorithm in conjunction with Multilayer Perceptron (MLP) as the base learner. The capability of the propose ensemble model was evaluated using basic performance measurements such as accuracy, sensitivity and specificity. (TP, TN, FP and FN representing true positive, true negative, false positive and false negative respectively).

Specificity=TN/ (TN+FP)*100

Sensitivity=TP/ (TP+FN)*100

Precision=TP/ (TP+FP)*100

Matthews correlation coefficient = (((TP*TN)-(FP*FN)))/($\sqrt{}$(TP+FP)*(TP+FN)*(TN+FP)*(TN+FN))

Accuracy= TP+TN/ (TP+TN+FP+TN)

The results of 5-fold cross validation indicate that our proposed model avoids over-fitting. The results are shown in Table2.

Table 2. Performance of proposed algorithm

Method	Specificity	Sensitivity	Precision	MCC	5-fold CV
Adaboost.M1 + MLP	96.82	97.76	97.58	0.94	97.36

In addition to MLP as a base classifier we further compared the effectiveness of the proposed algorithm with 5 weak base classifiers including SVM with radial basis function kernel, RBF network, Random Forest, Naïve Bayes and ADTree. The Results are shown in Table3.

Table 3. Classification performance comparison among multiple classifiers

Method	Specificity	Sensitivity	Precision	MCC	5-fold CV
Adaboost.M1+SVM	85.13	91.62	89.00	0.77	88.81
Adaboost.M1+RBF Network	68.54	82.72	77.55	0.51	76.59
Adaboost.M1+Random Forest	80.07	91.73	85.81	0. 72	86.71
Adaboost.M1+Naïve Bayes	69.54	82.78	78.12	0.52	77.06
Adaboost.M1+ADTree	75.65	86.88	82.41	0.63	82.02

We also computed F-measure which is the harmonic mean of sensitivity and specificity for both coding and non-coding regions. The results are given in Fig.1.

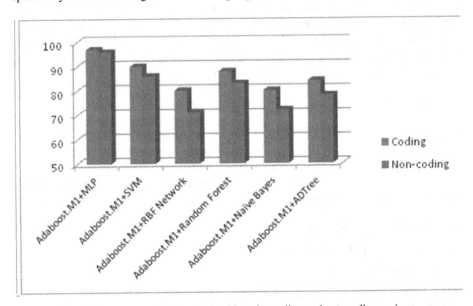

Fig. 1. F-measures of different classifiers for coding and non coding regions

Several alternative measures were used to evaluate the success of proposed ensemble method including mean absolute error, root mean squared error, relative absolute error and root relative squared error. These measures can be computed using the following formulas. The predicted values are noted by $p_1, p_2...p_n$ and the actual values are noted by $a_1, a_2...a_n$.

Mean absolute error = $\left(\left|p_1 - a_1\right|^2 + ... + \left(p_n - a_n\right)^2\right/n$

Root mean squared error = $\sqrt{\dfrac{(p_1 - a_1)^2 + ...(p_n - a_n)^2}{n}}$

Relative absolute error = $\left|p_1 - a_1\right| + ... + \left|p_n - a_n\right|\left/\left|a_1 - \overline{a}\right| + ... + \left|a_n - \overline{a}\right|\right.$

Root relative squared error = $\sqrt{\dfrac{(p_1 - a_1)^2 + ...(p_n - a_n)^2}{(a_1 - \overline{a})^2 + ...(a_n - \overline{a})^2}}$

The obtained error rates of proposed model are shown in Table.4 and Fig.2.

Table 4. The obtained error rates comparison among multiple classifiers

Method	Mean absolute error (%)	Root mean squared error (%)	Relative absolute error (%)	Root relative squared error (%)
Adaboost.M1+MLP	0.03	0.12	7.02	30.85
Adaboost.M1+SVM	0.13	0.29	27.83	60.40
Adaboost.M1+RBF Network	0.30	0.40	60.23	81.81
Adaboost.M1+Random Forest	0.13	0.35	27.25	72.37
Adaboost.M1+Naïve Bayes	0.30	0.40	61.89	82.34
Adaboost.M1+ADTree	0.23	0.36	48.01	73.04

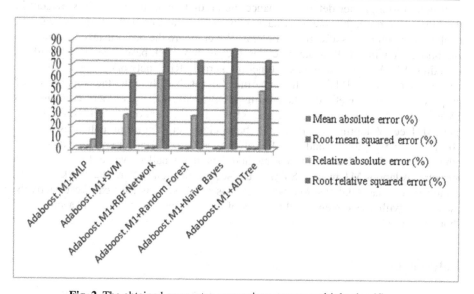

Fig. 2. The obtained error rates comparison among multiple classifiers

Self-consistency test (shown in Table 5) was also performed to evaluate the prediction model. Self-consistency test reflects the consistency of the developed model. It is an evaluation method to estimate the level of fitness of data in a developed method. In self-consistency test, observations of training datasets are predicted with decision rules acquired from the same dataset. The accuracy of self-consistency reveals the fitting ability of the rules obtained from the features of training sets. Since the prediction system parameters obtained by the self-consistency test are from the training dataset, the success rate is high. However poor result of self-consistency test reflects the inefficiency of classification method.

Table 5. Accuracy of proposed classifier for self-consistency test

Method	Specificity	Sensitivity	Precision	MCC	5-fold CV
Adaboost+MLP	98.59	99.90	98.94	0,98	99.34

All experiments were performed using algorithms implemented in data mining toolkit Weka version 3.6.6 [22] on a PC with a 2.13 GHz Intel CPU and 4 GB RAM, using Windows 7. Default setting was used for all methods.

4   Discussion

In this paper, we have introduced a novel coding feature along with a powerful ensemble method for prediction of short genes. Our results demonstrate that the proposed Physiochemical properties are able to bring out underlying property distribution to a greater detail to enhance the prediction accuracy. As it is evident in the literatures, ensemble methods have very large influence both on theoretical and applied research in classification problems. We compare the performance measures of Adaboost.M1 with MLP as the base classifier with 5 weak base classifier algorithms including SVM with radial basis function kernel, RBF network, Random Forest, Naïve Bayes and ADTree. The experimental results illustrate the superiority of proposed ensemble method, when using MLP as the base classifiers compared to SVM with radial basis function kernel, RBF network, Random Forest, Naïve Bayes and ADTree. Adopting MLP as the base classifier algorithm enhances 8.55%, 20.77%, 10.65%, 20.3% and 15.34% of proposed model accuracy compared with other 5 classifiers. Our proposed method has the smallest value for each error measure. Adaboos.M1 in with SVM is the second best classifier. Our work strongly demonstrates that Adaboost.M1 with MLP as the base classifier can be effectively used along with above mentioned features of the sequence to enhance the accuracy of short gene prediction.

References

1. Cutting, S., Anderson, M., Lysenko, E., Page, A., Tomoyasu, T., Tatematsu, K., Tatsuta, T., Kroos, L., Ogura, T.: SpoVM, a small protein essential to development in Bacillus subtilis, interacts with the ATP-dependent protease FtsH. Journal of Bacteriology 179, 5534–5542 (1997)

2. Schneider, D., Volkmer, T., Rogner, M.: PetG and PetN, but not PetL, are essential subunits of the cytochrome b6f complex from Synechocystis PCC 6803. Research in Microbiology 158, 45–50 (2007)

3. Yanofsky, C.: Transcription attenuation: once viewed as a novel regulatory strategy. Journal of Bacteriology 182, 1–8 (2000)

4. Brent, M.R., Guigo, R.: Recent advances in gene structure prediction. Current Opinion in Structural Biology 14, 264–272 (2004)

5. Fickett, J.W., Tung, C.S.: Assessment of protein coding measures. Nucleic Acids Research 20, 6441–6450 (1992)

6. Mathe, C., Sagot, M.F., Schiex, T., Rouze, P.: Current methods of gene prediction, their strengths and weaknesses. Nucleic Acids Research 30, 4103–4117 (2002)

7. Wang, Z., Chen, Y., Li, Y.: A brief review of computational gene prediction methods. Genomics, Proteomics & Bioinformatics / Beijing Genomics Institute 2, 216–221 (2004)

8. Do, J.H., Choi, D.K.: Computational approaches to gene prediction. Journal of Microbiology 44, 137–144 (2006)

9. Tramontano, A., Macchiato, M.F.: Probability of coding of a DNA sequence: an algorithm to predict translated reading frames from their thermodynamic characteristics. Nucleic Acids Research 14, 127–135 (1986)

10. Zhang, C.T., Wang, J.: Recognition of protein coding genes in the yeast genome at better than 95% accuracy based on the Z curve. Nucleic Acids Research 28, 2804–2814 (2000)

11. Zhou, L.Q., Yu, Z.G., Deng, J.Q., Anh, V., Long, S.C.: A fractal method to distinguish coding and non-coding sequences in a complete genome based on a number sequence representation. Journal of Theoretical Biology 232, 559–567 (2005)

12. Hutchinson, G.B., Hayden, M.R.: The prediction of exons through an analysis of spliceable open reading frames. Nucleic Acids Research 20, 3453–3462 (1992)

13. Zhu, H., Hu, G.Q., Yang, Y.F., Wang, J., She, Z.S.: MED: a new non-supervised gene prediction algorithm for bacterial and archaeal genomes. BMC Bioinformatics 8, 97 (2007)

14. Hyatt, D., Chen, G.L., Locascio, P.F., Land, M.L., Larimer, F.W., Hauser, L.J.: Prodigal: prokaryotic gene recognition and translation initiation site identification. BMC Bioinformatics 11, 119 (2010)

15. Markowitz, V.M., Korzeniewski, F., Palaniappan, K., Szeto, E., Werner, G., Padki, A., Zhao, X., Dubchak, I., Hugenholtz, P., Anderson, I., Lykidis, A., Mavromatis, K., Ivanova, N., Kyrpides, N.C.: The integrated microbial genomes (IMG) system. Nucleic Acids Research 34, D344–D348 (2006)

16. Dietterich, T.: Machine-learning research: four current directions. AI Magazine 18, 97–136 (1997)

17. Yang, P., Hwa Yang, Y., Zhou, B., Zomaya, A.Y.: A Review of Ensemble Methods in Bioinformatics. Current Bioinformatics 5, 296–308 (2010)

18. Dietterich, T.G.: Ensemble Methods in Machine Learning. In: Kittler, J., Roli, F. (eds.) MCS 2000. LNCS, vol. 1857, pp. 1–15. Springer, Heidelberg (2000)

19. Breiman, L.: Bagging Predictors. Machine Learning 24, 123–140 (1996)

20. Freund, Y., Schapire, R.E.: A Decision-Theoretic Generalization of Online Learning and an Application to Boosting. In: Vitányi, P.M.B. (ed.) EuroCOLT 1995. LNCS, vol. 904, pp. 23–37. Springer, Heidelberg (1995)

21. Freund, Y., Schapire, R.: Experiments with a New Boosting Algorithm. In: International Conference on Machine Learning, pp. 148–156 (1996)

Hall, M., Frank, E., Holmes, G., Pfahringer, B., Reutemann, P., Witten, I.H.: The WEKA data mining software: an update. SIGKDD Explor. Newsl. 11, 10–18 (2009)

Fault Resilient Galois Field Multiplier Design in Emerging Technologies

Mahesh Poolakkaparambil[1], Jimson Mathew[2], and Abusaleh Jabir[1]

[1] Oxford Brookes University, Department of Computing and Communication Technologies,
Wheatley Campus, Wheatley, Oxford, UK
[2] University of Bristol Department of Computer Science, UK

Abstract. Transient error is a critical issue in nano scale devices. The potential replacement technologies for CMOS circuits, such as Carbon Nano-Tube Field Effect Transistor (CNTFET) based circuits and Quantum Cellular Automata (QCA), are further scaled down below 20nm. This results in computations performed in lower energy levels than CMOS, making them more vulnerable to transient errors. This paper therefore investigates the performance of inevitable error mitigating schemes such as the Concurrent Error Detection (CED) over binary Galois Fields (GF) in CNTFETs and QCA. The results are then compared with their CMOS equivalents which are believed to be the first reported attempt to the best of the authors' knowledge. A QCA based GF multiplier layout has been presented as well. The detailed experimental analysis of CMOS with CNTFET design proves that the emerging technologies perform better for error tolerant designs in terms of area, power, and delay as compared to its CMOS equivalent.

Keywords: Galois Field (GF), Concurrent Error Detection (CED), Carbon Nano-Tube Field Effect Transistor (CNTFET), Quantum Cellular Automata (QCA), Transient Error.

1 Introduction

Modern day computing hardware requires much more processing power to perform complex computations efficiently and quickly than ever before. According to ITRS-2009 surveys, it is evident that further scaling in CMOS devices is limited by the adverse performance of the devices beyond 20nm geometry. However, the emerging technologies such as CNTFET's and QCA based digital circuits may be the potential replacement for the classical CMOS technology to continue with this trend of integration. As these devices are scaled down beyond certain nano meters, the energy used to convey information is further reduced which in turn makes the computation more vulnerable to transients and various manufacturing faults. This is a serious issue for, among other applications, stand alone cryptography hardware for example. Several decades ago, Intel reported faulty operations in their chips due to interference of radiation particles from the packaging materials [6]. Further research also proved that data processed within an integrated circuit can be analyzed and decoded using radiation effects. The impact of the resulting transient faults, either natural or malicious, in nano scale devices made up of CNTFETs and QCA cells can be critical especially when they are used in designing crypto arithmetic circuits [6].

J. Mathew et al. (Eds.): ICECCS 2012, CCIS 305, pp. 230–238, 2012.

Nowadays, crypto processors find critical applications in bank transactions, digital rights management, TV set top box, smart cards, mobile communications, etc. [5]. The faults are either natural or malicious and can result in multiple bit errors at the output. In either case, the end result maybe catastrophic. The primary component of an arithmetic logic block in a crypto processor is a GF multiplier [7], which is quite large compared to the other units. This makes the multiplier circuits more prone to exposure from radiation particles. Hence, the multipliers are considered to be particularly vulnerable to radiation induced tampering for leaking out secret information, and continue to be the main focus of attacks in the emerging technology based designs. During normal operations energy particle can come in contact with the nano scale devices causing upsets in the actual computed results. Based on this, a malicious attacker can subject a crypto chip to controlled radiations in a laboratory environment thus inducing random events in the hardware. By observing the chips behavioral changes due to the transient faults, the attacker can gain a good understanding of the internal of the chip. The main advantage of doing this is that, the attacker is not permanently tampering or damaging the chip but only making it perform faulty in the presence of radiation [8].

This paper investigates the performance figures such as power and delay of multiple error detecting schemes over bit parallel GF multiplier implemented using emerging technologies such as CNTFET and QCA.

The remainder of the paper is organized as follows. Section 2 explains the related recent research and other radiation tolerant techniques. Section 3 presents the basics of GF arithmetic and fundamentals of GF multiplier designs. Section 4 introduces the promising emerging technologies that may be replacements for existing CMOS technology. A hamming code based concurrent multiple error detection scheme is explained in Section 5. Section 6 presents experimental results of the CNTFET and QCA based designs and their performance compared to the CMOS counterpart. The conclusions and the future extensions of the proposed research is presented in Section 7.

2 Prior Research

This section details the state of the art multiple error correction schemes in GF multipliers and other GF circuits mainly targeted for CMOS technology. We also briefly review the fault tolerant research in CNTFET and QCA circuits.

Most of the existing approaches are based on replicating the functional block multiple times and checking for the correctness of the operations with the help of a voter. One example of such a space redundant scheme is the Triple Modular Redundancy (TMR). In TMR, the actual functional block is replicated three times and the output is compared for correctness with a voter [9]. If two out of three circuits agree to one result to be correct, the voter considers that as the correct result of the circuit output. The major drawback of TMR is that, the hardware overhead is at least 200%. Another drawback is that the entire reliability depends on the voter, which is not triplicated; as well as the assumption that the errors happen only in one functional block out of three. The design complexity of the voter is also non-trivial.

Another well known approach for error detection, termed as Concurrent Error Detection (CED), is based on time redundancy [10]. In CED, an additional error monitoring block is attached to the actual circuit that flags the occurrence of an error. Once the error flag is active, the functional block will roll back and recompute. This introduces a high delay penalty to the calculation that maybe unsuitable for many applications. There are also approaches reported for double error detection and single error correction known as the SEC/DED schemes. The SEC/DED schemes are based on hamming or LDPC codes that can correct only single bit errors in the calculations [12]. But analysis shows that transient error occurring at a critical node can cause multiple output errors due to large fan-out in most practical systems.

There has been little research done on faults and fault tolerant designs for QCA. The technique of [1,2] reports some of the causes of faults and fault tolerance in QCA based circuits. The primary cause of faults and errors in QCA seems to be due to the cell displacements and unwanted inversion during propagation. But to the best of our knowledge, this is the first effort that has been made to analyze the classical Hamming code based CED schemes in both CNTFET and QCA based designs.

3   Multiple Error in Galois Field Multiplier Circuits

This section presents the effects of transient errors and manufacturing faults on GF multiplier circuits arguably the most vital part of a crypto processor. The GF arithmetic circuits are AND-XOR logic with inner product nodes often shared between the output XOR gates.

Any element $A \in GF(2^m)$ is represented using the elements in the PB. The polynomial basis multiplication of $A(x)$ and $B(x)$ over $GF(2^m)$ is defined using the following expression: $C(x) = A(x) \cdot B(x) \mod P(x)$, where $A, B \in GF(2^m)$.

It can be shown that, the total number of inner product AND gates that are shared between the outputs can be calculated using the following formula,

$$N(AND_{Shared}) = \sum_{i=1}^{m-1} (m - i). \tag{1}$$

Fig. 1 shows example of a 3-bit GF multiplier with shared inner product terms. In this example $N(AND_{Shared}) = 3$, which is calculated using Eq. (1). It is noted in this example that the critical inner product terms a_2b_2, a_1b_2 and a_2b_1 are shared between the outputs c_0, c_1 and c_2. These nodes are said to be critical because an error at any one of these nodes propagates as multiple bit-errors at the output.

The number of times the AND gates are shared will depend on the primitive polynomial used to construct the fields. Clearly, traditional single bit-error detecting or correcting schemes are not sufficient to mitigate such multiple error propagating scenario. The effects of transient errors can be far worse in much smaller geometries such as those based on CNTFET and QCA. Hence in this paper, a Hamming code based multiple error detection scheme has been investigated over bit-parallel GF multipliers using both CNTFET and QCA based technologies. For better understanding of the performance, they are compared with CMOS equivalent circuits.

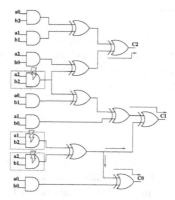

Fig. 1. Error Propagation in a bit-parallel GF multiplier

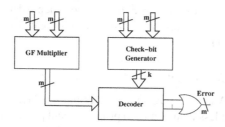

Fig. 2. General block diagram of CED based multiple error detection

Fig. 2 shows the basic block diagram of the Hamming CED scheme that is used to detect multiple errors in the test bench GF multiplier circuits. In this design, we have used an additional parity bit in order to increase the Hamming distance to detect up to 3-bit errors.

Fig. 3 shows the details of the decoder block that is used in the CED technique.

Due to the limitations of the available present day EDA tools for synthesis of CNT-FET and QCA circuits, the implementation results have been limited to circuits of smaller sizes and complexities. However, theoretically, the designs can be extended to effective multiple error correcting architectures, e.g. in [5].

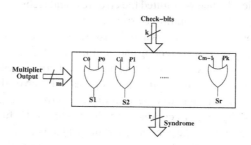

Fig. 3. General block diagram of CED decoder

4   Emerging Technologies

This section explores the two potential technologies that are considered to be the future replacement for the CMOS technology. The primary candidates seem to be CNTFETs and QCA. They are predominantly considered over other technologies due to their capabilities of maintaining high integration density, lower power consumptions, and lower chip area requirements.

4.1   CNTFET

CNTFET based circuits are the tough competent to QCA as a CMOS alternative. The CNTFET devices are proffered due to many reasons. One among the reason is the minute NRE cost in the fabrication. This is because of the fact that, CNTFETs are similar to that of MOSFETs in physical implementation except only with the channel material. In CNTFETs, the bulk silicon channel material of the MOSFET is replaced by a single carbon nano-tube or by an array of tubes. The in depth detail of CNTFET device properties are not discussed in this manuscript. The physical properties and features of CNTFETs are explained in [13].

4.2   Quantum Dot Cellular Automata

Quantum Dot Cellular Automata (QCA) is another emerging technology that uses quantum cells (with cell size less than 20nm) to propagate and process information. In QCA the interconnection between the QCA logic gates is done by quantum wires that are again realized using QCA cells as compared to the metallic wires in CNTFET and CMOS technologies.

In QCA the logic is propagated because of the coulombic interaction between the driver QCA cell and its neighboring cells. The QCA logic and basic working principle can be referred from [1].

5   Concurrent Error Detection in Emerging Technologies

From Fig. 1, it is obvious that a fault at any critical node can cause multiple bit errors at the output. In certain applications such as cryptography for example, detecting and correcting the errors in real time and in parallel is very important. The motivation of this paper is to investigate the performance of the error detection schemes in the potential emerging technologies. Hence we limited the error resilient technique used in this paper to simple Hamming code based multiple error detection.

The Hamming codes are well known and easy to implement error detecting codes generally known as single error correcting and double error detecting codes (SEC/DED). However, the Hamming codes can also detect an extra bit error if we increase the Hamming distance by adding an extra parity. In this paper 3-bit error detecting Hamming codes are considered. In practice, to detect multiple bit errors, we generate check bits (parity) from the primary input to compute the checksum for the functional block (GF multiplier) as shown in Fig. 2. The generated parities and the multiplier functional block outputs are then passed on to the decoder to generate syndromes that detect the occurrence of an error.

6   Experimental Results

For analysis of the performance of multiple error resilient scheme, we have implemented CED with various GF multiplier test bench circuits of different sizes. The implementation is done using the HSPICE simulator, 45nm CNTFET Stanford libraries
and CMOS models for fair comparison of performance. A 2-bit GF multiplier circuit
has been implemented using QCA Designer and a detailed layout has been achieved
with CED technique.

The multiplier sizes considered for comparison are 2, 3, 4 and 5-bit bit-parallel classical GF multipliers. The designs are implemented and simulated for power and delay
using the HSPICE simulator. Fig. 4 shows the power consumption of various GF multipliers in both 45nm CNTFET and CMOS technology. It is evident that the CNTFET
has much lower power signatures as compared to its CMOS counterpart.

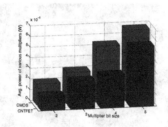

Fig. 4. Average power dissipation comparison of GF multipliers

The designs are then extended by coupling with the multiple error detecting CED
schemes. The power dissipation of the GF multiplier with CED is shown in Fig. 5. It
clearly shows that the CNTFET based technology is significantly superior to the CMOS
based implementation with lower power requirements.

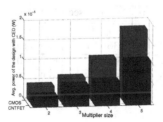

Fig. 5. Average power dissipation comparison of GF multiplier with CED circuits

For analysis, a QCA version of the 2-bit GF multiplier has been designed as shown
in Fig. 6. The implementation is achieved using AND-XOR logic based on the QCA
majority gates.

The multiplier designs are then extended to multiple error detectable versions, as
shown in Fig. 7. The various colors in the layout represent the various clocking zones
of the QCA.

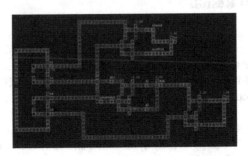

Fig. 6. 2-bit bit parallel multiplier using QCA

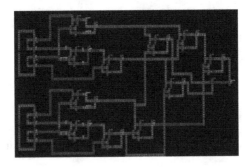

Fig. 7. 2-bit bit parallel multiplier with CED using QCA

Table 1. Delay information of various GF multipliers

No. of bits	CNTFET (sec)	CMOS (sec)
2	$1.324 * 10^{-11}$	$5.537 * 10^{-10}$
3	$1.36 * 10^{-11}$	$5.58 * 10^{-10}$
4	$1.362 * 10^{-11}$	$6.67 * 10^{-10}$
5	$1.38 * 10^{-11}$	$6.9 * 10^{-10}$

Table 2. Delay information of various GF multipliers with CED

No. of bits	CNTFET (sec)	CMOS (sec)
2	$3.18 * 10^{-11}$	$1.72 * 10^{-9}$
3	$3.6 * 10^{-11}$	$1.8 * 10^{-9}$
4	$4.14 * 10^{-11}$	$2.3 * 10^{-9}$
5	$5.06 * 10^{-11}$	$2.7 * 10^{-9}$

The propagation delay of the multipliers implemented in both CMOS and CNTFET technologies are compared in Table 1 and Table 2 with and without CED. From the tables it is clear that the CNTFETs based designs offer much less delay as compared to the CMOS based implementations.

7 Conclusions

This paper investigates the multiple error detection and fault tolerance, based on a concurrent error detection and correction scheme, in emerging technologies, namely, based on CNTFETs and QCAs. The motivation was two folds: Firstly, owing to the substantially reduced geometries, the possibilities of inherent errors or imperfections could be much higher in these technologies than in the existing CMOS technology, and secondly, these devices could be much more susceptible to soft errors due to radiation particles, either from natural causes or imparted with malicious intents. Hence, error mitigation is a very important way forward for continuous in field reliability. To this end the paper explored error detection with CED in GF multipliers, which are the primary components in crypto processors, designed over the 45nm CNTFET technology, and compared the findings in terms of power, area overhead, and delay with conventional 45nm CMOS technology. These metrics were also projected over the CED complexities in terms of various multiplier sizes. The scheme has also been implemented over the QCA technology to evaluate the logic performance. Due to the limitations of the available present day EDA tools for synthesis of CNTFET and QCA circuits, the implementations over CNTFET and QCA have been limited to circuits of smaller sizes and complexities. Our future work include extensions of the presented design to low complexity multiple error correction schemes, e.g. based on Low Density Parity Check (LDPC), Multiple Hamming and BCH codes, over these emerging technologies.

References

1. Crocker, M., Sharon Hu, X., Niemier, M.: Defects and Faults in QCA-Based PLAs. ACM Journal on Emerging Technologies in Computing Systems 5(2), Article 8 (2009)
2. Ma, X., Lombardi, F.: Fault Tolerant Schemes for QCA Systems. In: IEEE International Symposium on Defect and Fault Tolerance of VLSI Systems, pp. 236–244. IEEE (2008)
3. Dalui, M., Sen, B., Sikdar, B.K.: Fault Tolerant QCA Logic Design With Coupled Majority-Minority Gate. International Journal of Computer Applications 1(29), 90–96 (2010)
4. Pradhan, D.K.: A Theory of Galois Switching Functions. IEEE Trans. Computers 27(3), 239–248 (1978)
5. Poolakkaparambil, M., Mathew, J., Jabir, A., Pradhan, D.K.: BCH Code Based Multiple Bit Error Correction in Finite Field Multiplier Circuits. In: Proc. IEEE/ACM Int. Symp. Quality Electronic Design (ISQED 2011), pp. 1–6 (2011)
6. Alves, N.: State-of-the-art techniques for Detecting Transient Errors in Electrical Circuits. IEEE Potentials, 30–35 (2011)
7. Masoleh, A.R., Hasan, M.A.: Low Complexity Bit Parallel Architectures for Polynomial Basis Multiplication over GF(2^m). IEEE Trans. Computers 53(8), 45–959 (2004)
8. Ratnapal, G.B., Williams, R.D., Blalock, T.N.: An On-Chip Signal Suppression Countermeasure to Power Analysis Attacks. IEEE Trans. Dependable Sec. Comput. 1(3), 179–189 (2004)

9. Wakerly, J.F.: Microcomputer reliability improvement using triple-modular redundancy. IEEE Proceedings 64, 889–895 (1976)
10. Wu, K., Kari, R., Kuznetsov, G., Gossel, M.: Low Cost Concurrent Error Detection for the Advanced Encryption Standard. In: Proceedings of the International Test Conference, pp. 1242–1248 (2004)
11. Keren, O.: One to Many: Context Oriented Code for Concurrent Error Detection. Journal of Electronic Testing 26(3), 337–353 (2010)
12. Mathew, J., Jabir, A.M., Rahman, H., Pradhan, D.K.: Single Error Correctable Bit Parallel Multipliers Over GF(2^m). IET Comput. Digit. Tech. 3(3), 281–288 (2008)
13. Lin, S., Kim, Y.B., Lombardi, F.: CNTFET-Based Design of Ternary Logic Gates and Arithmetic Circuits. IEEE Trans. Nanotechnology 10(2), 217–225 (2011)

A Novel Security Architecture for Biometric Templates Using Visual Cryptography and Chaotic Image Encryption

Divya James and Mintu Philip

Department of Computer Science
Rajagiri School of Engineering & Technology, Rajagiri Valley, Cochin, India
{divyajames,mintu.philip}@gmail.com

Abstract. Protection of biometric data and templates is a crucial issue for the security of biometric systems This paper proposes new security architecture for biometric templates using visual cryptography and chaotic image encryption. The use of visual cryptography is explored to preserve the privacy of biometric image by decomposing the original image into two images (known as sheets) such that the original image can be revealed only when both images are simultaneously available; the individual sheet images do not reveal the identity of the original image. The algorithm ensures protection as well as privacy for image using a fast encryption algorithm based on chaotic encryption. Chaos based cryptography is applied on to individual shares. Using this one can cross verify his identity along with the protection and privacy of the image.

Keywords: visual cryptography, security, chaotic, Arnold map.

1 Introduction

Biometric based personal identification techniques that use physiological or behavioral characteristics are becoming increasingly popular compared to traditional token-based or knowledge based techniques such as identification cards (ID), passwords, etc. One of the main reasons for this popularity is the ability of the biometrics technology to differentiate between an authorized person and an impostor who fraudulently acquires the access privilege of an authorized person [1]. Among various commercially available biometric techniques such as face, voice, fingerprint, iris, etc., fingerprint-based and face recognition techniques are the most extensively studied and the most frequently deployed. While biometric techniques have inherent advantages over traditional personal identification techniques, the problem of ensuring the security and integrity of the biometric data is critical.

Researchers have proposed conventional algorithms such as AES, DES for the protection of biometric templates. However all the conventional algorithms take high processing power and high encryption/decryption time. The distinct properties of chaos, such as ergodicity, quasi-randomness, sensitivity dependence on initial conditions and system parameters, have granted chaotic dynamics as a promising alternative for the conventional cryptographic algorithms.

J. Mathew et al. (Eds.): ICECCS 2012, CCIS 305, pp. 239–246, 2012.

So we introduce a new method for ensuring the security and integrity of biometric data which utilizes both visual cryptography and chaos-based cryptographic scheme. The model is achieved by using visual cryptography for obtaining shares of biometric images. Chaos based system is used for encryption. This algorithm encrypts image shares pixel by pixel taking consideration the values of previously encrypted pixels.

This paper is organized as follows: Section 2 deals with the related work and Section 3 deals with 3Dchaotic map. Section 4 presents the proposed system and Section 5 deals with results and analysis. Section 6 contains the conclusion.

2   Related Work

2.1   Visual Cryptography

One of the best known techniques to protect data is cryptography. It is the art of sending and receiving encrypted messages that can be decrypted only by the sender or the receiver. Encryption and decryption are accomplished by using mathematical algorithms in such a way that no one but the intended recipient can decrypt and read the message. Naor and Shamir [2] introduced the visual cryptography scheme (VCS) as a simple and secure way to allow the secret sharing of images without any cryptographic computations. Visual cryptography schemes were independently introduced by Shamir [3] and Blakely [4], their original motivation was to safeguard cryptographic keys from loss. These schemes also have been widely employed in the construction of several types of cryptographic protocols [5]. A segment-based visual cryptography suggested by Borchert [6] can be used only to encrypt the messages containing symbols, especially numbers like bank account number, amount etc. The VCS proposed by Wei-Qi Yan et al., [7] can be applied only for printed text or images. A recursive VC method proposed by Monoth et al., [8] is computationally complex as the encoded shares are further encoded into number of sub-shares recursively. Similarly a technique proposed by Kim et al., [9] also suffers from computational complexity, though it avoids dithering of the pixels. Most of the previous research work on VC focused on improving two parameters: pixel expansion and contrast[10] [11],[12]. In these cases all participants who hold shares are assumed to be honest, that is, they will not present false or fake shares during the phase of recovering the secret image. Thus, the image shown on the stacking of shares is considered as the real secrete image. But, this may not be true always. So cheating prevention methodologies are introduced by Yan et al., [13], Horng et al., [14] and Hu et al., [15]. But, it is observed in all these methodologies, there is no facility of authentication testing.

2.2   Chaos Based Cryptography

In recent years, chaotic maps have been employed for image encryption. Most chaotic image encryptions (or encryption systems) use the permutation-substitution architecture. These two processes are repeated for several rounds, to obtain the final encrypted image. For example Fridrich suggested a chaotic image encryption method composed of permutation and substitution [16]. All the pixels are moved using a 2D chaotic map. The new pixels moved to the current position are taken as a permutation of the original pixels. In the substitution process, the pixel values are altered sequentially. Chen et al. employed a three-dimensional (3D) Arnold cat map[17] and a 3D Baker map in the permutation stage[18]. Guan et al. used a 2D cat map for pixel position permutation and the discretized Chen's chaotic system for pixel value masking[19]. Lian et al. used a chaotic standard map in the permutation stage and a quantized logistic map in the substitution stage[20].

3   The 3D Chaotic Cat Map

Arnold cat map is one of the most studied 2D invertible chaotic maps [17], particularly in image encryption, watermarking and steganographic algorithms. The generalized form of a 3D cat map is described as follows

$$
X_{i+1} = \begin{pmatrix} x_{i+1} \\ y_{i+1} \\ z_{i+1} \end{pmatrix} = A \begin{pmatrix} x_i \\ y_i \\ z_i \end{pmatrix} (\mathrm{mod}\,1), \tag{1}
$$

where A is a matrix which provides chaotic behavior. A family of such matrices is given by [17]

$$
A = \begin{pmatrix} 1 + a_x a_z b_y & a_z & a_y + a_x a_z + a_x a_y a_z b_y \\ b_z + a_x b_y + a_x a_z b_y b_z & a_z b_z + 1 & a_y b_z + a_x a_y a_z b_y b_z + a_x a_z b_z + a_x a_y b_y + a_x \\ a_x b_x b_y + b_y & b_x & a_x a_y b_x b_y + a_x b_x + a_y b_y + 1 \end{pmatrix} \tag{2}
$$

with $a_x, a_y, a_z, b_x, b_y, b_z$ all being positive integers.

As a special case, one can let $a_x = a_y = a_z = 1$ and $b_x = b_y = b_z = 2$. Thus, the map of Eq.(1) becomes

$$
X_{i+1} = \begin{pmatrix} x_{i+1} \\ y_{i+1} \\ z_{i+1} \end{pmatrix} = \begin{pmatrix} 3 & 1 & 4 \\ 8 & 3 & 11 \\ 6 & 2 & 9 \end{pmatrix} \begin{pmatrix} x_i \\ y_i \\ z_i \end{pmatrix} (\mathrm{mod}\,1) \tag{3}
$$

It is easy to verify that the determinant of A is equal to one and the eigen values of A are 14.3789,0.4745,0.1466.Since the leading eigenvalue is greater than 1,the map of Eq.(3) has chaotic behavior. Therefore it preserves all the features of chaos.

4   Proposed Security Architecture for Biometric Templates

This paper combines the concepts of visual cryptography as well as chaotic encryption to implement a two-tier cryptographic system for ensuring the security of biometric templates. The proposed architecture ensures that the time and computation power for decryption procedure is minimal when compared to conventional technologies.

4.1   Generation of Visual Cryptographic Shares

In this module the input image is a binary $m \times n$ image which is further decomposed into $m \times 2n$ shares by using the principle of pixel expansion mechanism of visual cryptography. Here a 2 out of 2 visual cryptography scheme is implemented. In the case of (2, 2) visual cryptography scheme, each pixel P in the original image is encrypted into two sub pixels called shares. There are two choices available for each pixel. For white and black pixels the following collection of matrices C_0 and C_1 are defined:

$$C_0 = \left\{ \text{all the matrices obtained by permuting the columns of } \begin{bmatrix} 1100 \\ 1100 \end{bmatrix} \right\}$$

$$C_1 = \left\{ \text{all the matrices obtained by permuting the columns of } \begin{bmatrix} 1100 \\ 0011 \end{bmatrix} \right\}$$

The choice of shares for a white and black pixel is randomly determined. When the two shares are overlaid using XOR operation the value of the original pixel P can be determined. Once two shares are obtained from the original image, for the protection and privacy of the shares which are stored in different database servers, a fast image encryption algorithm based on chaos are applied on to individual shares. Here we propose an image encryption algorithm based on a 3d chaotic map.

4.2   Chaotic Encryption for Image Shares

In order to impart more security for the shares without sacrificing processing power and decryption time we can apply a chaotic encryption technique based on 3d cat map. For the protection and privacy of images we resolve to apply a two level chaotic encryption. In the first level we scramble the whole image using the generated chaotic sequences. In the second level the already scrambled image is further segmented into $s \times s$ segments. From the $s \times s$ segments, n segments are randomly selected, where $n < s^2$ by iterating a logistic map. The selected segments are further encrypted using a new set of chaotic sequences generated from the 3d chaotic map. Generation and encryption of shares are depicted in Fig.1.

4.3   Chaotic Decryption and Re-combination of Image Shares

In this module the shares are recovered using suitable chaotic decryption algorithm. At first the shares are segmented into $s \times s$ segments. From these $s \times s$ segments,n segments are selected, where $n < s^2$ by iterating a logistic map using suitable keys. The selected segments are decrypted using the set of chaotic sequences generated from the 3d chaotic map. All the segments are combined and the whole image is decrypted once again done using the chaotic sequences from 3d cat map to retrieve the visual cryptographic shares. Once the decrypted shares are obtained, the two shares are overlaid using the XOR operator so that the loss in contrast in target images can be considerably reduced.

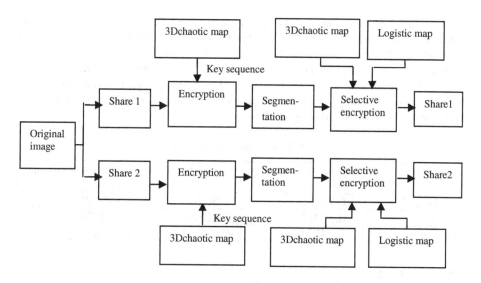

Fig. 1. Generation and encryption of shares

5   Statistical Analysis

The encrypted images should possess certain random properties in order to resist the statistical attack. In this section, we demonstrate the high security level of our suggested algorithm by showcasing its performance according to statistical analysis such as adjacent pixel correlation analysis, key sensitivity analysis and single bit test.

5.1   Adjacent Pixel Correlation

To investigate the quality of encryption, digital simulations are performed on MATLAB.The input image chosen for encryption is a fingerprint image. Fig.2. shows the original image, image share and encrypted image shares.

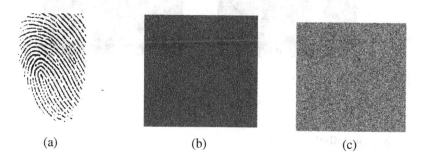

      (a)   (b)   (c)

Fig. 2. (a) Original image   (b) an image share   (c) encrypted image share

Since the encrypted image looks random, one would expect its adjacent pixels to have almost no correlation. On the other hand, adjacent pixel correlations are a means of measuring the performance of the algorithm. Each number in the table represents the correlation coefficient[17] of adjacent values x_i and y_i selected at random.

It is observed that in Table 1 that while in the original-image, adjacent pixels have high correlation, there is very little correlation between pairs of adjacent pixels in the image shares and encrypted image shares.

Table 1. Adjacent pixel correlation

	Original image	Image share	Encrypted Image share
Horizontal	0.73155	-0.50243	-0.00367
Vertical	0.70463	0.00226	0.00762
Diagonal	0.74781	-0.00268	0.00124

5.2   Single Bit Test

Since the histogram analysis may not be suitable for binary images where the pixel value is only zero and one we have adapted one bit testing for ensuring that the distribution of zero's and one's is almost uniform in the resulting image. In this test we count the number of 0's and 1's for each column of the image and plot the result. Fig.3(a). shows the plot for original image. It shows that black pixels are limited to a small portion of the image. Meanwhile in Fig.3(b) the bit counts of the encrypted image shares are uniformly distributed across all columns and also the number of one's are almost equal to the number of zero's in each column. This clearly shows the random nature of encrypted image.

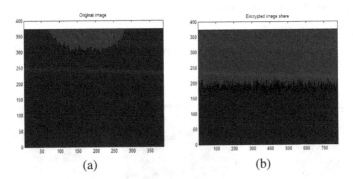

Fig. 3. (a) original image (b) an encrypted image share

5.3   Key Sensitivity Analysis

The proposed algorithm verifies the sensitivity of the keys by encrypting the image with a triplet of keys(k_{11}, k_{12}, k_{13}) and the same image was encrypted with a new triplet of keys(k_{21}, k_{22}, k_{23}).Here the difference between k11 and k21 is

0.000001.For the two encrypted images it was seen that the percentage of difference between the pixel values is 50.09% and thus ensuring an efficient key sensitivity. Analysis was carried out for another set of encrypted images, here also the percentage of difference between the pixel values was found to be 50.01.

6   Conclusion

In this paper, we use visual cryptography scheme and a 3D chaotic map to design novel security architecture for biometric templates. Applying Visual Cryptography scheme alone on biometric images does not lead to a complete solution for privacy as well as protection. Moreover, even if the shares have been obtained the original biometric image should not be revealed. A 3D chaotic cat map makes a suitable candidate for this purpose. Simulation results presented in this paper demonstrate that the suggested algorithm is efficient and possesses a high level of security against statistical attacks. Also a simple implementation of image encryption achieves high encryption rate on general purpose computer.

References

1. Jain, A., Uludag, U.: Hiding biometric data. IEEE Trans. Pattern Anal. Mach. Intell. 25(11), 1494–1498 (2003)
2. Naor, M., Shamir, A.: Visual Cryptography. In: De Santis, A. (ed.) EUROCRYPT 1994. LNCS, vol. 950, pp. 1–12. Springer, Heidelberg (1995)
3. Shamir, A.: How to Share a Secret. ACM Communication 22, 612–613 (1979)
4. Blakley, G.R.: Safeguarding Cryptographic Keys. In: Proceedings of AFIPS Conference, vol. 48, pp. 313–317 (1970)
5. Menezes, A., Van Oorschot, P., Vanstone, S.: Handbook of Applied Cryptography. CRC Press, Boca Raton (1997)
6. Borchert, B.: Segment Based Visual Cryptography. WSI Press, Germany (2007)
7. Yan, W.-Q., Jin, D., Kanakanahalli, M.S.: Visual Cryptography for Print and Scan Applications, pp. 572–575. IEEE (2004)
8. Monoth, T., Babu, A.P.: Recursive Visual Cryptography Using Random Basis Column Pixel Expansion. In: Proceedings of IEEE International Conference on Information Technology, pp. 41–43 (2007)
9. Kim, H.J., Sachnev, V., Choi, S.J., Xiang, S.: An Innocuous Visual Cryptography Scheme. In: Proceedings of IEEE-8th International Workshop on Image Analysis for Multimedia Interactive Services (2007)
10. Blundo, C., De Santis, A.: On the contrast in Visual Cryptography Schemes. Journal on Cryptography 12, 261–289 (1999)
11. Eisen, P.A., Stinson, D.R.: Threshold Visual Cryptography with speci_ed Whiteness Levels of Reconstructed Pixels. Designs, Codes, Cryptography 25(1), 15–61 (2002)
12. Verheul, E.R., Van Tilborg, H.C.A.: Constructions and Properties of k out of n Visual Secret Sharing Schemes. Designs, Codes, Cryptography 11(2), 179–196 (1997)
13. Yan, H., Gan, Z., Chen, K.: A Cheater Detectable Visual Cryptography Scheme. Journal of Shanghai Jiaotong University 38(1) (2004)
14. Horng, G.B., Chen, T.G., Tsai, D.S.: Cheating in Visua Cryptography. Designs, Codes, Cryptography 38(2), 219–236 (2006)

15. Hu, C.M., Tzeng, W.G.: Cheating Prevention in Visual Cryptography. IEEE Transaction on Image Processing 16(1), 36–45 (2007)
16. Fridrich, J.: Symmetric Ciphers Based on Two-dimensional Chaotic Maps. Int. J. Bifurcat. Chaos 8(6), 1259–1284 (1998)
17. Chen, G., Mao, Y.B., Chui, C.K.: A symmetric image encryption scheme based on 3D chaotic cat maps. Chaos, Solitons & Fractals 12, 749–761 (2004)
18. Mao, Y.B., Chen, G., Lian, S.G.: A novel fast image encryption scheme based on the 3D chaotic baker map. Int. J. Bifurcat. Chaos 14(10), 3613–3624 (2004)
19. Guan, Z.H., Huang, F.J., Guan, W.: Chaos-based image encryption algorithm. Phys. Lett. A 346, 153–157 (2005)
20. Lian, S.G., Sun, J., Wang, Z.: A block cipher based on a suitable use of chaotic standard map. Chaos, Solitons and Fractals 26(1), 117–129 (2005)

A New Fault-Tolerant Routing Algorithm for MALN-2

Nitin[1] and Durg Singh Chauhan[2]

[1] Jaypee University of Information Technology,
P.O. Waknaghat, Solan-173234, Himachal Pradesh, India
[2] Uttarakhand Technical University,
P.O. Chandanwadi, Prem Nagar, Sudhowala,
Dehradun-248007, Uttarakhand, India

Abstract. In this paper we have presented a new fault-tolerant Irregular Modified Alpha Multi-stage Interconnection Network (MALN-2). MALN-2 is a double switch fault-tolerant interconnection network with more alternate paths with full accessibility. It is better than existing Modified Alpha Network. The paper also presents the fault-tolerant Routing Algorithm for the proposed irregular Interconnection Network.

1 Introduction and Motivation

A brief survey of interconnection networks (INs) [1-2]and a survey of the fault-tolerant attributes of multistage interconnection networks (MINs) are found in [1-4]. A MIN in particular is an IN but consists of cascade of switching stages, contains switching elements (SEs) [1-21]. They are widely used for broadband switching technology and for multiprocessor systems. Besides this, they offer an enthusiastic way of implementing switches used in data communication networks. With the performance requirement of the switches exceeding several terabits/sec and teraflops/sec, it becomes imperative to make them dynamic and fault-tolerant [9-15].

The modest cost of unique paths MINs makes them attractive for large multiprocessors systems, but their lack of fault-tolerance, is a major drawback. To mitigate this problem, three hardware options are available [9-15, 21]:

- Replicate the entire network,
- Add extra stages and /or,
- Add chaining links.

Various researchers already done sufficient work on regular MINs and irregular topologies were out of limelight. Therefore, we have decided to work on irregular fault-tolerant interconnection networks. Any network is said to be irregular if the number of switching elements (SEs) in each stage of the network are different [9-15].

The rest of the paper is organized as follows: Section 2 provides the basic network description of Modified Alpha Network. Section 3 provides the network description of Modified Alpha Network-2 along with its fault-tolerant routing algorithm and cost analysis followed by conclusion and references.

J. Mathew et al. (Eds.): ICECCS 2012, CCIS 305, pp. 247–254, 2012.
© Springer-Verlag Berlin Heidelberg 2012

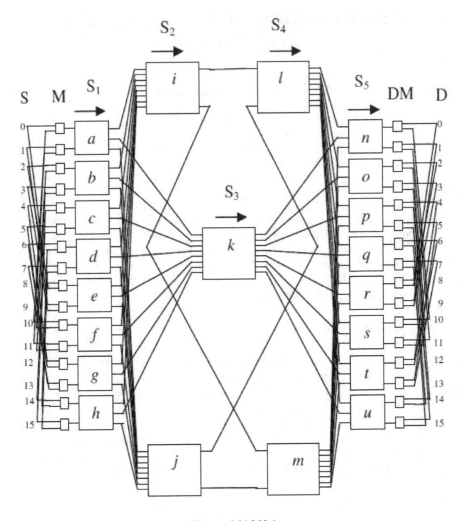

Fig. 1. MALN-2

2   Related Work

2.1   Modified Alpha Network

Modified Alpha Network (MALN) [21] is a single switch fault-tolerant interconnection network. This network has four stages. It has two groups G0 and G1.
G0 and G1 are connected with the 16 sources and 16 destinations through multiplexers and demultiplexers respectively. The first, second and third stage of
this interconnection network has SEs of size 3x3 whereas last stage has SEs of
size 2x1. It uses multiplexers of size 3x3 and demultiplexers of size 1x2. The
SEs of middle stage (stage 2 and stage 3) are connected with each other through
auxiliary links of each group. The main drawback of this network is that it is

a single switch fault-tolerant interconnection network with limited number of alternate paths, with less accessibility [21].

2.2 Proposed Interconnection Network

Modified Alpha Network-2 (MALN-2) is an irregular fault-tolerant interconnection network. It has (log2N)+1 stages hence, it has one extra stage as compare to MALN network. This network has 16 sources and 16 destinations. In the fig. 1, S, D, M and DM represents the source address, destination address, multiplexers and demultiplexers respectively. In the same way S1, S2, S3, S4 and S5 represents the first stage, second stage, third stage, fourth stage and fifth stage respectively. The first and last stage of network has N/2 SEs in each stage.

The second and fourth stage consists of N/8 SEs in each stage whereas third stage has only a single SE. This network has 16 multiplexers of size 2x1 at source side and 16 demultiplexers of size 1x2 at destination side. In first stage, the size of each SE is 2x3 and each SE of this stage is connected with 2 multiplexers, 2 SEs of second stage and 1 SE of third stage. In second stage, the size of each SE is 8x2 and each SE of this stage is connected with the SEs of first stage and fourth stage. In the same way, the third stage consists of SE of size 8x8 and this switch is connected with the SEs of first stage and fifth stage. In the fourth stage the size of each SE is 8x2 and each switch of this stage is connected with the SEs of second stage and fifth stage. Finally in the last stage the size of SE is 3x2 and each SE of this stage is connected with the SEs of third and fourth stage and demultiplexers. The MALN-2 is a two switch fault-tolerant interconnection network.

Routing Algorithm of MALN-2. In the first step, the source and destination address is obtained. Now send the request to the SE of first stage which is non faulty. In next step, if SE k is non faulty then send request to SE k. It will directly send the connection request to appropriate (non faulty) SE of fifth stage. If k is faulty, in this case connection request will be received by any one of non faulty SEs of stage 2. The SE of second stage send the request to appropriate SE of fourth stage. In the same way SE of fourth stage will send the appropriate SE of fifth stage and this SE will send the request to its destination address through demultiplexer.

1. Begin.
2. Get the source and its corresponding destination address sequentially.
3. Send the request to the appropriate SE of first stage through multiplexer and go to step 4.
4. SE of first stage gets the request on any of its two input links. If this SE is busy or faulty then request will be received by second SE of first stage and if the second SE is also busy or faulty then request will be received by third SE of first stage through multiplexer. If it is also busy or faulty then drop the request otherwise send the request to the SE of third stage and go to step 5.

5. SE of third stage gets the request on any of its eight input links from the SE of first stage, if this SE is busy or faulty then SE of first stage will send the request to the appropriate SE of second stage and go to step 6 otherwise send the request to the appropriate SE of fifth stage and go to step 8.

6. SE of second stage gets the request on any of its eight input links. If the required SE is busy or faulty then request will be received by second SE of stage2. If the second SE is also busy or faulty then drop the request otherwise send the request to appropriate SE of fourth stage and go to step 7.

7. SE of fourth stage gets the request on any of its two input links from the SE of second stage. If the required SE is busy or faulty then request will be received by second SE of stage4. If the second SE is also busy or faulty then drop the request otherwise send the request to appropriate SE of fifth stage and go to step 8.

8. SE of fifth stage gets the request on any of its three input links from the SE of third stage or from any SE of fourth stage. If this SE is busy or faulty then request will be received by second SE of fifth stage and if the second SE is also busy or faulty then request will be received by third SE of fifth stage. If it is also busy or faulty then drop the request otherwise go to step 9.

9. Send the request to its destination address through the demultiplexer.

10. End.

Proof of Correctness. Example: Let the source and destination addresses are 2 and 9 respectively. In this example we will consider two cases which are as follows:

– Case1: When SEs are not busy and fault free in every stage.

1. Begin.
2. According to the algorithm, we have source address 2 and destination address 9.
3. Now we will send the request to the appropriate SE of first stage i.e. to b through multiplexer and follow step 4 of algorithm.
4. In this step, SE b will get the request. Now, send the request to the SE of third stage and follow step 5 of algorithm.
5. SE k will get the request on its second input link from the SE b and it sends the request to SE r and go to step 8 of algorithm.
6. Now SE r of fifth stage gets the request on its second input link from the SE k and follow step 9 of algorithm.
7. In this step, send the request to its destination address i.e. to 9 through demultiplexer.
8. End.

In fig. 2 green lines show that how we can send the connection request from source 2 to destination 9 through SEs b, k and r. This is non faulty condition.

– Case2: When one SE is busy or faulty in second, third and fourth stage.

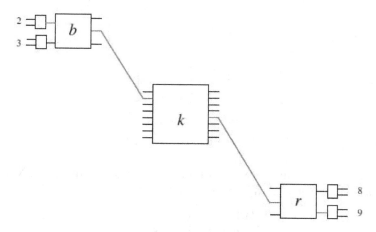

Fig. 2. Sending Request from 2 to 9

Let us suppose k, i, and l are faulty or busy SEs.

1. Begin.
2. According to the algorithm, we have source address 2 and destination address 9.
3. Now we will send the request to the appropriate SE of first stage i.e. to b through multiplexer and follow step 4 of algorithm.
4. In this step, SE b will get the request. Now, send the request to the SE of third stage and follow step 5 of algorithm.
5. SE k is busy or faulty then SE b of first stage will send the request to the SE i of second stage and follow step 6 of algorithm.
6. SE i is busy or faulty then request will be received by SE j of stage2. It send the request to appropriate SE of fourth stage and follow step 7 of algorithm.
7. SE l is busy or faulty then request will be received by SE m of stage4 and follow step 8 of algorithm.
8. Now SE r of fifth stage gets the request on its second input link from the SE m and follow step 9 of algorithm.
9. In this step, send the request to its destination address i.e. to 9 through demultiplexer.
10. End.

In fig. 3 green lines show that how we can send the connection request from source 2 to destination 9 through SEs b, j, m and r. This is faulty condition. The explanation of example1 shows that the proposed interconnection network is better than MALN. MALN-2 can tolerate multiple faults at a time and can pass the connection request from any source to any destination.

Theorem. There are at least five paths existing between every source and destination and MALN-2 is a double switch fault tolerant network.

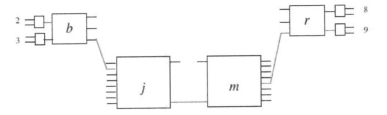

Fig. 3. Sending Request from 2 to 9

Proof: Let us suppose source address is 1 and destination address is 10. Now the possible paths between these addresses are as follows:

Path 1: $a \rightarrow k \rightarrow s$
Path 2: $a \rightarrow i \rightarrow l \rightarrow s$
Path 3: $a \rightarrow i \rightarrow m \rightarrow s$
Path 4: $a \rightarrow j \rightarrow l \rightarrow s$
Path 5: $a \rightarrow j \rightarrow m \rightarrow s$

It proves that MALN-2 has five paths from selected source to destination. Now suppose SEs i, j, l and m are busy or faulty, it means network is having two faulty SEs in second and fourth stage, then data will be passed through SE k. It proves that MALN-2 is a double switch fault-tolerant network.

Cost Analysis of MALN-2. Cost of the proposed interconnection network can be calculated as follows:

Total number of 23 SEs = 8, cost = 48
Total number of 82 SEs = 2, cost = 32
Total number of 88 SEs = 1, cost = 64
Total number of 28 SEs = 2, cost = 32
Total number of 32 SEs = 8, cost = 48
Total number of 2:1 multiplexers = 16, cost = 32
Total number of 1:2 demultiplexers = 16, cost = 32

The total cost of MALN-2 is 288 units and the cost of MALN is 240 units. The proposed network is little more costly as compare to MALN however, it is double switch fault-tolerant and every source address of this network is able to send its request to any destination in faulty situations.

3   Conclusion

In this paper we have presented a new fault-tolerant Irregular Modified Alpha Multi-stage Interconnection Network and have compared the same with exisitng MALN. MALN is a single switch fault-tolerant interconnection network with limited number of paths between every source and destination whereas MALN-2 is a double switch fault-tolerant interconnection network with more alternate paths with full accessibility.

References

1. Siegel, H.J.: Interconnection Networks for Large-scale Parallel Processing: Theory and Case Studies, 2nd edn. McGraw-Hill, Inc., New York (1990) ISBN: 0-07-057561-4
2. Hwang, K.: Advanced Computer Architecture: Parallelism, Scalability, Programmability. Tata McGraw-Hill, India (2000) ISBN: 0-07-053070-X
3. Duato, J., Yalamanchili, S., Ni, L.M.: Interconnection Networks: An Engineering Approach. Morgan Kaufmann (2003) ISBN: 1-55860-852-4
4. Dally, W., Towles, B.: Principles and Practices of Interconnection Networks. Morgan Kaufmann, San Francisco (2004) ISBN: 978-0-12-200751-4
5. Arabnia, H.R., Oliver, M.A.: Arbitrary Rotation of Raster Images with SIMD Machine Architectures. International Journal of Eurographics Association (Computer Graphics Forum) 6(1), 3–12 (1987)
6. Bhandarkar, S.M., Arabnia, H.R., Smith, J.W.: A Reconfigurable Architecture For Image Processing and Computer Vision. International Journal of Pattern Recognition And Artificial Intelligence 9(2), 201–229 (1995)
7. Bhandarkar, S.M., Arabnia, H.R.: The Hough Transform on a Reconfigurable Multi-Ring Network. Journal of Parallel and Distributed Computing 24(1), 107–114 (1995)
8. Wani, M.A., Arabnia, H.R.: Parallel Edge-Region-Based Segmentation Algorithm Targeted at Reconfigurable Multi-Ring Network. The Journal of Supercomputing 25(1), 43–63 (2003)
9. Nitin, Subramanian, A.: Efficient Algorithms to Solve Dynamic MINs Stability Problems using Stable Matching with Complete TIES. Journal of Discrete Algorithms 6(3), 353–380 (2008)
10. Nitin, Vaish, R., Srivastava, U.: On a Deadlock and Performance Analysis of ALBR and DAR Algorithm on X-Torus Topology by Optimal Utilization of Cross Links and Minimal Lookups. Journal of Supercomputing 59(3), 1252–1288 (2012)
11. Nitin, Chauhan, D.S.: Stochastic Communication for Application Specific Networks-on-Chip. Journal of Supercomputing 59(2), 779–810 (2012)
12. Nitin, Chauhan, D.S.: Comparative Analysis of Traffic Patterns on k-ary n-tree using Adaptive Algorithms based on Burton Normal Form. Journal of Supercomputing 59(2), 569–588 (2012)
13. Nitin, Garhwal, S., Srivastava, N.: Designing a Fault-tolerant Fully-chained Combining Switches Multi-stage Interconnection Network with Disjoint Paths. The Journal of Supercomputing 55(3), 400–431 (2009)
14. Nitin, Sehgal, V.K., Bansal, P.K.: On MTTF analysis of a Fault-tolerant Hybrid MINs. WSEAS Transactions on Computer Research 2(2), 130–138 (2007) ISSN: 1991-8755
15. Nitin: Component Level Reliability analysis of Fault-tolerant Hybrid MINs. WSEAS Transactions on Computers 5(9), 1851–1859 (2006) ISSN: 1109-2750
16. Rastogi, R., Nitin, Chauhan, D.S.: 3-Disjoint Paths Fault-tolerant Omega Multistage Interconnection Network with Reachable Sets and Coloring Scheme. In: Proceedings of the 13th IEEE International Conference on Computer Modeling and Simulation (IEEE UKSim), UK (2011)
17. Rastogi, R., Nitin: Fast Interconnections: A case tool for Developing Fault-tolerant Multi-stage Interconnection Networks. International Journal of Advancements in Computing Technology 2(5), 13–24 (2010) ISSN: 2005-8039

18. Fan, C.C., Bruck, J.: Tolerating Multiple Faults in Multistage Interconnection Networks with Minimal Extra Stages. IEEE Transactions on Computers 49(9) (2000)

19. Bhardwaj, V.P., Nitin: A New Fault Tolerant Routing Algorithm for Advance Irregular Augmented Shuffle Exchange Network. In: Proceedings of the 14th IEEE International Conference on Computer Modeling and Simulation (IEEE UKSIM), Emmanuel College, Cambridge, UK (2012)

20. Bhardwaj, V.P., Nitin: A New Fault Tolerant Routing Algorithm for Advance Irregular Alpha Multistage Interconnection Network. AISC, pp. 1867–5662. Springer (2012)

21. Gupta, A., Bansal, P.K.: Fault Tolerant Irregular Modified Alpha Network and Evaluation of Performance Parameters. International Journal of Computer Applications 4(1) (2010)

Multilanguage Block Ciphering Using 3D Array

C. Nelson Kennnedy Babu[1], M. Rajendiran[2], B. Syed Ibrahim[2], and R. Pratheesh[2]

[1] Sree Sowdambiga College of Engineering,
Virudhunagar Tamilnadu, India
cnkbabu63@yahoo.in
[2] Department of ECE, Chettinad College of Engineering & Technology,
Tamilnadu, Karur, India
{mrajendiran,ibzz82,pratheeshr.nair}@gmail.com

Abstract. Different types of coding techniques have been introduced in different languages in order to change the characters and information so that it is difficult for others to understand them. But these coding techniques correspond only to that particular language and they do not support the other languages. Also it was not possible to do Multi language encryption by applying one particular algorithm. In MULET [1] type encryption, the cipher text which is derived by applying the Multilanguage contains not only a shorter multi language characters but also two types of Replacement method are explained in them. Hence there are two shortcomings in this process But the MB3D method, which is now introduced here is free from the above two types of shortcomings and also the cipher text which we get at the end is also a Block cipher, which is a notable feature. This apart, the mapping array used here is made up with the help of Three Dimensional Multi language method and hence it is very difficult for others to understand the information even by attempting the Brute force attack or any similar types of attack. With this MB3D method, it is possible to do the encryption and decryption process for any language in the world which have Unicode.

1 Introduction

The cryptography method of encryption was being introduced for certain languages of the world. The encryption of information for the purpose of encoding, processing and sharing by using the cryptography method [5] was introduced only for a few languages of the world. It is also difficult to write cryptography by using only one algorithm. Though different methods of cryptography were introduced / used by different countries at various periods of time in the past, it is quite interesting to know that in India the cryptography method has been in use since 1600 BC[7]. In India 1.20 Billion people speak in 337 of the 348 languages and the rest 11 languages went out of usage by the passage of time. Around 8 million people speak in Tamil language, which is an oldest South Indian language. Tamil is the Administrative language in Tamil Nadu, Singapore, Malaysia and Sri Lanka. In this language, during the periods of 1600 BC [8], a cryptography method known as 'Porulkoal' (பொருள்கோள் in Tamil) was in existence. In the Ancient and Medieval Tamil grammar books [9] 'Tholkappiam'

J. Mathew et al. (Eds.): ICECCS 2012, CCIS 305, pp. 255–261, 2012.

(தொல்காப்பியம் *in Tamil*)) and 'Nannool' (நன்னூல் *in Tamil*), we can understand the meaning only by applying the 'Porulkoal' system. In Thirukkural, which is an ancient treasure of wisdom and also regarded as a Common Veda for the world by the Tamil people, also we can understand the meaning by using the 'Porulkoal' method. This 'Thirukkural' consists of 1330 couplets and over 100 couplets can be understood with the help of 'Porulkoal' method. 'Porulkoal' explains how the meaning of a poem or a literary work, with a secretly changed words, should be obtained. There are eight types of 'porulkoal'. By using the 'Idamaatru porulkoal'(இட மாற்று பொருள்கோள் *in Tamil*) [4], which is one among the above 8 types, we can find out the correct meaning of a sentence formed with secret changes in the words. If we directly consider the words, it shall give different meaning altogether.

After these periods, Caesar introduced the world famous character level cryptography method. By applying the shifting of words system, a sentence or a word can be secretly changed so that other do not understand the meaning. The Quantum Cryptography method [6], which came into use afterwards, has several limitations. During 20th century period a number of cryptography methods were developed by using ASCII code system. But since the ASCII [2] numbers are made up of 64 to 127, which is shorter in length, one can easily understand them by applying the Brute force method. This method is developed by using only the English word and characters. Hence, no encryption can be made for other languages by applying this method.

The MB3D encryption method, which is introduced here, is made up with Unicode and hence it is totally impossible for others to understand them. Through this method all languages of the world which have Unicode can be processed with encryption. For example, let us assume that each of the 337 Indian languages are having, on an average, 100 letters in them. Now the total number of letters are 337 x 100 = 33700. If we compare this combination with that of the 26 alphabet English encryption, it will be very clearly known that by applying Brute force, one cannot understand the Multilanguage encryption. That is why undoubtedly MB3D method is not only the relevant and exact method of encryption for multi linguistic nature like Indian languages but also for all the languages of the world which have Unicode.

2   Choice of Unicode Standard

In order to store letters and characters in a computer, they have to be assigned with separate numbers. The computer understands processes and shares the information through this encoding function. In yester years so many encoding methods were thus introduced. But each of these encoding methods encompassed only one language. For example, the ASCII code and EDCDIC code which were introduced earlier encompassed only the English language. Other languages cannot be encoded and processed with the ASCII and EDCDIC codes[3]. It was very difficult to represent the characters of all the world languages with the help of a single encoding method. Evidently many types of encoding methods were used to encode the languages of the European Union. The encoding methods that came into existence afterwards were used for languages which are being written from left to right. For languages which are written from Right to left, for example Arabic and Hebrew, there was no solution from the above encoding method. But through a single encoding system known as 'Unicode' method, we can make encoding, processing and sharing of information of all classical languages of the world. Normally it is very difficult to represent a language with 256 characters, like Japanese, with the previous methods of encryption. But with the help of Unicode method it has now become easier for us to encrypt these languages also. We can represent any of the classical languages of the world (i.e. we can encode, process and share the information) through this Unicode method.

The Unicode method [10] is constructed with four important design principles. They are: Universal, Efficient, Uniform and Unambiguous. It also provides unique numbers for every character. That is why there is no problem relating to Language, Platform and Character Length Limit in the Unicode method. The encryption method introduced here is based on the Unicode. Hence, we can easily encrypt any classical language of the world by using the Unicode. So Unicode is/has become a Successful, & Standard Multilingual Character set.

3   Existing Method

In MULET, the plain text is first converted to Unicode. Then it is converted as cipher text by using the Multilanguage character and Multilanguage numerical table. The cipher we get as above is known as stream cipher. The length of the cipher we got as above is too lengthier and also requires two types of Replacement method. The step by step procedure of MULET is clearly explained in Table 1.

Table 1. Characterwise Encryption Using One Dimensional Array Substitution

Plain Text	G	O	D	I	S	G	R	E	A	T
Unicode value (U)	71	79	68	73	83	71	82	69	65	84
Mapping Constant (M)	3	3	3	3	3	3	3	3	3	3
Quotient value(Q)	23	26	22	24	27	23	27	23	21	28
Remainder Value(R)	2	1	2	1	2	2	1	0	1	0
Stream Cipher	ई	आ	ई	आ	ई	ऽ	आ	अ	आ	अ

For the plain text English letters – 'God is great', first we have to write the equivalent Unicode. Assuming that the mapping constant M = 3, then we have to divide all the Unicode value by M. The stream cipher is made up with the resultant Quotient and Remainder value after the Division. By using Table 2, the Multilanguage character equivalent for the Remainder is written. Suppose a Remainder is repeated for one or more time, it is replaced with the help of Table 3.

Table 2. Mapping Array With M=3

Index	0	1	2
Character (Multilanguage)	आ	आ	ई

Table 3. Multilanguage Set

Index	0	1	2	3	4	5	6	7	8	9
Numerical (Multilanguage)	ŭ	ಕ	౨	ſh	♂	౫	ನ	e	ʊ	ε

In Table 1, since the remainder 2 for the letters 'S' and 'G' comes immediately after one another, the first '2' is replaced with the help of Table 2 and the second '2' is replaced with the help of Table 3. On the whole, the stream cipher is finally constructed through two Replacements. The cipher is also too lengthier compared to the cipher which we get through the proposed method. The mapping array in Table 2 is constructed with the help of Hindi language. This apart, Multilanguage numerical is used in Table 3.

Here, it is easier to know the information because the number of Multilanguage characters used as mapping array in Table 2 is very less. When compared with the MB3D method, the MULET method is seen as an unsecure method because of the fact that is requires more than one Replacement method and the lengthier stream cipher.

4   Proposed Method

The MB3D encryption method is clearly explained in Table 4. According to this method, the plain text – "God is great" is changed as Unicode. Let us assume the same mapping constant (i.e. M = 3). But, here the mapping constant array is created of a Three Dimensional value. The Unicode value is divided with M = 3 and then we get the Remainder and the Quotient value. The Remainder value is then grouped as x y z (3 x 3 groups). Then, for each group a Three Dimensional multi language character from Table 5 is written. The cipher we get here is known as Block cipher. The Three Dimensional mapping array in Table 5 is created with the help of Multi language characters. The x here denotes the Table number. y and z denote the Column and Row.

Table 4. MB3D Encryption

Plain Text in English	G	O	D	I	S	G	R	E	A	T		
Multilanguage Characters Unicode	71	79	68	73	83	71	82	69	65	84		
Three Dimensional Index value = M	3	3	3	3	3	3	3	3	3	3		
Quotient =Q=U/M	23	26	22	24	27	23	27	23	21	28		
Reminder =R=U mod M	2	1	2	1	2	2	1	0	1	0	0	0
X,Y,Z		2,1,2			1,2,2			1,0,1			0,0,0	
Cipher Text		ಲ			ౠ			ౘ			౮	

For example the mapping array constant for Hindi, Tamil and Telugu languages are given in Mapping Table 0, Mapping Table 1 and Mapping Table 3 respectively. Likewise all the languages of the world for which Unicode is available can be created with the mapping array constants. Since the last letter 'T' in Table 4 is short/ less by two pairs, two zero padding are created for it. That is why there shall be no difficulty in the process of encryption and decryption. Table 6 clearly explains the MB3D decryption process.

Table 5. Three Dimensional Mapping Array with M=3

Row/Column	Mapping Table 0				Mapping Table 1				Mapping Table 2			X
	0	1	2		0	1	2		0	1	2	Z
0	ए	ओ	ॐ	0	ஏ	ஊ	ஊ	0	ಖ	ಜ	ನ	Y
1	व	ड	ल	1	ஒ	ஈ	ஸ	1	ಎ	ಘ	ಒ	
2	र	अ	आ	2	ஜ	அ	ஆ	2	ಠ	ಆ	ಊ	

Table 6. MB3D Encryption

Cipher Text	ಒ			ಅ			ஊ			ए		
X,Y,Z	2,1,2			1,2,2			1,0,1			0,0,0		
Reminder=R	2	1	2	1	2	2	1	0	1	0	0	0
Quotient=Q	23	26	22	24	27	23	27	23	21	28	28	28
Three dimensional index value=M	3	3	3	3	3	3	3	3	3	3	3	3
U=M*Q + R	71	79	68	73	83	71	82	69	65	84	84	84
Plain text	G	O	D	I	S	G	R	E	A	T		

The MULET process gives the cipher which is lengthier when compared to the Block cipher derived from the MB3D process. Apart from this, the other notable feature is that one cannot understand the information even by attempting the Brute force method because the mapping array used in the MB3D process is of Three Dimensional multi language character.

4.1   Proposed Encryption Algorithm

Input : *Plain text*
Output: *Block Cipher*

Step	Description
1.	*Getting the plain text character by character.*
2.	*Getting Unicode for character (U).*
3.	*Getting the Mapping Constant (M).*
4.	*Calculating Quotient for each character $Q = U/M$.*
5.	*Getting Remainder value for each character $R=U\%M$.*
6.	*Split the 'R' value into three groups group*
7.	*In each group the first value states Table number (X), the second value states row (Y) a and the third value states column (Z). If there is no pair for the last character, pad zero a at the end*
8.	*By using x y z and referring to the mapping array table, the Block cipher is arrived at.*
9.	*Repeat the step 7 and Step 8 till the process is completed.*
	Comment: *The computation is completed.*

5   Performance Analysis

A detailed comparison between the MULET and MB3D process methods are well explained with the help of the graph given above. It shows how the cipher in the MB3D method is shorter and how the cipher in the MULET method is lengthier. The Quotient value repeated in the MULET method is re-arranged with the help of Table 4. But there is no need for such types of replacements or re-arrangements in the MB3D method. Since the number of mapping array constants are higher in MB3D method when compared to the MULET method, the MB3D method is proved to be more secure.

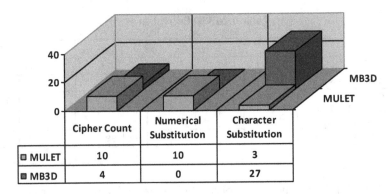

	Cipher Count	Numerical Substitution	Character Substitution
MULET	10	10	3
MB3D	4	0	27

Fig. 1. MULET Vs MB3D

6   Conclusion

The MB3D encryption method is created with the help of the Unicode and hence is very easier for encrypting the world languages which have Unicode. That is why this method can be called as Universal Cryptographic Algorithm. This method also ensures better security, since the encryption is made by applying the Three Dimensional mapping array. Hence it is a very difficult task for others to understand the information. In order to decrypt the Block cipher with due process, the 'Q' value becomes an important factor. Hence, by using the advanced stenography method, the information can be safely passed on to others.

In future, it shall become very easier to convert the cipher text in the MB3D process to Binary code.

References

1. Praveen Kumar, G., Arjun Kumar, M.: B. Parajuli, P. Choudhury.: MULET: A Multi-language Encryption Technique. In: IEEE Seventh International Conference on Information Technology, pp. 779–782 (2010)
2. Shankar, T.N.: Sahoo, G.: Cryptography by Karatsuba Multiplier with ASCII Codes. International Journal on Computer Applications, 53–60 (2010)

3. Sweetman, W.: The Bibliotheca Malabarica- An 18th century library of Tamil Literature. In: World Classical Tamil Conference (WCTC), Part 10, Coimbatore (2010)
4. Shanmugam, S.V.: Tholkappiyar concept of Semantics. In: Published in World Classical Tamil Conference (WCTC), Part 9, Coimbatore (2010)
5. Paar, C., Pelzl, J., Preneel, B.: Understanding Cryptography.: A Text book for student and Practitioners. Springer (2010)
6. Sakthi Vignesh, R., Sudharssun, S., Jegadish Kumar, K.J.: Limitations of Quantum & The Versatility of Classical Cryptography:Comparative Study. In: Second International Conference on Environmental and Computer Science, pp. 333–337 (2009)
7. Farmer, S., Sproat, R., Witzel, M.: The collapse of the Indus-script thesis: The myth of a literate Harappan civilization. Electronic Journal of Vedic Studies 11, 19 (2004)
8. Krishnamurti, B.: The Dravidian languages. Cambridge University Press (2003)
9. Parpola, A.: Syntactic methods in the study of Indus script. In: Proceedings of the Nordic South Asia Conference, Helsinki, pp. 125–136 (1980)
10. Unicode Character form, http://www.unicode.org

A Novel Invisible Watermarking Based on Cascaded PVD Integrated LSB Technique

J.K. Mandal and Debashis Das

Department of Computer Science and Engineering,
University of Kalyani, Kalyani,
West Bengal, India
{jkm.cse,debashisitnsec}@gmail.com

Abstract. To enhance the payload of secret data a scheme on the basis of least significant bit (LSB) substitution cascaded with pixel value differencing (PVD) has been proposed in this paper. Firstly secret bits are embedded using PVD method where smooth areas obtain small number of bits and higher bits in the edged area. Then bits are also embedded in the same pixel block by applying LSB replacement method as a cascading manner. The technique ensures more hiding capacity along with security as two methods are applied as cascaded manner on the same pixel block. Experimental result shows that proposed method can hide nearly double amount of secret data, compared to PVD method with good visual quality of stego-images.

Keywords: Steganography, Hiding capacity, Pixel-value differencing, Stego-image.

1   Introduction

Information security has become very much necessary for secret data transmission in the age of digital communication via Internet. To protect secret messages generally two popular schemes are used. One is called message encryption, where original text is changed to another unreadable form. Here secret key(s) are needed for changing it from one to another form. DES, RSA, AES are some popular encryption scheme. Another way of message protection is steganography, where secret message is hidden into some cover media like- text, image, audio, video etc. If a digital image is used as a cover media, is called image steganography and the image with hidden data is called stego-image. Steganographic techniques are mainly used in military, commercials, anti-criminal and various applications.

There are several Steganographic methods used for data hiding. The most common and simplest method is least-significant bits (LSB) substitution where LSB bits of cover image pixels are replaced by secret data bits. Hiding capacity and stego-image quality are the benchmarks needed for evaluation of hiding performance. Therefore Wang et al proposed a LSB substitution method to improve stego-image quality[7]. Furthermore, Cheng et al. proposed an efficient method using dynamic programming

J. Mathew et al. (Eds.): ICECCS 2012, CCIS 305, pp. 262–268, 2012.

strategy[4].Chan and Cheng proposed a method of data hiding using LSB replacement with an optimal pixel adjustment process (OPAP)[5]. Fu and Au proposed a technique which can embed large amount of data with good visual quality of stego-image[6]. An adaptive method of LSB substitution using variable amount of bits is proposed for increasing hiding capacity[8]. Wu and Tsai proposed a technique of data hiding by using the difference of two consecutive pixels of cover image, called pixel value differencing (PVD) method, can provide high embedding capacity and outstanding stego-image quality[1]. Therefore, based on PVD method various ideas have been proposed[2,3].

In this paper, a steganographic approach by using both pixel-value differencing and LSB replacement is proposed. Proposed method will provide double the hiding capacity as given by Wu-Tsai's PVD method and also maintain acceptable quality of stego-image.

The rest of this paper is as follows. Section 2 describes review of Wu-Tsai method. Proposed scheme is given in section 3. Experimental results are illustrated in section 4 and section 5 contains the discussion and analysis of the result. Conclusions are drawn in section 6 and references are given at end.

2   Review of Wu and Tsai's Method

In Wu-Tsai's PVD method[1], a gray scale image is used as a cover image which is partitioned into non-overlapping blocks of two consecutive pixels, p_i and p_{i+1}. The difference value d_i is calculated by subtracting p_i from p_{i+1} from each block. The set of all difference values can be ranged from -255 to 255. Therefore, $|d_i|$ ranges from 0 to 255. The blocks with small difference values considered as smooth area and can hide less bits where block with large difference values are the sharp edged area, can embed more bits. As because human eyes can tolerate more changes in sharp-edged area than smooth areas. Therefore, in PVD method a range table has been designed with n contiguous ranges R_k (where k=1,2,...,n) where the range is 0 to 255. The lower and the upper bound are denoted as l_k and u_k respectively, then $R_k \in [l_k, u_k]$. The width of R_k is calculated as $W_k = u_k - l_k + 1$. W_k decides how many bits can be embedded in a pixel block. R_k is kept as a variable for security purpose; as a result, original range table is required to extract the hidden information. The embedding algorithm is given as follows.

2.1   Algorithm

1. Calculate the difference value d_i of two consecutive pixels p_i and p_{i+1} for each block in the cover image. This is determined by $d_i = | p_{i+1} - p_i |$.
2. Compute the optimal range where the difference d_i lies in the range table. This is calculated as $R_i = \min(u_k - d_i)$, where $u_k \geq d_i$ for all $1 \leq k \leq n$.
3. Compute the number of bits 't' can be hidden in a pixel block, defined as $t = \lfloor \log_2 w_i \rfloor$. where w_i is the width of the specified range for the calculated d_i
4. Read 't' bits from binary secret data and convert it into its corresponding decimal value b. For instance if t=110, then b=6

5. Calculate the new difference value d_i' which is given by $d_i'=l_i +b$
6. Modify the values of p_i and p_{i+1} by the given formula:

$$(p_i', p'_{i+1}) = (p_i+ \lceil m/2 \rceil, p_{i+1}- \lfloor m/2 \rfloor), \text{ if } p_i \geq p_{i+1} \text{ and } d_i'>d_i$$

$$(p_i- \lfloor m/2 \rfloor, p_{i+1}+ \lceil m/2 \rceil), \text{ if } p_i<p_{i+1} \text{ and } d_i'>d_i$$

$$(p_i- \lceil m/2 \rceil), p_{i+1}+ \lfloor m/2 \rfloor), \text{ if } p_i \geq p_{i+1} \text{ and } d'_i \leq d_i$$

$$(p_i+ \lceil m/2 \rceil, p_{i+1}- \lfloor m/2 \rfloor), \text{ if } p_i<p_{i+1} \text{ and } d_i' \leq d_i$$

where $m=|d_i'-d_i|$. Now we obtain the pixel pair (p_i',p_{i+1}') after embedding the secret data into pixel pair (p_i,p_{i+1}). Repeat step 1-6 until all secret data are embedded into the cover image. Hence we get the stego-image.

When extracting the hidden information, original range table is required. For extraction, partition the stego-image into several pixel blocks of two consecutive non-overlapping pixels. Calculate the difference value for each block as $d_i'=| p_i'-p_{i+1}' |$. Then find the optimum range R_i for the d_i'. Then b' is obtained by subtracting l_i from d_i'. Convert b' into its corresponding binary of 't' bits, where $t= \lfloor \log_2 w_i \rfloor$. These bits are the hidden secret data obtained from the pixel block (p_i',p_{i+1}').

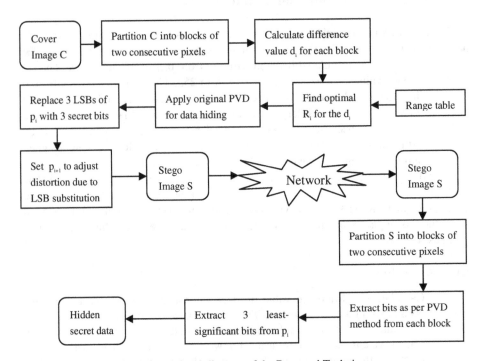

Fig. 1. Block level diagram of the Proposed Technique

3   The Proposed Method

In proposed method we have applied PVD and LSB replacement method in cascaded manner on each pixel block of the cover image. First we apply PVD method into each block and then 3 least significant bits (LSBs) of the first pixel is replaced by 3 secret bits. Now, distortion made in the first pixel for 3 bit replacement has been adjusted and reduced by setting the second pixel of the block in order to keep the difference, of two pixels, same as was after embedding data by PVD method. The embedding algorithm is given in 3.1. Figure 1 shows the embedding and extraction procedure.

3.1   Embedding Algorithm

1. Partition the cover image into separate pixel blocks which consists of two consecutive non-overlapping pixels and find difference values for each of them as $d_i = |p_i - p_{i+1}|$.
2. Find optimal range R_i for the d_i such that $R_i = \min(u_i - d_i)$, where $u_i \geq d_i$. Then $R_i \in [l_i, u_i]$ is the optimum range where the difference lies.
3. Compute the amount of secret data bits t from the width w_i of the optimum range, can be defined as $t = \lfloor \log_2 w_i \rfloor$.
4. Read t bits and convert it into a decimal value b. Then calculate the new difference value by the formula $d_i' = l_i + b$.
5. Now, calculate the pixel values after embedding t bits (p_i', p_{i+1}') by original PVD method.
6. Read 3 secret bits from secret bit-stream. Convert first pixel p_i into its equivalent binary form. Now replace 3 LSB positions of p_i with the 3 secret data bits. Suppose p_i is 20 (after applying PVD) whose binary form is 00010100 and secret bits are 110. Then after replacement, p_i becomes 00010110 which is 22.
7. Adjust the distortion, made from 3-bit replacement in p_i, by setting p_{i+1}. For instance, p_{i+1} is 50 after applying PVD. So, to keep the difference same, p_{i+1} has to be changed to 52 as p_i has increased by 2.

Repeat step 1-7 until all secret data are hidden into the cover image and hence obtain the stego-image.

3.2   Extraction Algorithm

The required steps for extracting the hidden secret data are as follows:

1. Partition the stego-image into pixel blocks and calculate the difference $d_i = |p_i - p_{i+1}|$.
2. Find optimal range from original range table and calculate the number of bits to extract as $t = \lfloor \log_2 w_i \rfloor$, where w_i is range width.
3. Find new difference value $d_i' = d_i + l_i$ and convert d_i' in t bits to get the secret hidden data.
4. Convert p_i in binary and extract 3 least significant bits. This 3 bits also the hidden secret bits.
5. Merge the bit-stream obtained from step 3 and 4 to get the total secret bits embedded in the pixel block.

Repeat step 1-5 until whole secret information is obtained.

4   Experimental Results

To illustrate the performance of the proposed method in terms of hiding capacity and stego-image quality, we have used C programming language for implementing our method and Wu-Tsai's PVD method. The standard digital images of size 512× 512 have been used as cover image and large bit-stream has been used as secret data. The range table widths are {8 , 8 , 16 , 32 , 64 , 128} used. Peak signal to noise ratio (PSNR) is calculated to evaluate the stego-image quality. The experimental results are depicted in figure 2 and figure 3. Finally the comparison in capacity and PSNR value between proposed technique and Wu-Tsai method is given as table 1. Each value given in the comparison table is the average value by executing 100 times where the secret bit-stream is different in each case.

(a)   (b)

Fig. 2. a) Cover image Lena b) Stego-image on embedding 100518 bytes, PSNR 36.34

(a)   (b)

Fig. 3. a) Cover image Elaine b) Stego-image on embedding 100219 bytes, PSNR 36.66

Table 1. Comparison of results of Wu-Tsai and proposed method

Cover image (512× 512)	Wu and Tsai's method		Our method	
	Capacity (Bytes)	PSNR (dB)	Capacity (Bytes)	PSNR (dB)
Lena	51370	41.07	100518	36.34
Baboon	57583	36.86	106732	34.43
Tank	50495	42.54	99644	36.80
Airplane	49735	42.31	98884	36.73
Truck	50061	43.05	99210	36.91
Elaine	51070	42.09	100219	36.66
Couple	51600	40.33	100749	36.12
Boat	52631	39.06	101780	35.54
Jet	51020	41.31	100168	36.42
Pepper	51107	40.55	100256	36.15

5   Analysis and Discussions

Table 1 shows that proposed method provide more hiding capacity of 1.5 bpB than Wu-Tsai's PVD method; however PSNR value reduced between 2.43 ~ 6.14 dB. We embed 3 extra bits in every pixel block, after embedding according to original PVD, which ensures high capacity in stego-image. On the other hand, pixel distortion can be increased by 7(maximum) due to LSB replacement up to 3 bits. That is why PSNR value of the stego-image decreased from original PVD.

6   Conclusions

The proposed steganographic technique using both PVD and LSB substitution method consecutively can hide more secret data into cover image than PVD method used alone. Experimental results show that the stego image quality is also imperceptible to human eyes. Furthermore, this approach has got superior hiding capacity than Wu-Tsai's method can be shown from experimental results. Also the extraction of hidden secret data can be done efficiently from the stego image.

References

1. Wu, D.C., Tsai, W.H.: A steganographic method for images by pixel-value differencing. Pattern Recognition Letters 24, 1613–1626 (2003)
2. Wu, H.C., Wu, N.I., Tsai, C.S., Hwang, M.S.: Image Steganographic scheme based on pixel-value differencing and LSB replacement method. IEEE Proceedings on Vision, Image and Signal Processing 152(5), 611–615 (2005)
3. Li, S.L., Leung, K.C., Cheng, L.M., Chan, C.K.: Data Hiding in Images by Adaptive LSB Substitution Based on Pixel-value Differencing. In: First International Conference on Innovative Computing, Information and Control (ICICIC 2006), vol. 3, pp. 58–61 (2006)

4. Chang, C.C., Hsiao, J.Y., Chan, C.S.: Finding Optimal least significant bit substitution in image hiding by dynamic programming strategy. Pattern Recognit. 36(7), 1583–1595 (2003)
5. Chan, C.K., Cheng, L.M.: Hiding data in images by simple LSB substitution. Pattern Recognition 37(3), 469–474 (2004)
6. Fu, M.S., Au, O.C.: 'Data hiding watermarking for halftone images'. IEEE Trans. Image Process. 147(3), 477–484 (2000)
7. Wang, R.Z., Lin, C.F., Lin, J.C.: Image hiding by optimal LSB substitution and genetic algorithm. Pattern Recognit. 34(3), 671–683 (2001)
8. Lie, W.N., Chang, L.C.: Data hiding in images with adaptive numbers of least significant bits based on the human visual system. In: IEEE International Conference on Image Processing, vol. 1, pp. 286–290 (1999)
9. Petitcolas, F.A.P., Anderson, R.J., Kuhn, M.G.: Information Hiding – a Survey. Proceedings of the IEEE 87, 1062–1078 (1999)
10. Bender, W., Gruhl, D., Morimoto, N., Lu, A.: Techniques for data hiding. IBM Systems Journal 35(3-4), 313–336 (1996)

Fault Tolerant High Performance Galois Field Arithmetic Processor

Vinu K. Narayanan, Rishad A. Shafik, Jimson Mathew, and Dhiraj K. Pradhan

Department of Computer Science, University of Bristol, Bristol, UK
jimson@cs.bris.ac.uk, csras@bris.ac.uk

Abstract. Reliability is an emerging design requirement for finite field processors used in cryptographic systems. However, reliable design of these systems is particularly challenging due to conflicting design requirements, including high performance and low power consumption. In this paper, we propose a novel design technique for reliable and low power Galois field (GF) arithmetic processor. The aim is to tolerate faults in the GF processor during on-line computation at reduced system costs, while maintaining high performance. The reduction in system costs is achieved through multiple parity prediction and comparison considering the trade-offs between performance and complexity. The effectiveness of the proposed technique is then validated using a case study of 163-bit digit serial multipliers using $90nm$ and $180nm$ technology nodes highlighting the resulting area, latency and power overheads. We show that up to 40 stuck-at faults can be tolerated during computation with reasonable system area and power costs.

1   Introduction

Finite field arithmetic is a coveted area of research due to its extensive usage in cryptographic applications. Using such arithmetic, complex cryptographic operations such as inversions and exponential operations can be efficiently performed using repeated computation over a finite field [1]. However, the implementation of such repeated or parallelized computations increase the underlying hardware complexities exponentially, often requiring integration of millions of gates. Hence, effective design technique of these systems is a prime design objective to ensure reduced overall system costs [2].

An emerging design objective of cryptographic systems is reliability against various types of faults: both permanent and transient faults [6]. Permanent faults manifest themselves in the form of defects due to design aberrations during manufacturing or due to wearout faults caused by dielectric or metal gate breakdown. On the other hand, transient errors are caused mainly by electromagnetic radiations, which generate large number of electron-hole pairs in transistor junctions to cause logic upset [16]. The importance of reliable cryptographic systems design in the presence of these faults has been studied extensively highlighting the various fault detection and correction techniques [11, 17]. Among these techniques, error detection

J. Mathew et al. (Eds.): ICECCS 2012, CCIS 305, pp. 269–281, 2012.

and correction coding schemes have been proposed by many researchers over the years. Parity prediction has been reported as one of the most efficient techniques for detecting single and multiple stuck-at faults [7, 13]. Using such technique in advanced encryption standards (AES) based cryptographic systems parity predicted for each byte is compared to the actual parity and verified at various stages of operation [12]. An effective parity prediction based detection and correction technique with high fault coverage is presented in [8]. The fault detection and correction in [8] is achieved at low hardware overheads as these are independent of the order of Galois field polynomial. A similar parity error detection and correction technique through scaling of the multiplier inputs and outputs by low-density parity check codes (LDPCs) is proposed by [9]. The use of such codes has been shown to be effective as high fault tolerance can be achieved with high performance for Galois processors. This is because LDPCs facilitate parallel processing at reasonable area overheads [10]. Furthermore, compared to traditional Hamming code based techniques for error correction decoding complexity of LDPCs is much lower. Underpinning the advantages of LDPCs, a multiple parity prediction circuit is proposed in [9] to detect and correct faults in GF processors. The complexity of the circuit in [9] is dependent only on the number of bits in the input and the number of parity bits to be predicted.

With continued technology scaling, the need for low-cost multiple fault detection and correction is becoming more obvious. However, currently proposed techniques, such as [7, 8, 9, 13, 19], have limited multiple fault correction abilities. Also, the overheads associated with such multiple fault correction can be significantly large, as reported by [13, 19]. Hence, more investigation is required into low-cost and low-complexity multiple fault detection and correction techniques, which is the aim of this paper. In this paper, a low-cost and fault tolerant design technique of Galois Field arithmetic processor is proposed. The fault tolerance using this technique is achieved through multiple parity prediction and correction based on low-density parity codes (LDPCs). The proposed multiple parity prediction technique features high fault coverage at low complexity. The effectiveness of the proposed design technique is validated using a case study of 163-bit digit serial multipliers. We show that the proposed design technique can achieve significant fault tolerance at low area and power overheads.

The rest of the paper is organized as follows. Section 2 outlines the preliminaries of Galois Field arithmetic, highlighting the need for fault tolerant implementation. Section 3 presents the proposed fault tolerant design technique, while Sections 4 shows the experimental results using 163-bit digit serial multipler as a case study. Finally, Section 5 concludes the paper.

2   Preliminaries

In this section, the Galois field arithmetic is briefly outlined highlighting the fault model used.

2.1   Arithmetic Operations over Fields

A field containing finite number of elements in it is called a finite field. It is also known as Galois field named after *Évariste Galois* [4]. The field, $GF(2^m)$ contains 2^m elements, where m is a prime number. Arithmetic operations like addition, subtraction and multiplication can be performed over this field. Further operations like inversion, exponentiation and division can be performed by repeated multiplication. In the following, addition and multiplication operations are briefly outlined.

A. Addition over Galois Field

Addition is the most basic operation in GF processor. However, it is also an integral part of many other complex operations, such as multiplication, inversion and exponentiation [3, 14]. In finite field arithmetic *addition* is implemented by *XOR* gates and *D* flip flops. Let the finite field polynomial $GF(2^m)$ be defined as $F(x) = x^m + f_{m-1}.x^{m-1} + ... + f_2.x^2 + f_1.x + 1$, where the co-efficient $f_i \in GF(2)$. The finite field generates a polynomial basis (PB) given by $1, \alpha, \alpha^2, \alpha^3 ... \alpha^{m-1}$, where α is a root of the equation $F(x)$. Now, let A and B be any two elements in $GF(2^m)$, which can be represented by the polynomial basis in the form of polynomials of degree $m-1$. Addition of A and B is given by Algorithm 1 as shown below.

Algorithm 1. Addition of $A, B \in GF(2^m)$
Require: $A = \sum_i a_i x^i$; *where i=0 to m-1 and $a_i \in GF(2) = \{0, 1\}$ and*
        $B = \sum_i b_i x^i$; *where i=0 to m-1 and $b_i \in GF(2)=\{0, 1\}$*
Ensure: $C \equiv A + B \bmod F(x) = \sum_i c_i x^i$; *where i=0 to m-1 and $c_i \in GF(2)=\{0, 1\}$*
    1. $C \leftarrow 0$
    2.   **for** $i = 0$ to $m-1$ do $\{ c_i = a_i \oplus b_i \}$ **end for**
    3.   return C

As can be seen in Algorithm 1, A and B are inputs, which belong to the finite field $GF(2^m)$. The output of addition is stored in C (step 2), which is essentially the bitwise XOR of A and B.

B. Multiplication over Galois Field

Multiplication is one of the important operations in GF processors. To understand GF implementation of multiplication, let $F(x) = x^m + f_{m-1}.x^{m-1} + ... + f_2.x^2 + f_1.x + 1$ be the degree m, which is an irreducible polynomial that defines the field $GF(2^m)$ and $f_i \in GF(2)$ for all the values of i from 0 to $m - 1$. If α is the root of the polynomial then we know that the polynomial basis is given by $1, \alpha, \alpha^2, \alpha^3, ... , \alpha^{m-1}$. Multiplication of any two elements A and B in $GF(2^m)$ is given by, $C \equiv A.B \bmod f(\alpha)$. This operation takes place in two steps, *viz.* (1) polynomial multiplication and (2) reduction modulo using the irreducible polynomial. In the following, these steps are briefly described.

Step 1: *Polynomial Multiplication:*
The product is a polynomial D of degree not less than $2m-1$ obtained by direct multiplication of the polynomial basis representations of the elements. [2, 15] i.e. $D = \sum_i (a_i \alpha^i) \sum_j (b_j \alpha^j) = \sum_k (d_k \alpha^k)$; where $i, j, k = 0$ to $m-1$. This can be obtained by a datapath containing a series of bitwise AND gates followed by bitwise XOR gates [2].

Step 2: Reduction modulo $F(x)$:

The aim of this step is to reduce the product polynomial of degree $2m-1$ using the monic polynomial of degree m. The aim is to reduce the degree of the final product to less than or equal to m. Such reduction modulo can be viewed as a linear mapping of all the $2m-1$ coordinates in D into m coordinates of C, where D is the product of the polynomial and C is the output of the multiplier [2]. This mapping can be represented by the following equation:

$$r_{i,j} = \begin{cases} f_j & ; j = 0 : m-1, i = 0 \\ r_{j-1,i-1} r_{m-1,i-1} r_{j,0} & ; j = 0 : m-1, i = 1 : m-2 \end{cases}$$

and $r_{j-1,i-1} = 0$, if j = 0 [2]. Algorithm 2 shows the classic multiplication operation in GF.

Algorithm 2. Classic Multiplication of A, B $\in$ GF(2^m)

Require: $A = \sum_i a_i x^i$; where $i=0$ to $m-1$ and $a_i \in GF(2) = \{0, 1\}$ and
         $B = \sum_i b_i x^i$; where $i=0$ to $m-1$ and $b_i \in GF(2) = \{0, 1\}$
Ensure: $C \equiv A . B \bmod f(\alpha)$
  1.   D = multiplication (A, B) and R = reduction (f)
  2.   for $j = 0$ to $m-1$ do   { $c(j) = d(j)$ }   end for
  3.   for $j = 0$ to $m-1$ do
    a.   **for** $i = 0$ to $m-2$ do
    b. $(j) = c(j) \oplus (R_{(j, i)} . d_{(m+i)})$
    c.   **end for**
  4.   **end for** and return C

As can be seen, the reduction operation is carried out on the output of the bitwise multiplication (steps 3), which is then stored in the final output C and returned (step 4).

2.2   Galois Field Processor

Galois field processors using arithmetic discussed in Section 2.1 can be designed using numerous techniques, such as [1, 2, 4, 8, 13, 18]. These techniques consider various design trade-offs in terms of performance, power, reliability and area. Due to its design complexity, achieving optimized design trade-off is extremely challenging in GF processors. In this work GF processor is demonstrated using a case study of a 163-bit digit serial multiplier. Such multiplier offers with an excellent trade-off between efficiency and complexity [4, 5] and hence is used in this work (see Section 4).

To illustrate the design considerations and complexity GF multipliers without any fault tolerance capability, Figure 1 shows a 4-bit digital serial single accumulator multiplier. As can be seen, the inputs A ($a_3 a_2 a_1 a_0$) and B($b_3 b_2 b_1 b_0$) are multiplexed serially and bitwise ANDed together. The outputs of bitwise AND is then accumulated into the accumulator unit ($ACC_4 ACC_3 ACC_2 ACC_1 ACC_0$) (Figure 1).

Upon accumulation, the final outcome C ($c_3 c_2 c_1 c_0$) is produced after modulo based reduction according to Algorithm 2 as

$$c[3] = ACC[0] \oplus ACC[4]$$
$$c[2] = ACC[1] \oplus ACC[4] \qquad \cdot$$
$$c[1] = ACC[2] \tag{1}$$
$$c[0] = ACC[3]$$

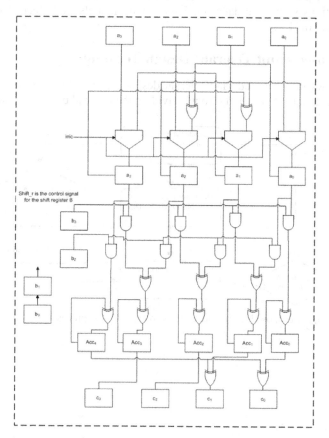

Fig. 1. A single accumulator 4-bit digit serial multiplier (with m = 4, D = 2 $F(x) = x^4 + x + 1$)

From Figure 1, it can be verified that for given m, k, and D values the number of AND gates, XOR gates and flip-flops (N_{AND}, N_{XOR} and N_{FF}) in GF multiplier can be given by:

$$N_{AND} = (m+k+1)D + (k+1)(D-1)$$
$$N_{XOR} = (m+k)D + (k+1)(D-1) \tag{2}$$
$$N_{FF} = 2(m+k) + (D-1) \qquad .$$

The resulting circuit delay will be a function of these gates, i.e. [delay (N_{AND}) + delay($\lceil \log_2(D) \rceil \Delta N_{XOR}$)]. Section 4 gives further insights into system area and latency.

2.3   Fault Model

In this work, we consider stuck-at-1 and stuck-at-0 fault models for permanent faults. For transient faults, similar model is used with an assumption of sufficient duration of fault occurrence. The faults are injected by perturbing the datapath nets in the hardware description of the circuit under test. Multiple stuck-at faults are injected in random locations within the datapath. Similar fault models and injection techniques have been used in [13].

3   Proposed Fault Tolerant Design Technique

Figure 2 shows the proposed fault tolerant design technique for GF processors. As shown, the technique is initiated through dividing the original design into manageable number of units.

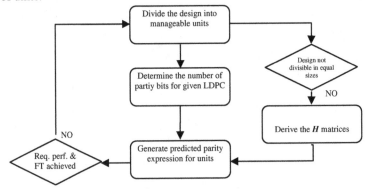

Fig. 2. Proposed technique for fault tolerant Galois processor design

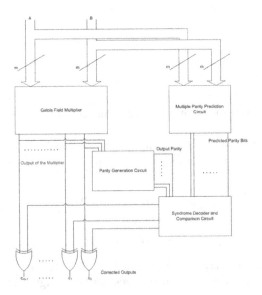

Fig. 3. Fault detection and correction architecture with LDPC codes proposed in this work

The term manageable refers to a number that gives required design trade-off between complexity and performance of the GF processor. Then, for each divided unit, the number of parity bits is determined for the given low-density parity codes (LDPCs). If the original design size cannot be divided into units of equal sizes, the corresponding H matrices from LDPC are derived. With the given H matrices for each unit, predicted parity codes are generated. The predicted parity codes are calculated using the parity derived from hamming matrices from the input data. The whole process of dividing the design and generating predicted parity is incrementally continued until the required performance and fault tolerance is achieved. Note that with more number of units, the number of multiple parity bits will increase but the overall system costs will also significantly increase.

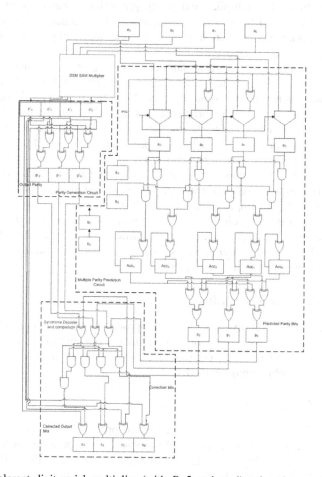

Fig. 4. Fault tolerant digit serial multiplier (with $D=2$ and $m=4$) using the proposed design technique

As an example demonstration of the how multiple parity codes are derived from an original design, consider a 163-bit digit serial multiplier, i.e. *m=163*. The number of parity bits required for *m = 163* is 8, since $2^8 > 163$. This indicates that only 8 bits are required to detect and correct single-bit faults 163-bit data. However, to provide with multiple stuck-at fault detection and correction, the multiplier is divided, for example, into fourteen 11-bit groups and one 9-bit group (14×11+9=163). Hence, since each group will require 4-bit parity bits (since $2^4 > 11$ or 9), a total of 60 parity bits will be required. These fifteen units provide with 15 stuck-at faults to be detected and corrected at once.

With the predicted LDPCs in Figure 2, these are then compared with the actual parity in the syndrome generator and comparator module. The parity prediction errors are identified and then corrected at the correction module as shown in Figure 3. As can be seen, both GF multiplier and parity prediction circuits operate in parallel, which is then followed by syndrome generator and correction. As a result of such parallel comparison and correction, highly efficient and multiple error detection and correction ability can be achieved.

With the given fault detection and correction using syndrome detector and comparator (Figure 3), required fault tolerance ability can be achieved. Figure 4 shows the implementation of a 4-bit digit serial multiplier (with *D=2* and *m=4*) as an example demonstration of the proposed design technique. As can be seen, the multiplier is followed by a parity generation circuit, while the actual serial inputs are followed by parity prediction circuits. The predicted parity and the generated parity are then both fed into the syndrome decoder and parity correction circuit. Clearly, the addition of parity prediction, generation and comparison adds to the system overheads. However, these overheads are far lower and less complex than that of a digit parallel multiplier implementation, which would require 2^m different datapaths and their corresponding parity prediction, generation and comparison units. In the following section, the experimental results carried out using a case study of fault tolerant 163-bit digit serial multiplier is presented highlighting the system overheads compared to non-fault tolerant design (Section 2.1).

4   Experimental Results

In this section, the effectiveness of the proposed design technique (Section 3) is demonstrated using a case study of 163-bit digit serial multiplier. In the following section, the area and power overheads and the worst case delay introduced by the proposed design technique is detailed.

4.1   Area Comparisons

To estimate the area overheads introduced by the proposed design technique (Section 3), first the number of AND gates, XOR gates and flip-flops of the non fault tolerant 163-bit digit serial multiplier design is determined systematically in the design (as discussed in Sections 2 and 3). Table 1 shows the number of different gates required for different digit size multipliers without fault tolerance capabilities using *m=163* with varied *D* values (Section 2.1).

Table 1. Number of gates for various digit size implementations

Digit Size, D	XOR gates	AND gates	Flip-Flops
2	330	332	329
4	662	666	331
6	994	1000	333
8	1326	1334	335

As expected, with increasing number of digit size, the number of gates increases substantially. For a given D and m, the required number of gates and fli-flops are found by (2).

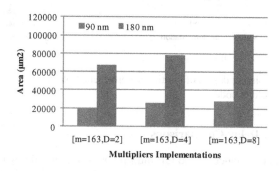

Fig. 5. Comparison of area of the multiplier implementations for various digit sizes

The 163-bit digit serial multiplier was then extended over the field GF (2^m) with various digit sizes. Figure 5 shows the resulting synthesized area results (obtained using Synopsys® design compiler) for both 90nm and 180nm technology nodes. As expected, with increased complexity of the GF multiplier, the area increases substantially.

To show comparative area overheads for introducing fault tolerance in the GF multiplier, Figure 6(a) and (b) show the synthesized area results of two163-bit digit serial multiplier implementations (90nm and 180nm technology nodes) using the proposed design technique (Section 3). The area results also highlighting the following component areas of the multipliers: digit serial multiplier (DSM), multiple parity prediction circuit (MPCC), parity generator (PG) and syndrome decoder and output generator (SDOG). From Figure 7, two main observations can be made. Firstly, it is evident that with increased parallelization (i.e. increasing D), the resulting multiplier areas increase for both 90nm and 180nm implementations. Secondly, in both implementations, parity prediction (MPCC) and multiplier units take up most of the area of the circuits. Comparing Figures 6 and 7, it can be seen that the proposed fault tolerance technique introduces an area overhead of upto 180% for multiplier implementation of $m=163$ and $D=8$. However, as expected, with low parallelization (i.e. lower D) the area overhead also decreases. For example for 90nm technology node multiplier implementation of $D=4$ shows a 145% area overhead compared to the non fault tolerant design (Section 2.1).

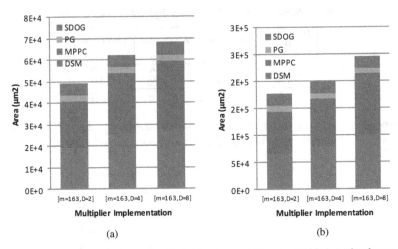

Fig. 6. Area of fault tolerant design of multiplier implementations in (a) 90nm technology node, and (b) 180nm technology node

4.2   Worst Case Delay Comparisons

To estimate the performance of the GF multiplier, worst case delay (and slack times) of the synthesized system is considered. Figure 7 shows the worst case slack times of various GF multiplier implementations without using any fault tolerance. As expected, the worst case delay reduces with higher parallelization (i.e. higher D). This is because, with higher delay, the system performance improves with higher utilization of the slack times.

To compare the comparative GF multiplier implementations between non fault tolerant version and the fault tolerant version using the proposed technique (Section 3), further experiments are carried out using Xilinx Design Suite^{TM}. Two Virtex-7 FPGA boards were used to estimate the worst case delay performance: XC7v1500T and XC7v200T (built on 28nm technology node) as shown in Table 2, which were chosen due to their optimization for high performance.

Table 2. Comparison of Worst Case delay of Multiplier implemented on FPGA devices

Multiplier	WCD on XC7v1500T	WCD on XC7v2000T	WCD on XC7v1500T (with FT)	WCD on XC7v2000T (with FT)
m=163, D=2	2.377nS	2.627nS	4.213nS	4.352nS
m=163, D=4	2.434nS	2.312nS	4.326nS	4.463nS
m=163, D=8	2.840nS	2.478nS	4.451nS	4.532nS

From Table 2, again two observations can be made. Firstly, with increased number of digit size, the worst case delay increases. This is because higher with higher digit size, the complexity of GF multiplier increases substantially. For example, it can be

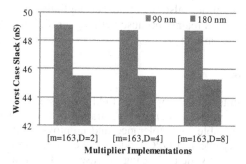

Fig. 7. Comparison of worst case slack times of various GF multiplier implementations (Section 2.1)

seen that upto 20% worst case delay increase when the digit size increases from *D=2* to *D=8*. Second observation is that when fault tolerance is introduced through proposed design technique (Section 3), the worst case delay increases, as expected. This is mainly due to extra hardware resources introduced by the proposed design technique in terms of multiple parity predictor circuit using LDPCs, parity generation circuits and also syndrome detector and correction circuits (Figure 4). It can be noted that up to 70% extra delay is introduced by the proposed design technique.

4.3   Power Comparisons

Figure 8 shows the comparative power consumptions of non fault tolerant and fault tolerant (using the proposed technique) designs using 90*nm* technology node for synthesis in Synopsys Power Compiler*TM*. The dynamic power consumptions are shown in Figure 8(a) and the leakapage consumptions are shown in Figure 8(b). As can be seen from Figure 8(a), due to higher resource usage in fault tolerant design the dynamic power is increased by upto 95%. Also, as expected the dynamic power increases for higher design complexity (i.e. higher *D*). From Figure 8(b), it can be seen that leakage power increases significantly due to addition of the extra hardware resources for fault tolerance (Section 3). Similar to the case of dynamic power, the

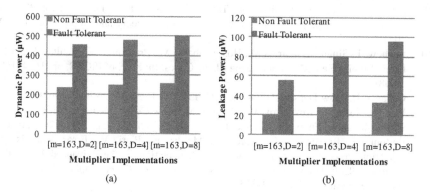

Fig. 8. Comparison of (a) dynamic and (b) leakage powers of non fault tolerant and fault tolerant version of various GF multiplier implementations

leakage power consumptions also increase substantially as digit size of the GF multiplier increases. This is mainly because with higher digit size the number of gates increases substantially (Table 1).

5   Conclusions

A fault tolerant design technique for Galois field processor has been presented in this paper. The design technique underpins the trade-off between design complexity, performance and achievable fault tolerance and employs multiple parity prediction using low density parity check (LDPC) codes. The effectiveness of the proposed fault tolerance technique was validated using a case study of a 163-bit digit serial multiplier used in crypto-processors. We showed that the proposed design technique achieves significant fault tolerance against up to 40 stuck-at faults at the cost of reasonable area, power and delay costs.

References

[1]   Mastrovito, E.D.: VLSI Architectures for Computation in Galois Fields. PhD thesis, Linkoping University, Linkoping, Sweden (1991)

[2]   Deschamps, J.-P., Imana, J.L., Sutter, G.D.: Hardware implementation of Finite Field Arithmetic. The McGraw-Hill Companies Inc. (2009)

[3]   Blake, I., Seroussi, G., Smart, N.P.: Elliptic curves in cryptography. London Mathematical Society Lecture Note Series. Cambridge University Press, Cambridge (1999)

[4]   Kumar, S., Wollinger, T., Paar, C.: Optimum Digit Serial $GF(2^m)$ Multipliers for Curve Based Cryptography. IEEE Transactions on Computers 55(10), 1306–1311 (2006)

[5]   Orlando, G., Paar, C.: A High-Performance Reconfigurable Elliptic Curve Processor for $GF(2^m)$. In: Paar, C., Koç, Ç.K. (eds.) CHES 2000. LNCS, vol. 1965, pp. 41–56. Springer, Heidelberg (2000)

[6]   Pradhan, D.K.: Fault-Tolerant Computer System Design, 1st edn. Prentice-Hall, NJ (1996)

[7]   Johnson, B.W.: Design and Analysis of Fault Tolerant Digital Systems. Addison-Wesley (1989)

[8]   Fenn, S., Gossel, M., Benaissa, M., Taylor, D.: Online Error Detection for Bit-serial Multipliers in $GF(2^m)$. Journal of Electronic Testing: Theory and Applications 13, 29–40 (1998)

[9]   Mathew, J., Singh, J., Jabir, A.M., Hosseinabady, M., Pradhan, D.K.: Fault Tolerant Bit Parallel Finite Field Multipliers using LDPC Codes. In: Proc. of ISCAS, pp. 1684–1687 (2008)

[10] Gallager, R.: Low-Density Parity-Check Codes. MIT Press, Cambridge (1963)

[11] Wu, K., Karri, R., Kuznetsov, G., Goessel, M.: Low Cost Concurrent Error Detection for the Advanced Encryption Standard. In: Proceedings of ITC, pp. 1242–1248 (2004)

[12] Bertoni, G., Breveglieri, L., Koren, I., Maistri, P., Piuri, V.: Error Analysis and Detection Procedures for a Hardware Implementation of the Advanced Encryption Standard. IEEE Trans. on Computers, Special issue on Cryptographic Hardware and Embedded Systems 52(4), 492–505 (2003)

[13] Reyhani-Masoleh, A., Hasan, M.A.: Fault Detection Architectures for Field Multiplication Using Polynomial Bases. IEEE Trans. on Computers 55(9) (September 2006)

[14] Meher, P.K.: On Efficient Implementation of Accumulation in Finite Field Over $GF(2^m)$ and Its Applications. IEEE Trans. on Very Large Scale Integration (VLSI) Systems 17(4) (2009)

[15] Reyhani-Masoleh, A., Hasan, M.A.: Low Complexity Bit Parallel Architectures for Polynomial Basis Multiplication over $GF(2^m)$. IEEE Trans. on Computers 53(8) (2004)

[16] Mitra, S., Seifert, N., Zhang, M., Shi, Q., Kim, K.: Robust System Design with Built-In Soft Error Resilience. IEEE Computer 38(2), 43–52 (2005)

[17] Boneh, D., DeMillo, R.A., Lipton, R.J.: On the Importance of Eliminating Errors in Cryptographic Computations. Journal of Cryptology 14, 1–119 (2001)

[18] Song, L., Parhi, K.K.: Low Energy Digit-Serial/Parallel Fnite Feld Multipliers. Journal of VLSI Signal Processing 19(2), 149–166 (1998)

[19] Gaubatz, G., Sunar, B.: Robust Fnite Feld Arithmetic for Fault Tolerant Public-key Cryptography. In: 2nd Workshop on Fault Tolerance and Diagnosis in Cryptography (FTDC), pp. 1–12 (2005)

Numerical Study on Separation of Analytes through Isotachophoresis

S. Bhattacharyya* and Partha P. Gopmandal

Department of Mathematics, Indian Institute of Technology Kharagpur
Kharagpur-721302, West Bengal, India
somnath@maths.iitkgp.ernet.in

Abstract. In this paper we present a high resolution numerical algorithm to capture the sharp interfaces in isotachophoretic transport of ions. We have considered a two-dimensional model of isotachophoresis (ITP). Both peak and plateau mode of separation is investigated in the present analysis. Our numerical algorithm is based on a finite volume method along with a second-order upwind scheme, QUICK. We have also presented an analytic solution for one-dimensional transport of two electrolytes where diffusion current is neglected. The formation of steady-state in a reference frame co-moving with ITP zones is analysed by providing the transient phase of the ITP separation. Results show that our numerical method can efficiently capture the sharp boundaries between adjacent anlaytes in a steady-state. The present numerical algorithm can handle dispersion of ITP due to pressure-driven convection of electroosmosis of ions.

1 Introduction

Under the influence of an applied electric field (E_0), ionic species i will move with a velocity v_i, as $v_i = \mu_i E_0$, where μ_i is the electrophoretic mobility of the ionic species i. In general different ionic species have different mobilities and therefore moves at different velocities in an electric field. Isotachophoreis (ITP) is separation technique which is based on this difference of migration speed of ionic species under the same electric field. In isotachophoretic separation, the analytes which is to be separated is placed between two electrolytes, namely the leading electrolyte (LE) and trailing electrolyte (TE). The LE is having the highest mobility and TE is of lowest mobility among the ionic species which are to be separated. When an electric field is applied, the ions of the analytes are arranged in order of their mobilities between the LE and TE. This consecutive stacking of ions in order of their mobilities is referred as isotachophoresis (ITP). A unique characteristic of ITP is that all species move at the same speed once steady-state has been reached. The cationic or anionic ITP corresponds to stacking of cations or anions, respectively. In ITP, the electric field adjust in such a way that each stack of ions migrate at constant speed U^{ITP} towards the highest mobility (cationic ITP) or lowest mobility (anionic ITP) zone. The electric field in each stack is constant. This leads to

---

* Corresponding author.

J. Mathew et al. (Eds.): ICECCS 2012, CCIS 305, pp. 282–292, 2012.

formation of a sharp interface between the stacks in which steep gradients in ionic concentration and electric field develop. This sharp interface region is referred as ITP transition zone. An accurate resolution of such sharp transition zones is a challenge in ITP modelling and has several technical applications [1]. In a recent review article by Gebauer et al. [2] provided a discussed about several potential applications of ITP.

Mathematical formulation of ITP was first proposed by Kohlarusch [3] through the introduction of a regulating function. The Kohlarusch regulating function relates the electric field across the interface of two different ions with the same counter ions. Subsequently, several theoretical studies on understanding the ITP and resolution of the transition zone has been reported in the literature. Bercovicci et al. [4] described the ITP transport as a similar phenomena as shock wave propagation in gas dynamics. In recent years, various high-resolution simulators for one-dimensional computation of ITP have been presented for analyzing electrophoresis separation problems, including SIMUL5, developed by Hruska et al. [5]. Yu et al [6] applied the space-time Conservation Element and Solution Element (CESE) method to solve 1-D transport equations to accurately resolve the sharp gradients in ITP. In the CESE method, like finite volume method, equations are cast into conservation law form and integrated over a space-time region. Chou and Yang [7] have provided a detailed discussion on several numerical schemes for simulating ITP and provided the ITP simulation based on flux conservative finite volume scheme in an adaptive grid system. Bercovici et al. [4, 8] developed an efficient numerical solver to resolve correctly the sharp ITP transition zones by using a sixth order compact difference scheme on an adaptive-grid system. Recently, in a review article, Thormann et al. [9] provided a state-of-art on the available computer simulation softwares for ITP separation.

In this paper we have computed the two-diemnsional transient ITP of ions within a rectangular channel. The transport of ions is governed by the Nernst-Planck equations. The electric field (or current density) is determined from the charge conservation equation along with the electro-neutrality condition. We have neglected the wall ζ-potential induced electroosmosis of ions. However, a uniform electroosmostic velocity does not create any dispersion in ITP. The governing equations are disretized through a finite volume method. The advective and electro-migration terms are discretized through an upwind algorithm, QUICK [12]. We have tested our algorithm for accuracy by comparing with analytic solutions for ITP of two electrolytes. Our numerical method can resolve efficiently the interface zones of ions. We present solution for the transient ITP as well as the steady ITP both for peak and plateau mode of separation. The stacking of sample ions and its dependence on mobilities and other parameters is investigated.

2   Mathematical Model

We assume the electrolytes to consists of monovalent leading electrolyte of cationic concentartion C_+ , monovalent trailing electrolyte (TE) of cationic concentration C_-, the cationic concentartion of mth sample species is C_{s_m} and a common anion of concentartion C_0. Here the solvent, i.e., water, is in excess and its ionization is assumed to be negligible. The mass conservation of the ionic species leads to the Nernst-Planck equation,

$$\frac{\partial C_i}{\partial t} + \nabla.N_i = 0 \tag{1}$$

where $i = +, s_m$ or$-$. The the molar flux of ith species is

$$\mathbf{N}_i = -D_i \nabla C_i + C_i(\mathbf{V} + z_i \mu_i \ \mathbf{E}) \tag{2}$$

where valance $z_i = +1$ for cations and $z_0 = -1$ for common anion. The mobility and diffusivity are related via Nernst-Einstein relation $\mu_i = D_i F/RT$, where F is Faraday constant, R is gas constant and T is absolute temperature. The electric field $\mathbf{E} = -\nabla \phi$ can be obtained from the charge conservation equation with electro-neutrality condition $(\rho_e = \sum_i z_i C_i = 0)$ as

$$\nabla.(\nu \mathbf{E}) = \nabla.(F \sum_i D_i z_i \nabla C_i) \tag{3}$$

where ionic conductivity is given by $\nu = (F \sum_i \mu_i z_i^2 C_i)$. The term in right-hand side of (3) is the diffusion current and its contribution is insignificant at all locations, except at the transition zones.

We imposed a fixed potential drop $\Phi = E_0 L$ along the channel. If E_i $(i = +, s_m$ or$-)$ are the strengths of electric field in each separated zone occupied by only one species, then using the relation $U^{ITP} = \mu_i E_i$ we get

$$E_i = E_0 \frac{1/\mu_i}{\sum_j (l_j/\mu_j)} \tag{4}$$

where the summation is taken over all the species and $l_i = L_i/L$ is the portion of the length L which is filled by i^{th} species. We can also express E_i in terms of the applied current density j_0 as follows

$$j_0 = F(\mu_- + \mu_0)C^\infty_- E_- = F(\mu_{s_m} + \mu_0)C^\infty_{s_m} E_{s_m} = F(\mu_+ + \mu_0)C^\infty_+ E_+ \tag{5}$$

The above relation leads to the relationships for the bulk concentration of TE, LE and all the sample species as

$$\frac{C^\infty_-}{C^\infty_+} = \frac{\mu_+ + \mu_0}{\mu_- + \mu_0} \frac{\mu_-}{\mu_+}, \quad \frac{C^\infty_{s_m}}{C^\infty_+} = \frac{\mu_+ + \mu_0}{\mu_s + \mu_0} \frac{\mu_{s_m}}{\mu_+} \tag{6}$$

The y-coordinate is non-dimensionalized by the height H of the channel and x-coordinate is scaled by ΔX, which is the maximum of all transition zone widths. The analytical formula for calculating the transition zone width is given latter. We non-dimensionalize the concentration by the bulk concentration of LE, C^∞_+, potential by $\phi_0(= RT/F)$ and time is scaled by $\tau = H/U^{ITP}$. We consider that the co-ordinate is moving with the average speed of the ions. This will allow us to consider a truncated computational domain. A variable transformation is introduced as

$$x = x - U^{ITP}t, \quad y = y, \quad t = t \tag{7}$$

With that the Nernst-Planck equations in non-dimensionalized form under no bulk fluid motion, can be written as

$$\frac{\partial c_i}{\partial t} - \epsilon_1 \frac{\partial c_i}{\partial x} - z_i \frac{D_i}{D_+} \frac{1}{Pe} \nabla.(c_i \nabla \phi) - \frac{D_i}{D_+} \frac{1}{Pe} \nabla^2 c_i = 0 \tag{8}$$

where D_i's are the diffusivities of the respective ions, Peclet number $Pe = U^{ITP} \Delta X / D_+$ and the aspect ratio $\epsilon_1 = H/\Delta X$. The charge conservation equation in nondimensional form is as follows

$$\nabla.\left[\left(c_+ + \sum_m \frac{D_{s_m} + D_0}{D_+ + D_0} c_{s_m} + \frac{D_- + D_0}{D_+ + D_0} c_- \right) \nabla \phi \right] = -\left(\frac{z_+ D_+ + z_0 D_0}{D_+ + D_0}\right)$$

$$\nabla^2 \left[c_+ + \sum_m \frac{z_{s_m} D_{s_m} + z_0 D_0}{z_+ D_+ + z_0 D_0} c_{s_m} + \frac{z_- D_- + z_0 D_0}{z_+ D_+ + z_0 D_0} c_- \right] \tag{9}$$

Both the left and right boundaries of the computational domain is placed sufficiently far away from the transition zones. The net flux through the channel walls are set to be zero i.e., $\mathbf{N}_i.\widehat{n} = 0$ for $i = -, s_m, +$, where $\widehat{n}$ is unit outward normal. The electric potential is subjected to insulating boundary conditions along the wall ($\nabla \phi.\widehat{n} = 0$). The concentration at both the end are governed by the Kohlrauch's conditions i.e., equation (6). As given in (4), upon application of constant voltage drop ϕ across the channel, due to the discontinuous conductivity, locally uniform but not globally, electric field appears at both TE and LE end. This allows for calculation of electric potential across the channel via charge conservation (9).

3   Numerical Methods

We have computed the equations for ion transport and electric field in a coupled manner through the finite volume method [13]. The computational domain is subdivided into a number of control volumes. In this method, the equation cast into conservative form, are integrated over each control volumes. The variables on the control volume interfaces are estimated by a linear interpolation between two neighbors to either sides of the control volume interfaces. This approach enables to compute the jump discontinuity as part of the solution. Thus, the finite volume discretization is more efficient than the finite difference scheme when a sharp gradient is expected. When the discretization based on central difference, the truncation error involves odd order derivatives, which leads to the dispersion error to the numerical scheme. The electric field across the transition zone suffers a sharp variation and hence the electromigration term i.e., $\nabla.(c_i \mathbf{E})$,

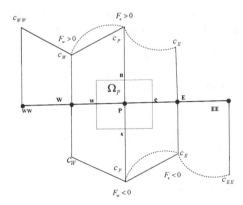

Fig. 1. Schematic diagram for control volume Ω_P and interpolation for a variable c based on QUICK scheme. Here, e, w, n and s are the cell faces of the cell centered at P.

appearing in the Nernst-Planck equation, leads to a hyperbolic characteristic of the PDE. Due to the hyperbolic nature of the ion transport equations, this error results into formation of wiggles near the sharpe variation of the variables. In order to resolve the sharpness of the variables, which occurs in ITP interface, we adopt the upwind scheme QUICK (Quardratic Upwind Interpolation Convection Kinematics) to discretize the electromigration and convection terms in the ion transport equations.

The ion transport equation when integrated over a cell Ω_P (Fig. 1) yields

$$\int\int_{\Omega_P} \frac{\partial c_i}{\partial t}\, d\Omega_P - \epsilon_1 \int\int_{\Omega_P} \frac{\partial c_i}{\partial x}\, d\Omega_P + \frac{D_i}{D_+}\frac{1}{Pe} \int\int_{\Omega_P} \left[-\frac{\partial}{\partial x}(\frac{\partial c_i}{\partial x}) + z_i(-\frac{\partial \phi}{\partial x}) \right]\, dxdy$$

$$+ \frac{D_i}{D_+}\frac{1}{Pe} \int\int_{\Omega_P} \left[-\frac{\partial}{\partial y}(\frac{\partial c_i}{\partial y}) + z_i(-\frac{\partial \phi}{\partial y}) \right]\, dxdy = 0 \tag{10}$$

The time derivatives are discretized through the second order accurate alternating direction implicit (ADI) scheme. In ADI scheme, we advance the solution from the time step k to the time step $k+1$ via an intermediate time step $k+1/2$. The derivatives with respect to x (or y) is obtained implicitly while advancing from t_k to $t_{k+1/2}$ and the direction of the implicit derivatives are reversed for computing the second step, i.e., from $t_{k+1/2}$ to t_{k+1}. In step-I where we advance the solution from t_k to $t_{k+1/2}$, the discretized equations are

$$\frac{c_P^{k+1/2} - c_P^k}{dt}\, d\Omega_P + (F_e c_e - F_w c_w)\big|^{k+1/2} dy_P + (G_n c_n - G_s c_s)\big|^k dx_P$$

$$- \left[\frac{\partial c}{\partial x}\Big|_e - \frac{\partial c}{\partial x}\Big|_w \right]^{k+1/2} dy_P - \left[\frac{\partial c}{\partial y}\Big|_n - \frac{\partial c}{\partial y}\Big|_s \right]^k dx_P = 0 \tag{11}$$

In step-II where we advance the solution from $(k+1/2)^{th}$ time step to $(k+1)^{th}$ and the discretized equations can be obtained in a similar manner. Here $F =$

$-\epsilon_1 - \frac{D_i}{D_+}\frac{1}{Pe}\frac{\partial\phi}{\partial x}$, $G = -\frac{\partial\phi}{\partial y}$. The spatial derivatives involve in the diffusion terms are evaluated as

$$\frac{\partial c}{\partial x}\bigg|_e = \frac{c_E - c_P}{0.5(dx_P + dx_E)}$$

Note that the big letter subscripts denote the cell centers in which variables are stored and small letter subscripts denotes the corresponding cell faces. The area of the rectangular control volume is $d\Omega_P$ with edges dx_P and dy_P. We have considered a variable grid size.

The variable c at the cell faces 'e' and 'w' are obtained as follows

$$c_e = \begin{cases} (\frac{3}{8}c_E + \frac{3}{4}c_P - \frac{1}{8}c_W), & F_e > 0; \\ (\frac{3}{4}c_E + \frac{3}{8}c_P - \frac{1}{8}c_{EE}), & F_e < 0, \end{cases}$$

$$c_w = \begin{cases} (\frac{3}{8}c_P + \frac{3}{4}c_W - \frac{1}{8}c_{WW}), & F_w > 0; \\ (\frac{3}{4}c_P + \frac{3}{8}c_W - \frac{1}{8}c_E), & F_w < 0, \end{cases}$$

Here $c = c_+$, c_- or c_{s_m}. At every time step we adopt an iterative method in which ϕ at every cell center is assumed to be known. In order to reduce the system of equations into a tri-diagonal form, the variables at the locations EE and WW are taken as the previous iterated values. We solve the ensuing system of linear algebraic equations through a block elimination method due to Varga [14].

At every iteration, the charge conservation equation is integrated over each control volume Ω_P through a finite volume method. The elliptic PDE for charge conservation is solved by a line-by-line iterative method along with the successive-over-relaxation (SOR) technique. Iteration at each time steps is continued till the desired order of accuracy is achieved.

4   Analytic Solution of 1-D Formulation for Two Electrolytes

When a steady istoachophoresis is formed in absence of external convection with electroneutrality assumption, the transition zones are planar and a one-dimensional analysis is sufficient for the concentration distribution of ionic species for two electrolytes (TE and LE) only. Nernst-Planck equation in one dimension form is given by

$$\frac{d}{dx}\left[-U^{ITP}C_i - D_i\frac{dC_i}{dx} + z_i\mu_i\frac{d}{dx}(C_i\mathbf{E}_x)\right] = 0 \tag{12}$$

where $\mathbf{E}_x$ is the axial potential gradient. Integrating (12) and using the fact that thermodynamic potential at TE and LE end remains constant along with constant values of concentrations at both the end, yields

$$\frac{d}{dx}\left[\ln\left(\frac{C_+}{C_-}\right)\right] = \frac{U^{ITP}}{\phi_0}\left[\frac{\mu_+ - \mu_-}{\mu_+\mu_-}\right] = \frac{E_- - E_+}{\phi_0} \tag{13}$$

Integrating above equation and considering the origin of the coordinate system where the concentration of both the species are equal, i.e., $C_+ = C_-$, then we get

$$\left[\ln\left(\frac{C_+}{C_-}\right)\right] = \frac{\Delta E}{\phi_0}x \tag{14}$$

where $\Delta E = E_- - E_+$, the difference in electric field strengths at TE and LE. We define the transition zone where the concentration ratio C_+/C_- changes from e^2 to $1/e^2$. The width of the transition zone where most of the concentration changes is given by

$$\Delta X = \frac{4\phi_0}{\Delta E} \tag{15}$$

Under the electro-neutrality condition, the axial electric field $\mathbf{E}_x$ can be obtained from the charge conservation equation (neglecting the diffusion current) as

$$\frac{d}{dx}(\nu\mathbf{E}_x) = 0 \tag{16}$$

Integrating (16) and using the relation given in (14), the axial electric field is given by

$$\frac{E_-}{E(x)} = \frac{C_-(x)}{C_-^\infty}[1 + \frac{\mu_+ + \mu_0}{\mu_- + \mu_0}e^{\frac{\Delta E}{\phi_0}}] \tag{17}$$

Using the above relation and integrating (13), concentration profile of TE can be expressed through the hypergeometric series (Goet et al. [15])

$$\frac{C_-(x)}{C_-^\infty} = F[1, \frac{E_-}{\Delta E}, 1 + \frac{E_-}{\Delta E}, \frac{\mu_+ + \mu_0}{\mu_- + \mu_0}e^{\frac{\Delta E}{\phi_0}x}] \tag{18}$$

and the concentration profile for LE can be obtained by using relation (14).

5   Results and Discussions

We first consider the ITP due to two electrolytes, LE and TE placed symmetrically in a channel. In order to check the time dependency of concentration profile for planer ITP, computed result for the logarithmic ratio of concentrations of LE and TE is compared with the analytical solution given by (14). The results show (Fig. 2a) that the concentration distribution becomes steady after a short transition. The LE-TE interface width as a function of electric E_0 is presented in Fig. 2b. The transition zone is the over-lapped region in which the concentration field of dissolved ions changes from their zone characteristic value to zero. The overlapping can be minimized by increasing the electric field. Our computer code has been validated by comparing the concentration profile and axial electric field for plane ITP with the 1-D solution as presented in Fig. 3(a) and (b). Our results agree well with the analytic solution based on the electroneutrality condition. The formation of a transition zone in the steady ITP is evident from the results. The distribution of an electric field along the channel is presented in Fig. 3(b) for

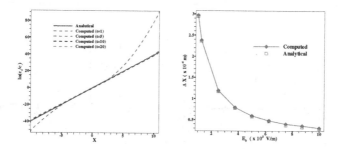

Fig. 2. Comparison of logarithmic ratio of TE and LE concentrations with analytical results to show how the steady state achieved. The electrophoretic mobility of TE and LE are taken as $\mu_- = 0.7\mu_+$ and $\mu_+ = 2.71 \times 10^{-8} m^2/Vs$ with $E_0 = 10^5 V/m$, $l_- = l_+ = 1/2$ and $j_0 = 2.87 \text{Amp}/m^2$. The bulk concentration of LE is taken as $C_+^\infty = 0.01M$. (b) Comparison of computed solution for the width of the transition zone in ideal ITP with the analytical results. All the parameters are taken same as (a) except the electric field is varied.

different mobility ratio $\mu_-/\mu+ = 0.6, 0.7$ and 0.8 . The electric field gradually decreased as we move from low conductivity TE to high conductivity LE. The jump in voltage between neighboring zones result to permanently sharp boundaries between the TE and LE and are typical for ITP (Fig. 3(a)). The sharp gradient in Ex which develops in the transition zone smear out as the mobility ratio of the buffer increases. Fig. 3(b) shows that low mobility ratio leads to a sharp transition zone, whereas a high mobility ratio widens the transition zone width and the lateral distribution of E_x is relatively smooth. Garcia-Schwartz et al. [1] have identified two modalities in which analytes can focus in ITP separation, namely 'plateau mode' and 'peak mode'. Occurrence of the modalities depends on the initial distribution of analytes. The plateau mode is referred to the case in which distinct boundaries of the separated zones appears, whereas,

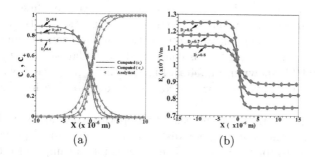

(a)   (b)

Fig. 3. Comparison of TE and LE concentration profiles and axial electric field with the results due to Goet et al. [15]. The electrophoretic mobility ratio are taken as $\mu_-/\mu_+ = 0.6, 0.7$ and 0.8 with $\mu_+ = 2.71 \times 10^{-8} m^2/Vs$ with $E_0 = 10^5 V/m$, $l_- = l_+ = 1/2$ and $j_0 = 2.87 \text{Amp}/m^2$. The bulk concentration of LE is taken as $C_+^\infty = 0.01M$.

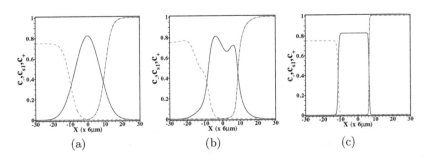

(a)   (b)   (c)

Fig. 4. Variation of concentration of TE, sample and LE at different time (a) $t = 0$; (b) $t = 10$ and (c) $t = 30$ when $l_- = l_{s_1} = l_+ = 1/3$ .The electrophoretic mobility of TE, sample and LE are respectively $\mu_- = 0.6\mu_+$, $\mu_{s_1} = 0.7\mu_+$ and $\mu_+ = 2.71 \times 10^{-8} m^2/Vs$ with $E_0 = 10^5 V/m$ and $j_0 = 2.55 \text{Amp}/m^2$. The bulk concentration of LE is $C_+^\infty = 0.01M$.

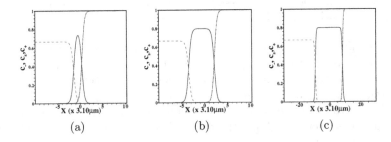

(a)   (b)   (c)

Fig. 5. Variation of concentration of ionic species $(-, s_1$ and $+$) for (a) $l_s = 0.1$; (b) $l_s = 0.3$ and (c) $l_s = 0.5$ and rest portion of the channel is filled symmetrically by the TE and LE. Here applied electric field $E_0 = 10^5 V/m$, current density $j_0 = 2.33 \text{Amp}/m^2$ with $C_+^\infty = 0.01M$. The electrophoretic mobility of TE, sample and LE are respectively $\mu_- = 0.5\mu_+$, $\mu_{s_1} = 0.666\mu_+$ and $\mu_+ = 2.71 \times 10^{-8} m^2/Vs$ with $l_- = l_{s_1} = l_+ = 1/3$.

in peak mode the separated zones appear as sharp peak. We first consider the ITP separation in which the analytes occupies $1/3^{rd}$ portions of the channel and the LE and TE are occupying the rest portion of the channel symmetrically. Initial distribution of the analytes is considered to be Gaussian, which is placed between LE and TE shown in Fig. 4(a). Fig.4(c) shows the steady plateau mode ITP separation. We have also presented the ion distribution for the transient phase in Fig. 4(b).

Figs. 5(a-c) shows that, by increasing the portion of the channel filled by sample electrolyte, the concentration distribution of the sample species, changes from 'peak mode' to 'plateau mode'. The penetration of sample electrolyte into LE or TE zone strongly depends on its mobility. Figs. 6(a-c) shows that as the mobility ratio between the sample and LE increases and consequently, the mobility ratio between TE and sample decreases, the sample penetrates more into the LE zone. The interface region between the LE-sample becomes sharp compared to the TE-sample interface as the sample mobility becomes close to

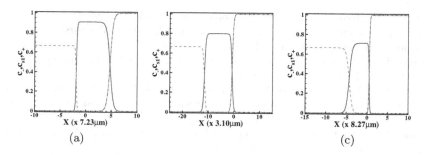

Fig. 6. Variation of concentration of ionic species $(-, s_1$ and $+$) when each species occupied $1/3^{rd}$ portion of the undertaken channel for (a) $\mu_- = 0.5\mu_+$, $\mu_{s_1} = 0.833\mu_+$; (b) $\mu_- = 0.5\mu_+$, $\mu_{s_1} = 0.666\mu_+$ and (c) $\mu_- = 0.5\mu_+$, $\mu_{s_1} = 0.555D_+$. Here applied electric field $E_0 = 10^5 V/m$ and electrophoretic mobility of LE is $\mu_+ = 2.71 \times 10^{-8} m^2/Vs$ with bulk concentration $C_+^\infty = 0.01M$.

the LE. Consequently, the TE-sample interface becomes smaller compared to the width of the LE-sample interface zone. Our results as presented in Fig. 6(a-c) agrees qualitatively with the observations made by Garcia-Schwarz et al. [1].

6   Conclusions

We developed an efficient numerical algorithm to accurately resolve the ITP interfaces involving step-jumps in electric field and concentration of ions. The hyperbolic characteristics of ion transport equations may lead to a numerical instability in the region where sharp change of variables occurs. Our numerical algorithm, based on the finite volume method along with a QUICK scheme, can capture the sharp interfaces between two electrolytes. Validation of our method is performed by comparing the plateau mode ITP of two electrolytes with the analytical solution. We have also presented the transient ITP characteristics in a plane micro-channel. The peak mode ITP due to presence of sufficiently low analytes concentration is investigated extensively.

The present numerical algorithm for solving the two-dimensional ITP transport of ions can effectively simulate ITP with imposed convection. In practical situation, convection of ions may arise due to unwanted pressure gradient and/or non-unifrom electroosmotic transport of ions.

References

1. Garcia-Schwartz, G., Bercovicci, M., Marshal, L.A., Santiago, J.G.: J. Fluid Mech. 679, 455–475 (2011)
2. Gebauer, P., Mala, Z., Bocek, P.: Electrophoresis 32, 83–89 (2011)
3. Kohlrausch, F.: Ann. Physik 62, 209–239 (1897)
4. Bercovicci, M., Lelea, S.K., Santiago, J.G.: Journal of Chromatography A 1217, 588–599 (2010)

5. Hruska, V., Jaros, M., Gas, B.: Electrophoresis 27, 984–991 (2006)
6. Yu, J.W., Chou, Y., Yang, R.J.: Electrophoresis 29, 1048–1057 (2008)
7. Chou, Y., Yang, R.J.: Journal of Chromatography A 1217, 394–404 (2010)
8. Bercovicia, M., Lelea, S.K., Santiago, J.G.: Journal of Chromatography A 1216, 1008–1018 (2009)
9. Thormann, W., Breadmore, M.C., Caslavska, J., Mosher, R.A.: Dynamic Computer Simulations of Electrophoresis: A versatile Research and Teaching Tool. Electrophoresis 31, 726–754 (2010)
10. Shim, J., Dutta, P., Ivory, C.F.: Numerical Heat Transfer. Part A: Applications 52, 441–461 (2007)
11. Choi, H., Jeon, Y., Cho, M., Lee, D., Shim, J.: Microsyst. Technol. 16, 1931–1938 (2010)
12. Leonard, B.P.: Comput. Meth. Appl. Mech. Eng 19, 59–98 (1979)
13. Fletcher, C.A.J.: Computation Technique for Fluid Dynamics, vol. 2. Springer, Berlin (1998)
14. Varga, R.S.: Matrix Iterative Analysis. Prentice-Hall, Englewood Cliffs (1962)
15. Goet, G., Baier, T., Hardt, S.: Biomicrofluidics 5, 014109(1-16) (2011)

Finite Dimensional Realization of a Guass-Newton Method for Ill-Posed Hammerstein Type Operator Equations

Monnanda Erappa Shobha and Santhosh George

Department of Mathematical and Computational Sciences
National Institute of Technology Karnataka
India-575 025
shobha.me@gmail.com, sgeorge@nitk.ac.in

Abstract. Finite dimensional realization of an iterative regularization method for approximately solving the non-linear ill-posed Hammerstein type operator equations $KF(x) = f$, is considered. The proposed method is a combination of the Tikhonov regularization and Guass-Newton method. The advantage of the proposed method is that, we use the Fréchet derivative of F only at one point in each iteration. We derive the error estimate under a general source condition and the regularization parameter is chosen according to balancing principle of Pereverzev and Schock (2005). The derived error estimate is of optimal order and the numerical example provided proves the efficiency of the proposed method.

Keywords: Newton's method, Tikhonov regularization, ill-posed Hammerstein operator, Balancing principle, Monotone operator, Regularization.

AMS Subject Classification : 47J06, 47A52, 65N20, 65J20.

1   Introduction

Let X be a real Hilbert space, $F : D(F) \subseteq X \to X$ be a monotone operator (i.e., $\langle F(x) - F(y), x - y \rangle \geq 0, \quad \forall x, y \in D(F)$) and $K : X \to Y$ be a bounded linear operator between the Hilbert spaces X and Y. Consider the ill-posed operator equation

$$KF(x) = f. \tag{1.1}$$

Equation (1.1) is called ill-posed Hammerstein type([1], [2], [3], [4]) operator equation. Throughout the paper, the domain of F is denoted by $D(F)$, the Fréchet derivative of F is denoted by $F'(.)$ and the inner product and norm in X and Y are denoted by $\langle .,. \rangle$ and $\|.\|$ respectively.

It is assumed that the available data is f^δ with $\|f - f^\delta\| \leq \delta$ and hence one has to consider the equation

$$KF(x) = f^\delta \tag{1.2}$$

J. Mathew et al. (Eds.): ICECCS 2012, CCIS 305, pp. 293–301, 2012.
© Springer-Verlag Berlin Heidelberg 2012

instead of (1.1). Since (1.1) is ill-posed, its solution is not depending continuously on the given data. Thus one has to use regularization method (see [1], [2], [3], [4], [6], [7] and [10]) for obtaining an approximation for $\hat{x}$.

Observe that the solution x of (1.2) can be obtained by first solving

$$Kz = f^\delta \qquad (1.3)$$

for z and then solving the non-linear problem

$$F(x) = z. \qquad (1.4)$$

This was exploited in [1], [2], [3], [4] and [5]. As in [4], we assume that the solution $\hat{x}$ of (1.1) satisfies

$$\|F(\hat{x}) - F(x_0)\| = \min\{\|F(x) - F(x_0)\| : KF(x) = f, x \in D(F)\}.$$

The prime motive of this study is to develop an iterative regularization method to obtain an approximation for $\hat{x}$ in the finite dimensional subspace of X. Precisely we considered Discretized Tikhonov regularization for solving (1.3) and Discretized Newton's method for solving (1.4).

This paper is organized as follows. Preliminaries are given in Section 2, Section 3 deals with the convergence of the proposed method. A numerical example is given in Section 4 and finally the paper ends with a conclusion in section 5.

2   Preliminaries

Let $\{P_h\}_{h>0}$ be a family of orthogonal projections on X, let $\varepsilon_h := \|K(I - P_h)\|$, $\tau_h := \|F'(x)(I - P_h)\|$, $\forall x \in D(F)$. Let $\{b_h : h > 0\}$ is such that $\lim\limits_{h\to 0} \frac{\|(I-P_h)x_0\|}{b_h} = 0$, $\lim\limits_{h\to 0} \frac{\|(I-P_h)F(x_0)\|}{b_h} = 0$ and $\lim\limits_{h\to 0} b_h = 0$. We assume that $\varepsilon_h \to 0$ and $\tau_h \to 0$ as $h \to 0$. The above assumption is satisfied if, $P_h \to I$ pointwise and if K and $F'(x)$ are compact operators. Further we assume that $\varepsilon_h < \varepsilon_0$, $\tau_h \le \tau_0$, $b_h \le b_0$ and $\delta \in (0, \delta_0]$.

In [5], the authors studied a two step newton method defined iteratively by

$$y_{n,\alpha_k}^{h,\delta} = x_{n,\alpha_k}^{h,\delta} - (P_h F'(x_{n,\alpha_k}^{h,\delta})P_h + \frac{\alpha_k}{c}P_h)^{-1}[F(x_{n,\alpha_k}^{h,\delta}) - z_{\alpha_k}^{h,\delta} + \frac{\alpha_k}{c}(x_{n,\alpha_k}^{h,\delta} - x_{0,\alpha_k}^{h,\delta})], \qquad (2.5)$$

$$x_{n+1,\alpha_k}^{h,\delta} = y_{n,\alpha_k}^{h,\delta} - (P_h F'(x_{n,\alpha_k}^{h,\delta})P_h + \frac{\alpha_k}{c}P_h)^{-1}[F(y_{n,\alpha_k}^{h,\delta}) - z_{\alpha_k}^{h,\delta} + \frac{\alpha_k}{c}(y_{n,\alpha_k}^{h,\delta} - x_{0,\alpha_k}^{h,\delta})], \qquad (2.6)$$

where $c \le \alpha_k$, $x_{0,\alpha_k}^{h,\delta} := P_h x_0$, the projection of initial guess x_0 and

$$z_{\alpha_k}^{h,\delta} = (P_h K^* K P_h + \alpha_k I)^{-1} P_h K^*[f^\delta - KF(x_0)] + P_h F(x_0), \qquad (2.7)$$

for obtaining an approximation for $\hat{x}$ in the finite dimensional subspace $R(P_h)$, the range of P_h, in X.

The main draw back of this approach was that the iterations (2.5) and (2.6) requires Fréchet derivative of F at each iteration $x_{n,\alpha_k}^{h,\delta}$.

In this paper we consider a modified form of (2.5) and (2.6), i.e.;

$$y_{n,\alpha_k}^{h,\delta} = x_{n,\alpha_k}^{h,\delta} - R(x_{0,\alpha_k}^{h,\delta})^{-1} P_h[F(x_{n,\alpha_k}^{h,\delta}) - z_{\alpha_k}^{h,\delta} + \frac{\alpha_k}{c}(x_{n,\alpha_k}^{h,\delta} - x_{0,\alpha_k}^{h,\delta})] \quad (2.8)$$

and

$$x_{n+1,\alpha_k}^{h,\delta} = y_{n,\alpha_k}^{h,\delta} - R(x_{0,\alpha_k}^{h,\delta})^{-1} P_h[F(y_{n,\alpha_k}^{h,\delta}) - z_{\alpha_k}^{h,\delta} + \frac{\alpha_k}{c}(y_{n,\alpha_k}^{h,\delta} - x_{0,\alpha_k}^{h,\delta})]. \quad (2.9)$$

We assume that F possess a uniformly bounded Fréchet derivative for all $x \in D(F)$ i.e., $\|F'(x)\| \le M$, for some $M > 0$ and $\|\hat{x} - x_0\| \le \rho$.

The advantage of the proposed method is that the Fréchet derivative of F is required only at $P_h(x_0)$ and the disadvantage of the proposed method is that it converges linearly where as the method in [5] converges cubically. The selection of the regularization parameter α is done according to the balancing principle considered by Pereverzev and Schock in [9] and the derived error estimate is of optimal order.

The following Assumption and Theorem from [5] are used in proving our results.

Assumption 1. *There exists a continuous, strictly monotonically increasing function $\varphi : (0, a] \to (0, \infty)$ with $a \ge \|K\|^2$ satisfying;*

- $\lim\limits_{\lambda \to 0} \varphi(\lambda) = 0,$

$$\sup_{\lambda \ge 0} \frac{\alpha \varphi(\lambda)}{\lambda + \alpha} \le \varphi(\alpha) \qquad \forall \lambda \in (0, a]$$

and

$$F(\hat{x}) - F(x_0) = \varphi(K^*K)w$$

for some $w \in X$ such that $\|w\| \le 1$.

Theorem 2. *(see [5], Theorem 2.4) Let $z_\alpha^{h,\delta}$ be as in (2.7). Further if $b_h \le \frac{\delta + \varepsilon_h}{\sqrt{\alpha}}$ and Assumption 1 holds. Then*

$$\|F(\hat{x}) - z_\alpha^{h,\delta}\| \le C(\varphi(\alpha) + (\frac{\delta + \varepsilon_h}{\sqrt{\alpha}})), \quad (2.10)$$

where $C = \max\{M\rho, 1\} + 1$

2.1 A Priori Choice of the Parameter

Let $\psi(\lambda) := \lambda\sqrt{\varphi^{-1}(\lambda)}, 0 < \lambda \le a$ and let $\delta + \varepsilon_h = \sqrt{\alpha(\delta, h)}\varphi(\alpha(\delta, h)) = \psi(\varphi(\alpha(\delta, h)))$. Then

$$\alpha(\delta, h) = \varphi^{-1}(\psi^{-1}(\delta + \varepsilon_h))$$

and this choice of $\alpha := \alpha(\delta, h)$ is of optimal order (see [5]). So the relation (2.10) leads to $\|F(\hat{x}) - z_\alpha^{h,\delta}\| \le 2C\psi^{-1}(\delta + \varepsilon_h)$.

2.2   An Adaptive Choice of the Parameter

We consider the balancing principle of Pereverzev and Shock [9] for choosing the regularization parameter α. Let

$$D_N = \{\alpha_i : 0 < \alpha_0 < \alpha_1 < \alpha_2 < \cdots < \alpha_N\}$$

be the set of possible values of the parameter α.

Let

$$l := \max\{i : \varphi(\alpha_i) \le \frac{\delta + \varepsilon_h}{\sqrt{\alpha_i}}\} < N, \tag{2.11}$$

$$k = \max\{i : \alpha_i \in D_N^+\} \tag{2.12}$$

where $D_N^+ = \{\alpha_i \in D_N : \|z_{\alpha_i}^\delta - z_{\alpha_j}^\delta\| \le \frac{4C(\delta+\varepsilon_h)}{\sqrt{\alpha_j}}, j = 0, 1, 2, ..., i - 1\}$.

We use the following theorem, proof of which is analogous to the proof of Theorem 4.3 in [3] for the error analysis.

Theorem 3. *(cf. [3], Theorem 4.3) Let l be as in (2.11), k be as in (2.12) and $z_{\alpha_k}^{h,\delta}$ be as in (2.7) with $\alpha = \alpha_k$. Then $l \le k$ and*

$$\|F(\hat{x}) - z_{\alpha_k}^{h,\delta}\| \le C(2 + \frac{4\mu}{\mu - 1})\mu\psi^{-1}(\delta + \varepsilon_h).$$

3   Discretized Newton's Method

First we consider the iterative scheme defined by (2.8) and (2.9) for approximating the zero $x_{c,\alpha_k}^{h,\delta}$ of

$$P_h(F(x) + \frac{\alpha_k}{c}(x - x_0)) = P_h z_{\alpha_k}^{h,\delta} \tag{3.13}$$

and then show that $x_{c,\alpha_k}^{h,\delta}$ is an approximation to the solution $\hat{x}$ of (1.1) where $c \le \alpha_k$.

The following Assumption is used throughout our analysis.

Assumption 4. *(cf. [11], Assumption 3 (A3)) There exists a constant $k_0 \ge 0$ such that for every $x, u \in D(F)$ and $v \in X$ there exists an element $\Phi(x, u, v) \in X$ such that $[F'(x) - F'(u)]v = F'(u)\Phi(x, u, v), \|\Phi(x, u, v)\| \le k_0\|v\|\|x - u\|$.*

One can observe that (see [5], equation 3.14)

$$\|R(x_{0,\alpha_k}^{h,\delta})^{-1} P_h F'(x_{0,\alpha_k}^{h,\delta})\| \le 1 + \tau_h \le 1 + \tau_0. \tag{3.14}$$

Let

$$e_{n,\alpha_k}^{h,\delta} := \|y_{n,\alpha_k}^{h,\delta} - x_{n,\alpha_k}^{h,\delta}\|, \qquad \forall n \ge 0 \tag{3.15}$$

and let $\delta_0 + \varepsilon_0 < (\frac{2}{2M+3})\sqrt{\alpha_0}$ and $\|\hat{x} - x_0\| \le \rho$, with

$$\rho < \frac{1}{M}(1 - (\frac{3}{2} + M)\frac{\delta_0 + \varepsilon_0}{\sqrt{\alpha_0}})$$

and
$$\gamma_\rho := M\rho + (\frac{3}{2} + M)(\frac{\varepsilon_0 + \delta_0}{\sqrt{\alpha_0}}).$$

Further let
$$\gamma_\rho < \frac{1}{4k_0(1 + \tau_0)},$$
$$r_1 = \frac{1 - \sqrt{1 - 4(1 + \tau_0)k_0\gamma_\rho}}{2(1 + \tau_0)k_0}$$

and
$$r_2 = min\{\frac{1}{(1 + \tau_0)k_0}, \frac{1 + \sqrt{1 - 4(1 + \tau_0)k_0\gamma_\rho}}{2(1 + \tau_0)k_0}\}.$$

For $r \in (r_1, r_2)$, let
$$q = (1 + \tau_0)k_0r, \tag{3.16}$$

then $q < 1$.

Further note that (see [5], Equation 3.24)
$$e_{0,\alpha_k}^{h,\delta} \le \gamma_\rho < 1. \tag{3.17}$$

Analogous to the proof of Theorem 3.4 and Theorem 3.5 in [5], one can prove the following theorem.

Theorem 5. *Let $e_{n,\alpha_k}^{h,\delta}$ and q be as in equation (3.15) and (3.16) respectively, $y_{n,\alpha_k}^{h,\delta}$ and $x_{n,\alpha_k}^{h,\delta}$ be as defined in (2.8) and (2.9) respectively with $\delta \in (0, \delta_0]$ and $\varepsilon_h \in (0, \varepsilon_0]$. Then under the Assumptions of Theorem 3, (3.14) and (3.17), $x_{n,\alpha_k}^{h,\delta}, y_{n,\alpha_k}^{h,\delta} \in B_r(P_hx_0)$ and the following estimates hold for all $n \ge 0$.*

(a) $\|x_{n,\alpha_k}^{h,\delta} - y_{n-1,\alpha_k}^{h,\delta}\| \le qe_{n-1,\alpha_k}^{h,\delta}$;
(b) $\|x_{n,\alpha_k}^{h,\delta} - x_{n-1,\alpha_k}^{h,\delta}\| \le (1 + q)e_{n-1,\alpha_k}^{h,\delta}$;
(c) $\|y_{n,\alpha_k}^{h,\delta} - x_{n,\alpha_k}^{h,\delta}\| \le q^2e_{n-1,\alpha_k}^{h,\delta}$;
(d) $e_{n,\alpha_k}^{h,\delta} \le q^{2n}\gamma_\rho, \quad \forall n \ge 0.$

The main result of this section is the following theorem.

Theorem 6. *Let $y_{n,\alpha_k}^{h,\delta}$ and $x_{n,\alpha_k}^{h,\delta}$ be as in (2.8) and (2.9) respectively and assumptions of Theorem 5 hold. Then $(x_{n,\alpha_k}^{h,\delta})$ is a Cauchy sequence in $B_r(P_hx_0)$ and converges, say to $x_{c,\alpha_k}^{h,\delta} \in \overline{B_r(P_hx_0)}$. Further $P_h[F(x_{c,\alpha_k}^{h,\delta}) + \frac{\alpha_k}{c}(x_{c,\alpha_k}^{h,\delta} - x_0)] = z_{\alpha_k}^{h,\delta}$ and $\|x_{n,\alpha_k}^{h,\delta} - x_{c,\alpha_k}^{h,\delta}\| \le C_0q^{2n}$ where $C_0 = \frac{\gamma_\rho}{1-q}$.*

Proof. Using the relation (b) and (c) of Theorem 5 and (3.17), we observe that

$$\|x_{n+m,\alpha_k}^{h,\delta} - x_{n,\alpha_k}^{h,\delta}\| \le \sum_{i=0}^{i=m-1} \|x_{n+i+1,\alpha_k}^{h,\delta} - x_{n+i,\alpha_k}^{h,\delta}\|$$

$$\le \sum_{i=0}^{i=m-1} (1 + q)q^{2(n+i)} e_{0,\alpha_k}^{h,\delta}$$

$$\le q^{2n}[\frac{1 - (q^2)^{m+1}}{1 - q}]\gamma_\rho \le C_0q^{2n}.$$

Thus $(x_{n,\alpha_k}^{h,\delta})$ is a Cauchy sequence in $B_r(P_h x_0)$ and hence it converges, say to $x_{c,\alpha_k}^{h,\delta} \in \overline{B_r(P_h x_0)}$. Observe that from (2.8)

$$\|P_h(F(x_{n,\alpha_k}^{h,\delta}) - z_{\alpha_k}^{h,\delta}) + \frac{\alpha_k}{c}(x_{n,\alpha_k}^{h,\delta} - P_h x_0)\| = \|R(x_{0,\alpha_k}^{h,\delta})(y_{n,\alpha_k}^{h,\delta} - x_{n,\alpha_k}^{h,\delta})\|$$

$$\leq (M + \frac{\alpha_k}{c})q^{2n}\gamma_\rho. \qquad (3.18)$$

Now by letting $n \to \infty$ in (3.18) we obtain $P_h F(x_{c,\alpha_k}^{h,\delta}) + \frac{\alpha_k}{c}(x_{c,\alpha_k}^{h,\delta} - P_h x_0) = z_{\alpha_k}^{h,\delta}$. This completes the proof.

Remark 7. *Note that* $0 < q < 1$ *and hence the sequence* $(x_{n,\alpha_k}^{h,\delta})$ *converges linearly to* $x_{c,\alpha_k}^{h,\delta}$.

Hereafter we assume that $r < \frac{1}{k_0}$ and $K_1 < \frac{1-k_0 r}{1-c}$. The following Assumption and Theorems (Theorem 9 and Theorem 10) from [5] are used to prove our results in this section.

Assumption 8. *There exists a continuous, strictly monotonically increasing function* $\varphi_1 : (0, b] \to (0, \infty)$ *with* $b \geq \|F'(x_0)\|$ *satisfying;*

$- \displaystyle\lim_{\lambda \to 0} \varphi_1(\lambda) = 0,$

$$\sup_{\lambda \geq 0} \frac{\alpha \varphi_1(\lambda)}{\lambda + \alpha} \leq \varphi_1(\alpha) \qquad \forall \lambda \in (0, b]$$

and

$-$ *there exists* $v \in X$ *with* $\|v\| \leq 1$ *(cf.[8]) such that*

$$x_0 - \hat{x} = \varphi_1(F'(x_0))v.$$

$-$ *for each* $x \in B_r(x_0) := \{x : \|x - x_0\| < r\}$ *there exists a bounded linear operator* $G(x, x_0)$ *(cf.[10]) such that* $F'(x) = F'(x_0)G(x, x_0)$ *with* $\|G(x, x_0)\| \leq K_1$.

Theorem 9. *(see [5], Theorem 3.7) Suppose* x_{c,α_k}^{δ} *is the solution of* $F(x) + \frac{\alpha_k}{c}(x - x_0) = z_{\alpha_k}^{\delta}$ *and Assumption 4 and 8 hold. Then*

$$\|\hat{x} - x_{c,\alpha_k}^{\delta}\| \leq \frac{\varphi_1(\alpha_k) + (2 + \frac{4\mu}{\mu-1})\mu\psi^{-1}(\delta + \varepsilon_h)}{1 - (1 - c)K_1 - k_0 r}.$$

Theorem 10. *(see [5], Theorem 3.8) Suppose* $x_{c,\alpha_k}^{h,\delta}$ *is the solution of (3.13) and Assumption 1 and Theorem 9 hold. In addition if* $\tau_0 < 1$, *then*

$$\|x_{c,\alpha_k}^{h,\delta} - x_{c,\alpha_k}^{\delta}\| \leq \frac{2}{1 - \tau_0}(\frac{\delta + \varepsilon_h}{\sqrt{\alpha_k}})$$

The following Theorem is a consequence of Theorem 6, Theorem 9 and Theorem 10.

Theorem 11. *Let $x_{n,\alpha_k}^{h,\delta}$ be as in (2.9), assumptions in Theorem 6, Theorem 9 and Theorem 10 hold. Then*

$$\|\hat{x} - x_{n,\alpha_k}^{h,\delta}\| \leq C_0 q^{2n} + \frac{\varphi_1(\alpha_k) + (2 + \frac{4\mu}{\mu-1})\mu\psi^{-1}(\delta + \varepsilon_h)}{1 - (1-c)K_1 - k_0 r} + \frac{2}{1 - \tau_0}\left(\frac{\delta + \varepsilon_h}{\sqrt{\alpha_k}}\right)$$

where C_0 is as in Theorem 6.

Theorem 12. *Let $x_{n,\alpha_k}^{h,\delta}$ be as in (2.9) and assumptions in Theorem 11 hold. Further let $\varphi_1(\alpha_k) \leq \varphi(\alpha_k)$ and $n_k := \min\{n : q^{2n} \leq \frac{\delta + \varepsilon_h}{\sqrt{\alpha_k}}\}$. Then*

$$\|\hat{x} - x_{n_k,\alpha_k}^{h,\delta}\| = O(\psi^{-1}(\delta + \varepsilon_h)).$$

4 Numerical Example

Let $KF : L^2(0,1) \longrightarrow L^2(0,1)$ with $K : L^2(0,1) \longrightarrow L^2(0,1)$ defined by

$$K(x)(t) = \int_0^1 k(t,s)x(s)ds$$

and $F : D(F) \subseteq L^2(0,1) \longrightarrow L^2(0,1)$ defined by

$$F(u) := \int_0^1 k(t,s)u^3(s)ds,$$

where

$$k(t,s) = \begin{cases} (1-t)s, 0 \leq s \leq t \leq 1 \\ (1-s)t, 0 \leq t \leq s \leq 1 \end{cases}.$$

Then F is monotone and the Fréchet derivative of F is given by

$$F'(u)w = 3\int_0^1 k(t,s)(u(s))^2 w(s)ds.$$

(see [11], section 4.3) Further observe that

$$[F(x_0) - F(\hat{x})](s) = \int_0^1 k(s,t)(x_0^3(s) - \hat{x}^3(s))ds$$

$$= K(x_0^3 - \hat{x}^3) \in R(\varphi(K^*K)^{\frac{1}{2}})$$

where $\varphi(\lambda) = \lambda^{\frac{1}{2}}$. In our computation, we take $y(t) = \frac{t^{13}}{17160} - \frac{279t^3}{12320} + \frac{t^4}{192} + \frac{t^7}{1120} + \frac{t^{10}}{3360} + \frac{15557t}{960960}$ and $y^\delta = y + \delta$. Then the exact solution

$$\hat{x}(t) = 0.5 + t^3.$$

We use $x_0(t) = 0.5 + t^3 - \frac{3t^8}{56} - \frac{3t^5}{20} - \frac{3t^2}{8} + \frac{81t}{140}$ as our initial guess, so that the function $x_0 - \hat{x}$ satisfies the source condition $x_0 - \hat{x} = \varphi_1(F'(x_0))1$ where $\varphi_1(\lambda) = \lambda$. Thus we expect to have an accuracy of order at least $O((\delta + \varepsilon_h)^{\frac{1}{2}})$.

We choose $\alpha_0 = (1.3)^2(\delta + \varepsilon_h)^2$, $\mu = 1.3$, $\delta = 0.0667 =: c$, $\varepsilon_h = O(n^{-2})$, the Lipschitz constant k_0 equals approximately 0.21 as in [11] and $r = 1$, $\tau_0 = \frac{1}{64}$, so that $q = (1 + \tau_0)k_0 r = 0.2133$. The results of the computation are presented in Table 1. The plots of the exact solution and the approximate solution $x_{n,\alpha}^{h,\delta}$ obtained for n=512 and n=1024 are given in Figure 1.

Table 1.

n	k	α_k	$\|x_k - \hat{x}\|$	$\frac{\|x_k - \hat{x}\|}{(\delta + \varepsilon_h)^{1/2}}$
8	4	0.0494	0.8157	3.1229
16	4	0.0477	0.8854	3.4192
32	4	0.0473	0.9188	3.5557
64	4	0.0472	0.9351	3.6208
128	4	0.0471	0.9431	3.6525
256	4	0.0471	0.9471	3.6682
512	4	0.0471	0.9491	3.6760
1024	4	0.0471	0.9501	3.6798

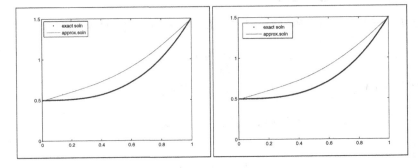

Fig. 1. Curves of the exact and approximate solutions for $x_{n,\alpha}^{h,\delta}$, when n=512 (left), n=1024 (right)

5   Conclusion

We presented a finite dimensional realization of Guass Newton method for the solution of non-linear ill-posed Hammerstein type operator equations $KF(x) = f$. Here $F : D(F) \subseteq X \to X$ is nonlinear monotone operator, $K : X \to Y$ is a bounded linear operator and the available data is f^δ with $\|f - f^\delta\| \leq \delta$. The proposed method combines the Tikhonov regularization and Guass Newton method. As the iterations involve the Fréchet derivative only at the Projection of the initial approximation, the method is simpler than the method considered in George and Shobha (2012). We have chosen the regularization parameter according to balancing principle of Pereverzev and Schock (2005). The error estimate is of optimal order and the proposed method leads to local linear convergence. The numerical example confirm the efficiency of our method.

Acknowledgement. Ms.Shobha, thanks National Institute of Technology Karnataka, India, for the financial support.

References

1. George, S.: Newton-Tikhonov regularization of ill-posed Hammerstein operator equation. J. Inverse and Ill-Posed Problems 2(14), 135–146 (2006)
2. George, S.: Newton-Lavrentiev regularization of ill-posed Hammerstein operator equation. J. Inverse and Ill-Posed Problems 6(14), 573–582 (2006)
3. George, S., Kunhanandan, M.: An iterative regularization method for Ill-posed Hammerstein type operator equation. J. Inv. Ill-Posed Problems 17, 831–844 (2009)
4. George, S., Nair, M.T.: A modified Newton-Lavrentiev regularization for nonlinear ill-posed Hammerstein-Type operator equation. Journal of Complexity 24, 228–240 (2008)
5. George, S., Shobha, M.E.: Two Step Newton Tikhonov Method for Hammerstein Type Equations: Finite Dimensional Realization. ISRN, Journal of Applied Mathematics (to appear-ID 783579)
6. Kaltenbacher, B., Neubauer, A., Scherzer, O.: Iterative regularisation methods for nolinear ill-posed porblems. de Gruyter, Berlin (2008)
7. Kelley, C.T.: Iterative Methods for Linear and Nonlinear Equations. SIAM, Philadelphia (1995)
8. Nair, M.T., Ravishankar, P.: Regularized versions of continuous newton's method and continuous modified newton's method under general source conditions. Numer. Funct. Anal. Optim. 29(9-10), 1140–1165 (2008)
9. Pereverzev, S., Schock, E.: On the adaptive selection of the parameter in regularization of ill-posed problems. SIAM J. Numer. Anal. 43(5), 2060–2076 (2005)
10. Ramm, A.G., Smirnova, A.B., Favini, A.: Continuous modified Newton's-type method for nonlinear operator equations. Ann. Mat. Pura Appl. 182, 37–52 (2003)
11. Semenova, E.V.: Lavrentiev regularization and balancing principle for solving ill-posed problems with monotone operators. Comput. Methods Appl. Math. 4, 444–454 (2010)

Projection Scheme for Newton-Type Iterative Method for Lavrentiev Regularization

Suresan Pareth and Santhosh George

Department of Mathematical and Computational Sciences,
National Institute of Technology Karnataka, Surathkal-575025
sureshpareth@rediffmail.com, sgeorge@nitk.ac.in

Abstract. In this paper we consider the finite dimensional realization of a Newton-type iterative method for obtaining an approximate solution to the nonlinear ill-posed operator equation $F(x) = f$, where $F : D(F) \subseteq X \to X$ is a nonlinear monotone operator defined on a real Hilbert space X. It is assumed that $F(\hat{x}) = f$ and that the only available data are f^δ with $\|f - f^\delta\| \leq \delta$. It is proved that the proposed method has a local convergence of order three. The regularization parameter α is chosen according to the balancing principle considered by Perverzev and Schock (2005) and obtained an optimal order error bounds under a general source condition on $x_0 - \hat{x}$ (here x_0 is the initial approximation). The test example provided endorses the reliability and effectiveness of our method.

Keywords: Newton Lavrentiev method, nonlinear ill-posed operator equation, nonlinear monotone operator, balancing principle, finite dimensional.

1 Introduction

Throughout this paper X is a real Hilbert space and $F : D(F) \subseteq X \to X$ is a monotone operator, i.e.,

$$\langle F(x) - F(y), x - y \rangle \geq 0, \quad \forall x, y \in D(F).$$

The inner product and the norm in X are denoted by $\langle .,. \rangle$ and $\|.\|$ respectively. We consider the problem of approximately solving the ill-posed operator equation

$$F(x) = f \tag{1}$$

in the finite dimensional setting.

Let $S := \{x : F(x) = f\}$. Then S is closed and convex if F is monotone and continuous (see, e.g., [10]) and hence has a unique element of minimal norm, denoted by $\hat{x}$ such that $F(\hat{x}) = f$.

We assume that F possesses a locally uniformly bounded, self adjoint Fréchet derivative $F'(.)$ (i.e., there exists some constant C_F such that $\|F'(x)\| \leq C_F$) in the domain $D(F)$ of F. Note that since F is monotone, $F'(.) \geq 0$, i.e., $F'(.)$ is

J. Mathew et al. (Eds.): ICECCS 2012, CCIS 305, pp. 302–310, 2012.
© Springer-Verlag Berlin Heidelberg 2012

a positive self adjoint operator and hence $(F'(.) + \alpha I)^{-1}$ exists for any $\alpha > 0$. In application, usually only noisy data f^δ are available, such that $\|f - f^\delta\| \leq \delta$. Since (1) is ill-posed, the regularization methods are used to obtain a stable approximate solution for (1).

The Lavrentiev regularization method (see [1,6,11,12]) is used for appropriately solving (1) when F is monotone. In this method the regularized approximation x_α^δ is obtained by solving the operator equation

$$F(x) + \alpha(x - x_0) = f^\delta \qquad (2)$$

where $\alpha > 0$ is the regularization parameter and $x_0 \in D(F)$ is a known initial approximation of the solution $\hat{x}$. From the general regularization theory it is known that the equation (2) has a unique solution x_α^δ for any $\alpha > 0$ and $x_\alpha^\delta \to \hat{x}$ as $\alpha \to 0, \delta \to 0$ provided α is chosen appropriately (see, [9] and [12]).

In [3], the authors considered a Two Step Newton Lavrentiev Method (TSNLM) for approximating the solution x_α^δ of (2). In this paper we consider the finite dimensional realization of the method considered in [3].

This paper is organized as follows. In section 2, we set up the method and analyze its convergence. The error analysis under a general source condition is considered in Section 3. The numerical example and the computational results are presented in section 4. Finally a conclusion is made in section 5.

2   The Method and Its Convergence

The purpose of this section is to obtain an approximate solution for the equation (2), in the finite dimensional subspace of X. Let $\{P_h\}_{h>0}$ be a family of orthogonal projections on X.

Let $\varepsilon_h := \|F'(x)(I - P_h)\|, \quad \forall x \in D(F)$ and $\{b_h : h > 0\}$ be such that $\lim_{h \to 0} \frac{\|(I - P_h)x_0\|}{b_h} = 0$ and $\lim_{h \to 0} b_h = 0$. We assume that $\varepsilon_h \to 0$ as $h \to 0$. The above assumption is satisfied if, $P_h \to I$ pointwise and if $F'(x)$ is a compact operator. Further we assume that $\varepsilon_h \leq \varepsilon_0$, $b_h \leq b_0$ and $\delta \in (0, \delta_0]$.

2.1   Projection Method

Let $x_{0,\alpha}^{h,\delta} := P_h x_0$ be the projection of the initial guess x_0 on to $R(P_h)$, the range of P_h and let $R_\alpha(x) := P_h F'(x) P_h + \alpha P_h$ with $\alpha > \alpha_0 > 0$. We define the iterative sequence as:

$$y_{n,\alpha}^{h,\delta} = x_{n,\alpha}^{h,\delta} - R_\alpha^{-1}(x_{n,\alpha}^{h,\delta}) P_h[F(x_{n,\alpha}^{h,\delta}) - f^\delta + \alpha(x_{n,\alpha}^{h,\delta} - x_0)] \qquad (3)$$

and

$$x_{n+1,\alpha}^{h,\delta} = y_{n,\alpha}^{h,\delta} - R_\alpha^{-1}(x_{n,\alpha}^{h,\delta}) P_h[F(y_{n,\alpha}^{h,\delta}) - f^\delta + \alpha(y_{n,\alpha}^{h,\delta} - x_0)]. \qquad (4)$$

Note that the iteration (3) and (4) are the finite dimensional realization of the iteration (3) and (4) in [3]. We will be selecting the parameter $\alpha = \alpha_i$ from some finite set $D_N = \{\alpha_i : 0 < \alpha_0 < \alpha_1 < \alpha_2 < \cdots < \alpha_N\}$ using the adaptive method considered by Perverzev and Schock in [9].

The following assumptions and Lemmas are used for proving our results.

Assumption 1. *(cf. [11], Assumption 3) There exists a constant $0 < k_0 \leq 1$ such that for every $x, u \in D(F)$ and $v \in X$, there exists an element $\Phi(x, u, v) \in X$ satisfying*

$$[F'(x) - F'(u)]v = F'(u)\Phi(x, u, v), \quad \|\Phi(x, u, v)\| \leq k_0 \|v\| \|x - u\|.$$

Assumption 2. *There exists a continuous, strictly monotonically increasing function $\varphi : (0, a] \to (0, \infty)$ with $a \geq \|F'(\hat{x})\|$ satisfying;*

(i) $\lim_{\lambda \to 0} \varphi(\lambda) = 0$,

(ii) $\sup_{\lambda \geq 0} \frac{\alpha \varphi(\lambda)}{\lambda + \alpha} \leq c_\varphi \varphi(\alpha) \quad \forall \lambda \in (0, a]$ *and*

(iii) *there exists $v \in X$ with $\|v\| \leq 1$ (cf. [8]) such that $x_0 - \hat{x} = \varphi(F'(\hat{x}))v$.*

Let

$$e_{n,\alpha}^{h,\delta} := \|y_{n,\alpha}^{h,\delta} - x_{n,\alpha}^{h,\delta}\|, \qquad \forall n \geq 0 \tag{5}$$

and for $0 < k_0 < min\{1, \dfrac{\sqrt{8}}{(1 + \frac{\varepsilon_0}{\alpha_0})\sqrt{4 + 3(1 + \frac{\varepsilon_0}{\alpha_0})}}\}$, let $g : (0, 1) \to (0, 1)$ be the function defined by

$$g(t) = \frac{k_0^2}{8}(1 + \frac{\varepsilon_0}{\alpha_0})^2 (4 + 3k_0(1 + \frac{\varepsilon_0}{\alpha_0})t)t^2 \qquad \forall t \in (0, 1). \tag{6}$$

Hereafter we assume that $b_0 < \dfrac{\sqrt{1 + \frac{2k_0}{(1 + \frac{\varepsilon_0}{\alpha_0})}(1 - \frac{\delta_0}{\alpha_0})} - 1}{k_0}$, $\delta_0 < \alpha_0$ for some $\alpha_0 > 0$ and $\|x_0 - \hat{x}\| \leq \rho$ where

$$\rho \leq \frac{\sqrt{1 + \frac{2k_0}{(1 + \frac{\varepsilon_0}{\alpha_0})}(1 - \frac{\delta_0}{\alpha_0})} - 1}{k_0} - b_0. \tag{7}$$

Let

$$\gamma_\rho := (1 + \frac{\varepsilon_0}{\alpha_0})[\frac{k_0}{2}(\rho + b_0)^2 + (\rho + b_0)] + \frac{\delta_0}{\alpha_0}. \tag{8}$$

The proof of the following lemmata and theorem are analogous to the proof of lemmata and theorem in [4].

Lemma 1. *(see [4], Lemma 1) Let $x \in D(F)$. Then $\|R_\alpha^{-1}(x)P_h F'(x)\| \leq (1 + \frac{\varepsilon_0}{\alpha_0})$.*

Lemma 2. *(see [4], Lemma 2) Let $e_0 = e_{0,\alpha}^{h,\delta}$ and γ_ρ be as in (8). Then $e_0 \leq \gamma_\rho < 1$.*

Lemma 3. *(cf. [4], Lemma 3) Let $y_{n,\alpha}^{h,\delta}$, $x_{n,\alpha}^{h,\delta}$ and $e_{n,\alpha}^{h,\delta}$ be as in (3), (4) and (5) respectively with $\delta \in (0, \delta_0]$. Then*

(a) $\|x_{n,\alpha}^{h,\delta} - y_{n-1,\alpha}^{h,\delta}\| \leq \frac{k_0}{2}(1 + \frac{\varepsilon_0}{\alpha_0})e_{n-1,\alpha}^{h,\delta}\|y_{n-1,\alpha}^{h,\delta} - x_{n-1,\alpha}^{h,\delta}\|$;

(b) $\|x_{n,\alpha}^{h,\delta} - x_{n-1,\alpha}^{h,\delta}\| \leq (1 + \frac{k_0}{2}(1 + \frac{\varepsilon_0}{\alpha_0})e_{n-1,\alpha}^{h,\delta})\|y_{n-1,\alpha}^{h,\delta} - x_{n-1,\alpha}^{h,\delta}\|$;

Theorem 1. *(cf. [4], Theorem 1) Let $e_{n,\alpha}^{h,\delta}$ and g be as in equation (5) and (6) respectively, $y_{n,\alpha}^{h,\delta}$ and $x_{n,\alpha}^{h,\delta}$ be as in (3) and (4) respectively with $\delta \in (0, \delta_0]$. Let the Assumption 1 holds. Then*

(a) $\|y_{n,\alpha}^{h,\delta} - x_{n,\alpha}^{h,\delta}\| \leq g(e_{n-1,\alpha}^{h,\delta})\|y_{n-1,\alpha}^{h,\delta} - x_{n-1,\alpha}^{h,\delta}\|$;

(b) $g(e_{n,\alpha}^{h,\delta}) \leq g(\gamma_\rho)^{3^n}$, $\qquad \forall n \geq 0$;

(c) $e_{n,\alpha}^{h,\delta} \leq g(\gamma_\rho)^{(3^n - 1)/2}\gamma_\rho \qquad \forall n \geq 0$.

Theorem 2. *Let $r = [\frac{1}{1-g(\gamma_\rho)} + \frac{k_0}{2}(1 + \frac{\varepsilon_0}{\alpha_0})\frac{\gamma_\rho}{1-g(\gamma_\rho)^2}]\gamma_\rho$ and the assumptions of Theorem 1 hold. Then $x_{n,\alpha}^{h,\delta}, y_{n,\alpha}^{h,\delta} \in B_r(P_h x_0)$, for all $n \geq 0$.*

Proof. Note that by Lemma 2 and (b) of the Lemma 3 we have

$$\|x_{1,\alpha}^{h,\delta} - P_h x_0\| = \|x_{1,\alpha}^{h,\delta} - x_{0,\alpha}^{h,\delta}\|$$

$$\leq [1 + \frac{k_0}{2}(1 + \frac{\varepsilon_0}{\alpha_0})e_0]e_0 \qquad (9)$$

$$\leq [1 + \frac{k_0}{2}(1 + \frac{\varepsilon_0}{\alpha_0})\gamma_\rho]\gamma_\rho < r$$

i.e., $x_{1,\alpha}^{h,\delta} \in B_r(P_h x_0)$. Again note that by (9), Lemma 2 and (a) of Theorem 1 we have

$$\|y_{1,\alpha}^{h,\delta} - P_h x_0\| \leq \|y_{1,\alpha}^{h,\delta} - x_{1,\alpha}^{h,\delta}\| + \|x_{1,\alpha}^{h,\delta} - P_h x_0\|$$

$$\leq [1 + g(e_0) + \frac{k_0}{2}(1 + \frac{\varepsilon_0}{\alpha_0})e_0]e_0$$

$$\leq [1 + g(\gamma_\rho) + \frac{k_0}{2}(1 + \frac{\varepsilon_0}{\alpha_0})\gamma_\rho]\gamma_\rho < r$$

i.e., $y_{1,\alpha}^{h,\delta} \in B_r(P_h x_0)$. Further by (9), Lemma 2 and (b) of Lemma 3 we have

$$\|x_{2,\alpha}^{h,\delta} - P_h x_0\| \leq \|x_{2,\alpha}^{h,\delta} - x_{1,\alpha}^{h,\delta}\| + \|x_{1,\alpha}^{h,\delta} - P_h x_0\|$$

$$\leq [1 + \frac{k_0}{2}(1 + \frac{\varepsilon_0}{\alpha_0})e_{1,\alpha}^{h,\delta}]e_{1,\alpha}^{h,\delta} + [1 + \frac{k_0}{2}(1 + \frac{\varepsilon_0}{\alpha_0})e_0]e_0$$

$$\leq [1 + \frac{k_0}{2}(1 + \frac{\varepsilon_0}{\alpha_0})g(e_0)e_0]g(e_0)e_0 + [1 + \frac{k_0}{2}(1 + \frac{\varepsilon_0}{\alpha_0})e_0]e_0$$

$$\leq [1 + g(e_0) + \frac{k_0}{2}(1 + \frac{\varepsilon_0}{\alpha_0})e_0(1 + g(e_0)^2)]e_0 \qquad (10)$$

$$\leq [1 + g(\gamma_\rho) + \frac{k_0}{2}(1 + \frac{\varepsilon_0}{\alpha_0})\gamma_\rho(1 + g(\gamma_\rho)^2)]\gamma_\rho < r$$

and by (10), Lemma 2 and (a) of Theorem 1 we have

$$\|y_{2,\alpha}^{h,\delta} - P_h x_0\| \leq \|y_{2,\alpha}^{h,\delta} - x_{2,\alpha}^{h,\delta}\| + \|x_{2,\alpha}^{h,\delta} - P_h x_0\|$$

$$\leq g(e_{1,\alpha}^{h,\delta})e_{1,\alpha}^{h,\delta} + [1 + g(e_0) + \frac{k_0}{2}(1 + \frac{\varepsilon_0}{\alpha_0})e_0(1 + g(e_0)^2)]e_0$$

$$\leq g(e_0)^4 e_0 + [1 + g(e_0) + \frac{k_0}{2}1 + \frac{\varepsilon_0}{\alpha_0})e_0(1 + g(e_0)^2)]e_0$$

$$\leq [1 + g(e_0) + g(e_0)^2 + \frac{k_0}{2}(1 + \frac{\varepsilon_0}{\alpha_0})e_0(1 + g(e_0)^2)]e_0$$

$$\leq [1 + g(\gamma_\rho) + g(\gamma_\rho)^2 + \frac{k_0}{2}(1 + \frac{\varepsilon_0}{\alpha_0})\gamma_\rho(1 + g(\gamma_\rho)^2)]\gamma_\rho < r$$

i.e., $x_{2,\alpha}^{h,\delta}, y_{2,\alpha}^{h,\delta} \in B_r(P_h x_0)$. Continuing this way one can prove that $x_{n,\alpha}^{h,\delta}, y_{n,\alpha}^{h,\delta} \in B_r(P_h x_0), \forall n \geq 0$. This completes the proof.

The main result of this section is the following Theorem.

Theorem 3. *Let $y_{n,\alpha}^{h,\delta}$ and $x_{n,\alpha}^{h,\delta}$ be as in (3) and (4) respectively with $\delta \in (0, \delta_0]$ and assumptions of Theorem 2 hold. Then $(x_{n,\alpha}^{h,\delta})$ is Cauchy sequence in $B_r(P_h x_0)$ and converges to $x_\alpha^{h,\delta} \in \overline{B_r(P_h x_0)}$. Further $P_h[F(x_\alpha^{h,\delta}) + \alpha(x_\alpha^{h,\delta} - x_0)] = P_h f^\delta$ and*

$$\|x_{n,\alpha}^{h,\delta} - x_\alpha^{h,\delta}\| \leq Ce^{-\gamma 3^n}$$

where $C = [\frac{1}{1-g(\gamma_\rho)^3} + \frac{k_0}{2}\gamma_\rho(1 + \frac{\varepsilon_0}{\alpha_0})\frac{1}{1-(g(\gamma_\rho)^2)^3}g(\gamma_\rho)^{3^n}]\gamma_\rho$ and $\gamma = -\log g(\gamma_\rho)$.

Proof. Using the relation (b) of Lemma 3 and (c) of Theorem 1, we obtain

$$\|x_{n+m,\alpha}^{h,\delta} - x_{n,\alpha}^{h,\delta}\| \leq \sum_{i=0}^{m-1} \|x_{n+i+1,\alpha}^{h,\delta} - x_{n+i,\alpha}^{h,\delta}\|$$

$$\leq \sum_{i=0}^{m-1}(1 + \frac{k_0}{2}(1 + \frac{\varepsilon_0}{\alpha_0})e_0 g(e_0)^{3^{n+i}})g(e_0)^{3^{n+i}}e_0$$

$$\leq [(1 + g(e_0)^3 + g(e_0)^{3^2} + \cdots + g(e_0)^{3^m}) + \frac{k_0}{2}(1 + \frac{\varepsilon_0}{\alpha_0})e_0(1 +$$

$$(g(e_0)^2)^3 + (g(e_0)^2)^{3^2} + \cdots + (g(e_0)^2)^{3^m})g(e_0)^{3^n}]g(e_0)^{3^n}e_0$$

$$\leq Cg(\gamma_\rho)^{3^n} \leq Ce^{-\gamma 3^n}.$$

Thus $x_{n,\alpha}^{h,\delta}$ is a Cauchy sequence in $B_r(P_h x_0)$ and hence it converges cubically, say to $x_\alpha^{h,\delta} \in \overline{B_r(P_h x_0)}$. Observe that

$$\|P_h(F(x_{n,\alpha}^{h,\delta}) - f^\delta + \alpha(x_{n,\alpha}^{h,\delta} - x_0))\| = \|R_\alpha(x_{n,\alpha}^{h,\delta})(x_{n,\alpha}^{h,\delta} - y_{n,\alpha}^{h,\delta})\|$$

$$\leq \|R_\alpha(x_{n,\alpha}^{h,\delta})\|\|(x_{n,\alpha}^{h,\delta} - y_{n,\alpha}^{h,\delta})\|$$

$$= \|(P_h F'(x_{n,\alpha}^{h,\delta})P_h + \alpha P_h)\|e_{n,\alpha}^{h,\delta}$$

$$\leq (C_F + \alpha)g(\gamma_\rho)^{\frac{3^n-1}{2}}\gamma_\rho. \qquad (11)$$

Now by letting $n \to \infty$ in (11) we obtain

$$P_h[F(x_\alpha^{h,\delta}) + \alpha(x_\alpha^{h,\delta} - x_0)] = P_h f^\delta. \qquad (12)$$

This completes the proof.

Hereafter we assume that $\rho \leq r$.

Remark 1. *Note that $\rho \leq r$, if $\frac{4(1-g(\gamma_\rho)^2)^2}{9\gamma_\rho^4}[2(1 - \frac{\delta_0}{\alpha_0}) - \frac{3\gamma_\rho^2}{1-g(\gamma_\rho)^2}] \leq k_0 \leq 1$. Further observe that $0 < g(\gamma_\rho) < 1$ and hence $\gamma > 0$.*

3   Error Bounds under Source Conditions

The following proposition and theorem from [4], are used to prove our main result of this section.

Proposition 1. *(cf. [4], Proposition 1) Let $F : D(F) \subseteq X \to X$ be a monotone operator in X. Let $x_\alpha^{h,\delta}$ be the solution of (12) and $x_\alpha^h := x_\alpha^{h,0}$. Then $\|x_\alpha^{h,\delta} - x_\alpha^h\| \le \frac{\delta}{\alpha}$.*

Theorem 4. *(cf. [4], Theorem 4) Let $\rho < \frac{2}{k_0(1+\frac{\varepsilon_0}{\alpha_0})}$ and $\hat{x} \in D(F)$ be a solution of (1). And let Assumption 1, Assumption 2 and the assumptions in Proposition 1 be satisfied. Then*

$$\|x_\alpha^h - \hat{x}\| \le \tilde{C}(\varphi(\alpha) + \frac{\varepsilon h}{\alpha})$$

where $\tilde{C} := \frac{\max\{1, \rho + \|\hat{x}\|\}}{1 - (1 + \frac{\varepsilon_0}{\alpha_0})\frac{k_0}{2}\rho}$.

Theorem 5. *Let $x_{n,\alpha}^{h,\delta}$ be as in (4). And the assumptions in Theorem 3 and Theorem 4 hold. Then*

$$\|x_{n,\alpha}^{h,\delta} - \hat{x}\| \le Ce^{-\gamma 3^n} + \max\{1, \tilde{C}\}(\varphi(\alpha) + \frac{\delta + \varepsilon h}{\alpha}).$$

Proof. Observe that,

$$\|x_{n,\alpha}^{h,\delta} - \hat{x}\| \le \|x_{n,\alpha}^{h,\delta} - x_\alpha^{h,\delta}\| + \|x_\alpha^{h,\delta} - x_\alpha^h\| + \|x_\alpha^h - \hat{x}\|$$

so, by Proposition 1, Theorem 3 and Theorem 4 we obtain,

$$\|x_{n,\alpha}^{h,\delta} - \hat{x}\| \le Ce^{-\gamma 3^n} + \frac{\delta}{\alpha} + \tilde{C}(\varphi(\alpha) + \frac{\varepsilon h}{\alpha})$$

$$\le Ce^{-\gamma 3^n} + \max\{1, \tilde{C}\}(\varphi(\alpha) + \frac{\delta + \varepsilon h}{\alpha}).$$

Let

$$n_\delta := \min\{n : e^{-\gamma 3^n} \le \frac{\delta + \varepsilon h}{\alpha}\} \tag{13}$$

and

$$C_0 = C + \max\{1, \tilde{C}\}. \tag{14}$$

Theorem 6. *Let n_δ and C_0 be as in (13) and (14) respectively. And let $x_{n_\delta, \alpha}^{h,\delta}$ be as in (4) and the assumptions in Theorem 5 be satisfied. Then*

$$\|x_{n_\delta, \alpha}^{h,\delta} - \hat{x}\| \le C_0(\varphi(\alpha) + \frac{\delta + \varepsilon h}{\alpha}). \tag{15}$$

3.1   A Priori Choice of the Parameter

Note that the error estimate $\varphi(\alpha) + \frac{\delta + \varepsilon h}{\alpha}$ in (15) is of optimal order if $\alpha_\delta := \alpha(\delta, h)$ satisfies, $\varphi(\alpha_\delta)\alpha_\delta = \delta + \varepsilon h$.

Now as in [2], using the function $\psi(\lambda) := \lambda\varphi^{-1}(\lambda), 0 < \lambda \leq a$ we have $\delta + \varepsilon_h = \alpha_\delta\varphi(\alpha_\delta) = \psi(\varphi(\alpha_\delta))$, so that $\alpha_\delta = \varphi^{-1}(\psi^{-1}(\delta + \varepsilon_h))$.

Theorem 7. *Let* $\psi(\lambda) := \lambda\varphi^{-1}(\lambda)$ *for* $0 < \lambda \leq a$ *and the assumptions in Theorem 6 hold. For* $\delta > 0$, *let* $\alpha_\delta = \varphi^{-1}(\psi^{-1}(\delta + \varepsilon_h))$ *and let* n_δ *be as in (13). Then*

$$\|x_{n_\delta,\alpha}^{h,\delta} - \hat{x}\| = O(\psi^{-1}(\delta + \varepsilon_h)).$$

3.2   An Adaptive Choice of the Parameter

In this subsection, we present a parameter choice rule based on the balancing principle studied in [7], [9]. In this method, the regularization parameter α is selected from some finite set $D_N(\alpha) := \{\alpha_i = \mu^i\alpha_0, i = 0, 1, \cdots, N\}$ where $\mu > 1$, $\alpha_0 > 0$ and let $n_i := min\{n : e^{-\gamma 3^n} \leq \frac{\delta + \varepsilon_h}{\alpha_i}\}$. Then for $i = 0, 1, \cdots, N$, we have

$$\|x_{n_i,\alpha_i}^{h,\delta} - x_{\alpha_i}^{h,\delta}\| \leq C\frac{\delta + \varepsilon_h}{\alpha_i}, \qquad \forall i = 0, 1, \cdots N.$$

Let $x_i := x_{n_i,\alpha_i}^{h,\delta}$, $i = 0, 1, \cdots, N$. Proof of the following theorem is analogous to the proof of Theorem 3.1 in [11].

Theorem 8. *(cf. [11], Theorem 3.1) Assume that there exists* $i \in \{0, 1, 2, \cdots, N\}$ *such that* $\varphi(\alpha_i) \leq \frac{\delta + \varepsilon_h}{\alpha_i}$. *Let the assumptions of Theorem 6 and Theorem 7 hold and let*

$$l := max\{i : \varphi(\alpha_i) \leq \frac{\delta + \varepsilon_h}{\alpha_i}\} < N,$$

$k := max\{i : \|x_i - x_j\| \leq 4C_0\frac{\delta + \varepsilon_h}{\alpha_j}, \quad j = 0, 1, 2, \cdots, i\}$. *Then* $l \leq k$ *and* $\|\hat{x} - x_k\| \leq c\psi^{-1}(\delta + \varepsilon_h)$ *where* $c = 6C_0\mu$.

4   Computation

Example 1. *(see [11], section 4.3) Let* $F : D(F) \subseteq L^2(0,1) \longrightarrow L^2(0,1)$ *defined by*

$$F(u) := \int_0^1 k(t,s)u^3(s)ds,$$

where

$$k(t,s) = \begin{cases} (1-t)s, & 0 \leq s \leq t \leq 1 \\ (1-s)t, & 0 \leq t \leq s \leq 1 \end{cases}.$$

The Fréchet derivative of F *is given by*

$$F'(u)w = 3\int_0^1 k(t,s)u^2(s)w(s)ds. \tag{16}$$

In our computation, we take $f(t) = \frac{t-t^{11}}{110}$ *and* $f^\delta = f + \delta$. *Then the exact solution is* $\hat{x}(t) = t^3$. *We use* $x_0(t) = t^3 + \frac{3}{56}(t - t^8)$ *as our initial guess, so*

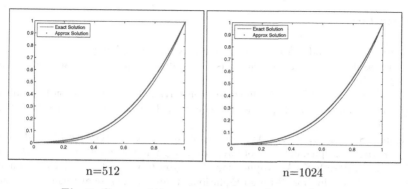

n=512   n=1024

Fig. 1. Curves of the exact and approximate solutions

Table 1. Iterations and corresponding error estimates

n	k	n_k	$\delta + \varepsilon_n$	α_k	$\|x_k - \hat{x}\|$	$\frac{\|x_k - \hat{x}\|}{(\delta + \varepsilon_n)^{1/2}}$
8	2	2	0.1016	0.3428	0.2634	0.8266
16	2	2	0.1004	0.3388	0.1962	0.6191
32	2	2	0.1001	0.3378	0.1429	0.4518
64	2	2	0.1000	0.3376	0.1036	0.3275
128	2	2	0.1000	0.3375	0.0755	0.2387
256	2	2	0.1000	0.3375	0.0560	0.1772
512	2	2	0.1000	0.3375	0.0430	0.1360
1024	2	2	0.1000	0.3375	0.0347	0.1096

that the function $x_0 - \hat{x}$ satisfies the source condition $x_0 - \hat{x} = \varphi(F'(\hat{x}))1$ where $\varphi(\lambda) = \lambda$. For the operator $F'(.)$ defined in (16), $\varepsilon_n = O(n^{-2})$ (cf. [5]). Thus we expect to obtain the rate of convergence $O((\delta + \varepsilon_n)^{\frac{1}{2}})$.

We choose $\alpha_0 = (1.1)(\delta + \varepsilon_n)$, $\mu = 1.1, \rho = 0.1, k_0 = 1, \gamma_\rho = 0.766$ and $g(\gamma_\rho) = 0.461$. The results of the computation are presented in Table 1. The plots of the exact solution and the approximate solution $x_{n,\alpha}^{h,\delta}$ obtained for $n = 2^i$, $i = 9, 10$ are given in Figure 1.

5   Conclusion

We considered a finite dimensional realization of a method considered in [3] for finding an approximate solution for a nonlinear ill-posed operator equation $F(x) = f$ where $F : D(F) \subseteq X \to X$ is a monotone operator defined on a real Hilbert space X, when the available data is f^δ with $\|f - f^\delta\| \leq \delta$. The regularization parameter α is chosen according to the balancing principle considered by Perverzev and Schock in [9]. The error bounds obtained under a general source condition on $x_0 - \hat{x}$ is order optimal. The computational results provided confirm the reliability and effectiveness of our method.

Acknowledgments. S.Pareth thanks NITK for the financial support.

References

1. George, S., Elmahdy, A.I.: A quadratic convergence yielding iterative method for nonlinear ill-posed operator equations. Comput. Methods Appl. Math. 12(1), 32–45 (2012)
2. George, S., Nair, M.T.: A modified Newton-Lavrentiev regularization for nonlinear ill-posed Hammerstein-Type operator equation. Journal of Complexity 24, 228–240 (2008)
3. George, S., Pareth, S.: An Application of Newton-type Iterative Method for the Approximate Implementation of Lavrentiev Regularization (2012) (communicated)
4. George, S., Pareth, S.: An Application of Newton-type Iterative Method for Lavrentiev Regularization for Ill-Posed Equations: Finite Dimensional Realization (2012) (communicated)
5. Groetsch, C.W., King, J.T., Murio, D.: Asymptotic analysis of a finite element method for Fredholm equations of the first kind. In: Baker, C.T.H., Miller, G.F. (eds.) Treatment of Integral Equations by Numerical Methods, pp. 1–11. Academic Press, London (1982)
6. Jaan, J., Tautenhahn, U.: On Lavrentiev regularization for ill-posed problems in Hilbert scales. Numer. Funct. Anal. Optim. 24(5-6), 531–555 (2003)
7. Mathe, P., Perverzev, S.V.: Geometry of linear ill-posed problems in variable Hilbert scales. Inverse Problems 19(3), 789–803 (2003)
8. Nair, M.T., Ravishankar, P.: Regularized versions of continuous Newton's method and continuous modified Newton's method under general source conditions. Numer. Funct. Anal. Optim. 29(9-10), 1140–1165 (2008)
9. Perverzev, S.V., Schock, E.: On the adaptive selection of the parameter in regularization of ill-posed problems. SIAM J. Numer. Anal. 43, 2060–2076 (2005)
10. Ramm, A.G.: Dynamical system method for solving operator equations. Elsevier, Amsterdam (2007)
11. Semenova, E.V.: Lavrentiev regularization and balancing principle for solving ill-posed problems with monotone operators. Comput. Methods Appl. Math. (4), 444–454 (2010)
12. Tautanhahn, U.: On the method of Lavrentiev regularization for nonlinear ill-posed problems. Inverse Problems 18, 191–207 (2002)

Modelling the Wind Speed Oscillation Dynamics

K. Asokan[1] and K. Satheesh Kumar[2]

[1] Department of Mathematics, College of Engineering, Thiruvananthapuram,
Kerala, India - 695 016
asokan.kc@gmail.com
[2] Department of Futures Studies, University of Kerala, Thiruvananthapuram,
Kerala, India - 695 034
kskumar@bsnl.in

Abstract. We present a detailed nonlinear time series analysis of the daily mean wind speed data measured at COCHIN/WILLINGDON (Latitude: +9.950, Longitude: +76.267 degrees, Elevation: 3 metres) from 2000 to 2010 using tools of non-linear dynamics. The results of the analysis strongly suggest that the underlying dynamics is deterministic, low-dimensional and chaotic indicating the possibility of accurate short term prediction. The chaotic behaviour of wind dynamics explains the presence of periodicities amidst random like fluctuations found in the wind speed data, which forced many researchers to model wind dynamics by stochastic models previously. While most of the chaotic systems reported in the literature are either confined to laboratories or theoretical models, this is another natural system showing chaotic behaviour.

1 Introduction

The importance of the role of surface wind in shaping the climate and weather system of the earth has been well recognized. Knowledge of the nature, direction and speed of wind is important in aiding agriculture, navigation, structural engineering calculations and reduction of atmospheric pollution. Of late, there has been an upsurge of interest in research related to all areas of wind energy due to its potential as a cheap alternate source of energy when the natural resources of earth are deteriorating (Lei et al. (2009); Sfetsos (2000) and references therein). Wind speed prediction is an important aspect of wind power generation, but is also one of the most difficult due to the complex interactions between pressure and temperature differences, rotation of earth and local topography. Among the various models currently used in wind speed prediction are those relying on statistical methods such as ARMA and ARIMA, models based on probability distribution of wind speed and recent models based on artificial neural networks(Lei et al., 2009; Sfetsos, 2000).

Most of the above models assume that wind speed data is stochastic in nature but the results are not very encouraging in terms of reducing prediction error (Sfetsos, 2000) or accounting for the persistent fluctuations found in wind speed data. A few researchers (see Martín et al. (1999) and references therein) have

J. Mathew et al. (Eds.): ICECCS 2012, CCIS 305, pp. 311–318, 2012.

reported the presence of periodic cycles of 1 year, 24 hour and 12 hour dura-
tions in the data–corresponding to natural earth cycles and high/low tides– but
since these cycles lay buried under persistent random fluctuations the models
used in the study are predominantly stochastic. One of the reasons why probabilistic models are advocated to study wind speed fluctuations is the influence
on wind energy of solar radiations which is believed to be stochastic in nature.
However, recent studies of the data of Total Electron Content (TEC), which is
strongly influenced by the solar radiation, shows strong evidence of deterministic low dimensional character of the underlying dynamics (Kumar et al., 2004).
The surface wind speed is a similar atmospheric parameter with solar influence
and it is quite plausible that the random like fluctuations in wind speed are a
result of the chaotic dynamics of the underlying system. The presence of various
periodicities amongst the apparent random fluctuations found in the data is also
an indication of the possibility of chaotic dynamics.

In this work we carry out a detailed systematic analysis of the time series of
daily mean wind speed (DMWS) measured at COCHIN/WILLINGDON (Latitude: +9.950, Longitude: +76.267 degrees Elevation: 3 metres) for the period
from year 2000 to 2010 using tools of nonlinear dynamics. We demonstrate that
the dynamics of wind speed is essentially deterministic with a low dimensional
chaotic character. The chaotic behaviour explains the presence of various periodicities along side the apparent random fluctuations found in the data and
also the reason why long term predictions are generally erroneous. On the other
hand, it must be possible to make better short term predictions using deterministic models than would be possible with the existing statistical and probabilistic
methods.

2   Analysis of the Data

Deterministic dynamical systems which evolve continuously over time is de-
scribed by a state vector $x(t)$ and an equation of motion $\dot{x} \equiv \frac{dx}{dt} = f(x)$. Such
systems are usually characterised by an attractor, which is a bounded subset of
the phase space reached asymptotically by a set of trajectories over an open set
of initial conditions as time $t \to \infty$. In some some dynamical systems, the trajectories on the attractor may exhibit *sensitive dependence on initial conditions*,
meaning that trajectories starting from neighbouring initial conditions may separate from each other at an exponential rate. *Chaos is the bounded aperiodic
behaviour in a deterministic system that shows sensitive dependence on initial
conditions* (Ott, 1993). The exponential divergence of trajectories confined to a
bounded region leads to complex dynamics on the attractor resulting in random
like behaviour and long term unpredictability, often making a chaotic system
indistinguishable from a truly stochastic system. In fact, many a system which
were earlier dubbed as stochastic were later shown to be chaotic (Ott, 1993).

Most of the time the state vector $x(t)$ is not measured directly but indirectly at
discrete time intervals τ using a scalar measurement function $y(t) = h(x(t))$ leading to a time series $y_i = y(i\tau)$. The embedding theorems (Ott et al., 1994)assert

that the dynamics of the state vector $x(t)$ is topologically identical to the dynamics of the 'delay vector' defined by $y(t) = (y(t), y(t+\tau), \cdots y(t+(m-1)\tau)$, for almost all values of time delay τ and suitable values of m. This means that the dynamical and geometrical characteristics of the original system are preserved in the reconstructed space and can be computed from the flow defined by $y(t)$ (Kantz and Schreiber, 1997; Ott et al., 1994).

The literature on the analysis of time series based on attractor reconstruction is extensive and a discussion of the basic tools with comprehensive references can be found in Hegger et al. (1999) and Kantz and Schreiber (1997). Proper choices of the delay τ and the embedding dimension m are very important for successful attractor reconstruction. A useful tool for finding an optimal delay is the method of time-delayed mutual information, which is based on computing the mutual information $I(\tau)$ (Hegger et al., 1999) for various delays as a measure of the predictability of $y(t + \tau)$ given $y(\tau)$. A good choice for time delay is then the value of τ at which the graph of mutual information exhibits a marked minimum (Hegger et al., 1999).

A practical method for choosing the right embedding dimension is to find the fraction of 'false neighbours' as a function of the embedding dimension. False neighbours arise when the current dimension is not large enough for the attractor to unfold its true geometry, leading to crossing of trajectories due to projection onto a smaller dimension. The method checks the neighbours in progressively higher dimensions until it finds only a negligible number of false neighbours in passing from dimension m to $m+1$. The first time the fraction of false neighbours attains a minimum indicates a suitable value for the embedding dimension.

A quantitative measure of the structure and self-similarity of the attractor is provided by various dimension estimates. The easiest to compute among them is the correlation dimension (Kantz and Schreiber, 1997) which is computed from the correlation sum $C(\epsilon, m)$ is given by

$$C(\epsilon, m) = \frac{1}{N_p} \sum_{j=m}^{N} \sum_{k<j-w}^{N} \Theta(\epsilon - \|y_j - y_k\|), \tag{1}$$

for a set of N data points of m-dimensional delay vectors y_i on the reconstructed attractor. Here $\Theta(a) = 1$ if $a > 0$, $\Theta(a) = 0$ if $a \leq 0$ and $N_p = (N - m - w)(N - m - w + 1)/2$ where w is the Theiler window which is approximately equal to the product of the time delay and embedding dimension. In practice, one computes the local slopes

$$D_2(\epsilon, m) = \frac{d \ln C(\epsilon, m)}{d \ln \epsilon} \tag{2}$$

and plots them as a function of ϵ for various m; the value corresponding to a plateau in the curves is identified as an approximation to D_2. A detailed discussion of the subtleties in the computation of correlation dimension can be found in Hegger et al. (1999).

The Lyapunov exponents quantify the average rate of divergence or convergence of nearby orbits in the principal directions, and the existence of a positive

Lyapunov exponent is one of the most striking signatures of chaos (Ott, 1993). Among the various algorithms for estimating the maximal Lyapunov exponent from time series, the most popular is the Kantz algorithm (Kantz and Schreiber, 1997), which proceeds by computing the sum

$$S(\Delta n) = \frac{1}{N} \sum_{n_0=1}^{N} \ln\left(\frac{1}{|U(\boldsymbol{y}_{n_0})|} \sum_{y_n \in U(y_{n_0})} \left|\boldsymbol{y}_{n_0+\Delta n} - \boldsymbol{y}_{n_0+\Delta n}\right|\right) \tag{3}$$

for a point $\boldsymbol{y}_{n_0}$ of the time series in the embedded space and over a neighbourhood $U(\boldsymbol{y}_{n_0})$ of $\boldsymbol{y}_{n_0}$ with diameter ϵ. If the plot of $S(\Delta n)$ against Δn is linear over small Δn and for a reasonable range of ϵ, and all have identical slope for sufficiently large values of the embedding dimension m, then that slope can be taken as an estimate of the maximum Lyapunov exponent (Kantz and Schreiber, 1997).

The time series of DMWS plotted in Fig. 1(a) exhibits persistent temporal fluctuations which, as mentioned previously, could indicate a noise driven stochastic system or a deterministic system with chaotic dynamics and calls for

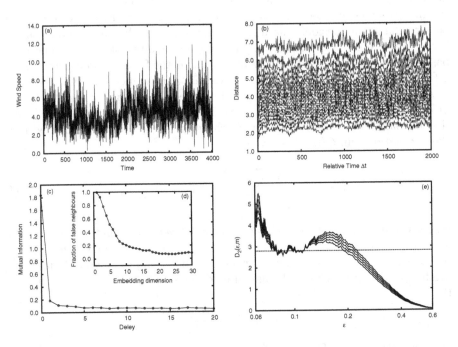

Fig. 1. (a) Time series of the measured daily mean wind speed (DMWS) in knots. (b) Space time separation plot for the time series for $\tau = 1$ and $m = 14$ (c) The mutual information. (d) The fraction of false nearest neighbours as a function of m with $\tau = 1, \omega = 20$. (e) The local slopes $D_2(\epsilon, m)$ for $m = 14, 15, 16, 17, 18$ with $\tau = 1, \omega = 20$ giving a plateau for small values of ϵ and giving an estimate of $D_2 = 2.801 \pm 0.006$.

a detailed analysis of the time series. But before doing this we must remove from the data the effect of measurement errors and averaging which are assumed to contribute an additive noise to the data. We have performed a simple noise reduction based on a locally constant approximation (Kantz and Schreiber, 1997), to remove the effect of this noise so that the resulting de-noised data can be regarded as a time series representing the true dynamics of wind speed. Further, the times series plot, Fig. 1(a) and the corresponding space time separation plot, Fig. 1(b), show the presence of annual cyclic variations in the data. We performed an *epoch analysis*(Kumar et al., 2004) to remove the effect of annual variations and its modulation effects on the nonlinear analysis. The resulting data, cleaned by noise reduction and epoch analysis, nonetheless retains the temporal fluctuations. We will now carry out a systematic analysis of the de-noised de-trended data, referred to as the DMWS-data hereafter, to investigate the nature of the underlying dynamics of these fluctuations.

The first step in nonlinear time series analysis is the theoretical reconstruction of the attractor of the system using delay co-ordinates. For the DMWS-data, plot of the mutual information as a function of delay is shown in Fig. 1(c) which attains a minimum at $\tau = 1$ indicating an optimum value of $\tau = 1$. Fig. 1(d) plots the fraction of false neighbours as a function of embedding dimension and it can be seen that an optimal choice of m must be higher than 14, since for $m \geq 14$ the fraction of false neighbours become negligibly small. We have chosen $m = 15$ for the further analysis.

To estimate the correlation dimension of the reconstructed attractor, we plot in Fig. 1(e) the local slopes $D_2(\epsilon, m)$ of the DMWS-data for embedding dimensions ranging from 14 to 18 using 20 as value for Theiler window ω. The curves exhibit convergence for larger m, an indication of low dimensionality of the attractor, and exhibits a plateau near $D_2(\epsilon, m) = 2.8$ for all the curves, suggesting a value $D_2 = 2.8$ for the correlation dimension. The low-dimensionality is an indication of determinism in the data and shows that while the original system may be affected by a myriad of factors, the eventual behaviour is determined by a few variables.

Fig. 2(a) which shows curves of $S(\Delta n)$, Eq. (3), for embedding dimensions $m = 14, 15, 16$. The curves increases linearly with Δn and gives $\lambda = 0.1815 \pm 0.0018$ as an estimate of the maximum Lyapunov exponent, being the common slope of the curves before settling down. A positive Lyapunov exponent is an indication of chaotic behaviour.

A colour noise time series can mimic many characteristics of a chaotic time series. To ensure that the above characteristics of the DMWS-data are not an artefact of colour noise, we compared the data with its phase randomized time series. The phase randomization of a chaotic signal can destroy its profile whereas a colour noise time series retains its profile. We calculated the local slopes $D_2(\epsilon, m)$ of correlation sum of for both the original time series and the phase randomized time series and are plotted in Fig. 2(b) and (c) respectively. These figures clearly shows that phase randomization destroys deterministic profile.

3   Comparison with Surrogate Data

The features exhibited by the the DMWS-data, revealed by the analysis in the previous section, are hallmarks of chaos if the underlying system is deterministic. However, it is possible that a stochastic system driven by noise may also show some of these features. The method of surrogate data (Hegger et al., 1999) is widely used as a tool for discriminating whether the source of random fluctuations in time series data is deterministic or stochastic. It is basically a statistical procedure to test the validity of a null hypothesis that the observed data is a linear noise process. In this method we first formulate a null hypothesis, which is usually an assumption that the observed data is random, and then generate an ensemble of time series of random numbers, called surrogate data, which satisfy the null hypothesis but are otherwise similar to the original data. Then one compares the values of some discriminating statistic, such as correlation dimension, computed from the given data to the distribution of values obtained from the surrogates. If the values differ significantly, then the null hypothesis may be rejected.

For the following analysis we generated from the given time series a set of 40 surrogates consistent with the null hypothesis that the observed time series is a linear Gaussian noise process. By construction, the surrogates are like realisations of a linear Gaussian noise process but are otherwise similar to the original data in terms of amplitude distribution, power spectrum and auto-correlation. Invariances such as fraction of false nearest neighbours, local slopes of the correlation sums and curves of $S(\Delta_n)$ are used as discriminating statistic for testing the null hypothesis. Each of the above measures are calculated for both the original data and the surrogate data and the value and the null hypothesis is accepted or rejected depending on the value of the significance of difference given by (Pavlos et al., 1999) $S = (\mu - \mu_{\text{orig}})/\sigma$ where μ and σ are the mean and standard deviation of the characteristic computed from the surrogates and μ_{orig} is the mean of the characteristic on the original data. It is estimated that (Pavlos et al., 1999) we may reject the null hypothesis with 95% confidence if $S > 2$, which means that the probability is 95% or more that the observed time series is not a realization of a Gaussian stochastic process.

A consistently good tool for discriminating non-linearity is the *prediction error* which computes the deviations of the values predicted using past data from the actual values in the trajectory. We calculated the prediction errors by using a locally constant approximation to predict future values and the root mean square prediction errors of each of the 40 surrogates and the original data were computed. The results, displayed in Fig. 2(d), how that the prediction errors are significantly lower for the original data than all the surrogates with $S = 8.13$. This is a strong case for the rejection of the null hypothesis.

Fig. 2(e) plots the mean values of fraction of false nearest neighbours of all the surrogates and values one standard deviation away from the mean alongside the values of fraction of false nearest neighbours of the DMWS-data. We can observe that the curves corresponding to the surrogates deviate significantly from that of the original data for a range of the embedding dimensions. As shown in Fig. 2(f)

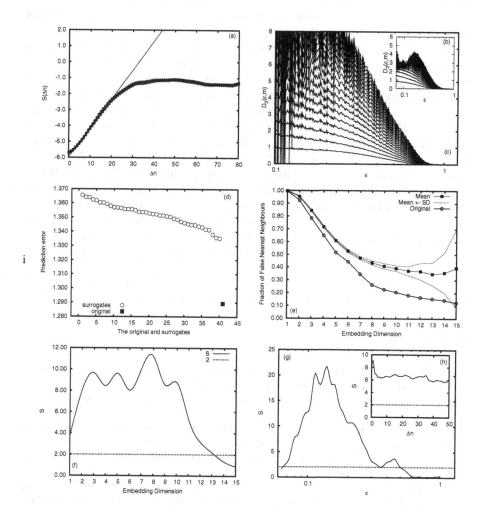

Fig. 2. (a) The curve of $S(\Delta n)$ for $m = 14, 15, 16$ with $\tau = 1, \omega = 20$. The maximum Lyapunov exponent λ is the slope of the dashed line 0.1815 ± 0.0018. (b) The local slopes $D_2(\epsilon, m)$ of original time series for the embedding dimensions $m = 1, \ldots, 24$ with $\tau = 1, \omega = 20$. (c) The same for the phase randomized time series (d) The plot of the prediction errors for $m = 15, \tau = 1$ for the surrogates (denoted by circles) and that of the original series (denoted by filled square) shows the determinism in the time series. The significance of difference $S = 8.13$. (e) The mean values of the fraction of false nearest neighbours of the surrogates with standard deviation. (f) Plot of the significance of difference S versus m. (g) Plot of S versus ϵ calculated for the mean values of the local slopes $D_2(\epsilon, 14)$. (h) Plot of the significance of difference S versus Δn calculated for $S(\Delta n)$.

the significance of difference S for the fraction of false nearest neighbours is larger than 2 for all values embedding dimension up to 13 indicating a strong case for rejection of the null hypothesis. We have repeated the surrogate data test further with the discriminating statistics local slopes $D_2(\epsilon, 14)$ (Eq. (2)) of the logarithms of correlations sums and $S(\Delta_n)$-values (Eq. (3)) and the corresponding corresponding significance of difference S is plotted in Fig. 2(g) and (h) respectively which clearly support rejection of the null hypothesis.

Based on the results of these statistical tests using surrogates we can conclude that the dynamical features exhibited by the data are characteristic of a deterministic system and not an artefact of a noise driven system. Since the dynamics is chaotic long term predictions are prone to errors, but it must be possible to make better short term predictions by using a non-linear prediction method suited to the data at hand. The purpose of this study was to investigate the suitability of a deterministic model to represent the wind speed fluctuations by employing tools of nonlinear dynamics. The analysis shows strong evidence for the existence of an underlying deterministic system which is chaotic and low dimensional indicating the possibility of accurate short term predictions.

Acknowledgment. The authors are grateful to Prof. Manoj Changat, Department of Futures Studies, Univ. of Kerala for his support and encouragement.

References

Hegger, R., Kantz, H., Schreiber, T.: Practical implementation of nonlinear time series methods:The TISEAN package. Chaos 9, 413–435 (1999)

Kantz, H., Schreiber, T.: Nonlinear Time Series Analysis. Cambridge University Press, Cambridge (1997)

Kumar, K.S., George, B., Renuka, G., Kumar, C.V.A., Venugopal, C.: Analysis of the fluctuations of the total electron content (TEC) measured at Goose Bay using the tools of nonlinear methods. J. Geophys. Res. 109, A02308 (2004), doi:10.1029/2002JA009768

Lei, M., Shiyan, L., Chuanwen, J., Hongling, L., Yan, Z.: A review on the forecasting of wind speed and generated power. Renewable and Sustainable Energy Reviews 13, 915–920 (2009)

Martín, M., Cremades, L.V., Santabárbara, J.M.: Analysis and modelling of time series of surface wind speed and direction. Int. J. Climatol. 19, 197–209 (1999)

Ott, E.: Chaos in Dynamical Systems. Cambridge University Press, Cambridge (1993)

Ott, E., Sauer, T., Yorke, J.A.: Coping with Chaos. Wiley, New York (1994)

Pavlos, G.P., Athanasiu, M.A., Diamantidis, D., Rigas, A.G., Sarris, E.T.: Nonlinear analysis of magnetospheric data Part I. Geometric characteristic of the AE index time series and comparison with nonlinear surrogate data. Nonlinear Proc. Geophys. 6, 51–65 (1999)

Sfetsos, A.: A comparison of various forecasting techniques applied to mean hourly wind speed time series. Renewable Energy 21, 23–35 (2000)

Some Modelling Aspects of Aggregation Kernels and the Aggregation Population Balance Equations

Nageswara Rao Narni[1], Gerald Warnecke[2], Jitendra Kumar[3], Mirko Peglow[4], and Stefan Heinrich[5]

[1] Department of Mathematics, Rajiv Gandhi University of Knowledge Technologies, Hyderabad-500 032, India
narninrao@gmail.com
[2] Institute for Analysis and Numerics, Otto-von-Guericke-University Magdeburg, Universitätsplatz 2, D-39106 Magdeburg, Germany
[3] Department of Mathematics, IIT Kharagpur, India
[4] Institute for Process Engineering, Otto-von-Guericke-University Magdeburg, Universitätsplatz 2, D-39106 Magdeburg, Germany
[5] Institute of Solids Process Engineering and Particle Technology, Hamburg University of Technology, Hamburg, Germany

Abstract. Aggregation is one of the important processes in chemical and process engineering. To model this process aggregation population balance equation is widely used. The aggregation kernel is one of the most effective parameter of this aggregation equation. This is usually defined as the product of the aggregation efficiency and collision frequency functions. Several authors attempted to model the aggregation kernel using different approaches like theoretical, experimental and empirical approaches. In the present paper we are giving some important modelling aspects of these kernels. One can use these modelling aspects in other variants of the population balance equations with breakage, nucleation, growth, ..., etc.

Keywords: Aggregation equation, aggregation kernel, collision frequency function, dimensional analysis.

1 Introduction

Population balance equations are widely used in the literature to model the granulation process in the fluidized beds on a macroscopic level. Kernels are the major parameters in these integro-partial differential equations. Aggregation equation is one of the important class of these equations. It describes the particle growth through aggregation of particles. The main parameter in this aggregation equation is the aggregation kernel.

Modelling of the aggregation kernel was done through theoretical, empirical and experiments approaches. Smoluchowski [13] derived the shear kernel and brownian kernel using theoretical considerations. Sastry [11], Kapur [7] obtained

J. Mathew et al. (Eds.): ICECCS 2012, CCIS 305, pp. 319–326, 2012.

the aggregation kernels by empirical approach. Aggregation kernels can be modelled by fitting the experimental data to the empirical kernels. Peglow [10] obtained an empirical kernel by fitting the experimental data to the empirical kernel proposed by Kapur [7].

Due to improvement in computational power, some authors try to extract the kernels by using computer simulations. Recently Gant et al. [6] obtained a multi dimensional aggregation kernel by simulating a granulator. Tan et al. [14] derived an aggregation kernel based on the principle of kinetic theory of granular flow to the fluidized bed melt granulation.

In the present paper we wish to present some modelling aspects of these kernels for population balance equations and a correction to the population balance equation for the fluidized bed. It is shown that when the number of particles increases the correction is negligible. The present modelling aspects can be applied to population balance equations with growth, breakage and nucleation.

2   Different Forms of Aggregation Equation

The aggregation equation was first derived by Smoluchowski [13]. He derived an infinite set of differential equations describing the particle aggregation. Later Mueller [8] derived a continuous form for these equations and is an integro-partial differential equation of nonlinear type. The analytic solutions of these equations exists only for a few types of kernels and can be found in [2].

2.1   Discrete Population Balance Equation

Disperse systems describes the solid or liquid particles suspended in a medium (usually in a gas). The disperse systems are usually unstable and changes with time due to the movement of the particles in the medium. The particles gets the movement by external fields like gas flow, stirrer, ...etc. As a result of this movement, the particles collides. The collision and coagulation in this system leads to the change in the particle size (Here we took particle volume as size) distribution. Population balance equations are very frequently used to study the particle size distribution in the disperse systems. Smoluchowski [13] derived an infinite set of nonlinear differential equations for his theory of rapid coagulation process with the following assumptions:

- The number of particles per unit volume of the fluid are sufficiently small (i.e. dilute system) with constant porosity
- Binary collisions are assumed to occur simultaneously
- Incompressible spherical particles are assumed to collide

Based on the above assumption he derived the following equations:

$$\frac{dn_i(t)}{dt} = \frac{1}{2}\sum_{j=1}^{i-1} K_{i-j,j}n_{i-j}(t)n_j(t) - n_i\sum_{j=1}^{\infty} K_{i,j}n_j(t) \qquad (1)$$

for $i = 1, 2, 3, ...$, with the initial conditions $n_i(0) = n_i^{(0)} \geq 0$.

The above equation (1) is known as the discrete coagulation equation. The complexity of the above system is dependent on the form of the kernel $K_{i,j}$. The function $K_{i,j}$ is known as the coagulation kernel. It describes the intensity of the particle interactions between particle classes i and j. The coagulation kernel is non-negative and symmetric, i.e. $K_{i,j} \geq 0$ and $K_{i,j} = K_{j,i}$. The unknown non-negative function $n_i(t)$ is the concentration of particles with size $i, i \geq 1$. The term $\frac{1}{2} \sum_{j=1}^{i-1} K_{i-j,j} n_{i-j}(t) n_j(t)$ is known as the birth term and $n_i \sum_{j=1}^{\infty} K_{i,j} n_j(t)$ is knows as the death term.

2.2   Continuous Population Balance Equation

Mueller [8] derived the continuous form for the above equation as an integro partial differential equation defined as:

$$\frac{\partial n(t, u)}{\partial t} = \frac{1}{2} \int_0^u K(t, u - v, v) n(t, u - v) n(t, v) dv$$
$$-n(t, u) \int_0^\infty K(t, u, v) n(t, v) dv, \tag{2}$$

with the initial condition $n(0, u) = n_0(u) \geq 0$. The Kernel $K(t, u - v, v)$ describes the coagulation of particles of size $u - v$ with particles of size v. The first term of the right hand side of equation(2) describes birth of the particles. The second term in the right hand side of the equation(2) describes the death of the particles.

3   Dimensional Analysis of the Equation

Dimensional analysis is one of the primitive modelling tool for Physicist, Engineers and Applied mathematicians for modelling a physical phenomena. It allows us to calculate the scale factors when using physical models. We are going to use it in calculating the scaling factor $K(u, v)$, which is known as aggregation kernel. To discuss the concepts of dimensional analysis, we need the following definitions:

- We defines the mass M as the mass of the particle with units Kilogram.
- We defines the volume for two physical quantities:
    1. Volume of fluid as V_{fluid} and has the dimension length L^3 for notational purpose we writes it as L_{fluid}, which is an external parameter corresponds to space coordinate with units Meter.
    2. Volume of particles as $V_{particles}$ and has the dimension length L^3 for notational purpose we writes it as $L_{particles}$, which is an internal parameter corresponds to particle property with units Meter.
- We defines time as T and has units Seconds.

3.1   Dimensional Analysis for Discrete Equation

- Particle concentration $n_i(t)$ is defined as the number of particles per unit volume of the fluid and had the dimension $M^0 L_{fluid}^{-3} T^0$
- The coagulation kernel $K_{i,j}$ is defined as the number of particle interactions in unit volume of fluid per unit time and had the dimension $M^0 L_{fluid}^3 T^{-1}$

3.2   Dimensional Analysis of the Continuous Equation

Case 1:

- Particle concentration $n(t, u)$ is defined as the number of particles per unit volume of fluid per unit particle volume and had the dimension $M^0 L_{fluid}^{-3} L_{particle}^{-3} T^0$
- Dimension of kernel $K(t, u, v)$ is $M^0 L_{fluid}^3 T^{-1}$

Case 2:

- Suppose for the continuous equation we have dimension of $n(t, u)$ as Number of particles per Unit volume of particles. That is we are considering the number density as function of time and material volume with the dimension $M^0 L_{particle}^{-3} T^0$.
- Dimension of the kernel $K(t, u, v)$ is $M^0 L_{particle}^0 T^{-1}$. This is usually called as collision frequency kernel in the literature [4].

3.3   Number Density Selection and Dimensional Analysis of the Kernel in Applications

When we are trying to use this aggregation equation for a particulate process, the dimensions of the kernels can be calculated as:

In the study of atmosphere sciences the particle volume is negligible compared to the volume of the air. So in such models discrete kernel is widely used by considering the number density as function of fluid and time. In this case the discrete kernel $K(t, u, v)$ has the dimension $L_{air}^3 T^{-1}$.

In a fluidized bed the volume of the bed is not constant. Therefore the porosity of the bed changes with the time. So in such situations it is necessary to consider both, the volume of fluid and volume of the particles in the calculation of particle concentration. Since the porosity is not constant, we consider the case 1 and the kernel has the dimension $L_{fluid}^3 T^{-1}$.

As we stated earlier, the number density can be a function of any material coordinate. In certain applications the mass of the particles is important compared to volume of the particles. In crystallization processes, particles are formed as a result of change in the thermodynamics of the material. So in such processes the number density solely depends on the amount of material present in the vessel. Therefore we can replace the number density as a function of material and time, and having the dimension $M^{-1} L^{-3}$. In this case the aggregation kernel has the dimension of $M L^3 T^{-1}$. Here L^3 represents the volume of the vessel.

4   Derivation of Collision Frequency Functions

Aggregation processes lead to a reduction in the total number of particles and an increase in the average size. Therefore agglomeration leads a major change in the system as the time changes. As a result, the aggregation rate will be effected, which is a function of collision frequency function and aggregation efficiency. Here we are deriving an expression for collision frequency function.

Let $N_{i,j}$ be the number of collisions occuring per unit time per unit volume between the two classes of particles of volumes v_i and v_j. All particles are assumed to be spherical in shape, and collisions are treated as binary hard sphere collisions. The collision frequency function $K_{i,j}$ can be written in terms of concentrations of particles of volumes v_i and v_j as:

$$N_{i,j} = K_{i,j} n_i n_j, \qquad 1 \le i, j \le I$$

where n_i, n_j are the number concentrations of classes i, j. The number I denotes the total number of classes. The parameter $K_{i,j}$ is the frequency function among different classes i and j. The collision frequency function $K_{i,j}$ is a function of flow properties, i.e. particle size, gas velocity, temperature, etc.

The number of collisions occuring during time t_{sim} in the fluidized bed of volume V_{bed} is given as

$$N_{i,j}^{tot} = K_{i,j} \frac{N_i}{V_{bed}} \frac{N_j}{V_{bed}} t_{sim} V_{bed}, \tag{3}$$

where N_i, N_j are the number of particles of class i, j present in the fluidized bed of volume V_{bed}, i.e. $n_i = \frac{N_i}{V_{bed}}, n_j = \frac{N_j}{V_{bed}}$.

From the Equation (3), the collision rate function is obtained as

$$K_{i,j} = \frac{N_{i,j}^{tot}}{N_i N_j} \cdot \frac{V_{bed}}{t_{sim}}. \tag{4}$$

In case of collisions within a size class, i.e. $i = j$, the collision rate function is obtained as

$$K_{i,i} = \frac{N_{i,i}^{tot}}{\frac{N_i(N_i-1)}{2}} \cdot \frac{V_{bed}}{t_{sim}} \tag{5}$$

which differs from the previous one due to the correction for self collisions of the particles, which was observed by Hu and Mei [5]. The authors observed a deviation of about 2 percent in their simulations. This correction is negligible in the case of continuous spectrum, because of the assumption of large concentrations, in which, we take the approximation $N_i(N_i - 1)/2 \approx N_i^2/2$. The details of the derivation of (5) by using a combinatorial approach can be found in Sastry [12].

5   Correction to Aggregation Equation

Let us consider the combinatorial approach of the particle collisions among different classes and within the same class of particles. Here classes are based on the volume of the particle, i.e, $V_0, 2V_0, 3V_0, ...$ and V_0 is the initial particle volume class.

The Figure 1 describes the number of possible collisions among two classes A and B which contain 2 and 3 particles respectively. The number of possible collisions between A and B are 6. The number of possible collisions among the particles of class B are 3.

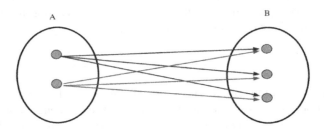

Fig. 1. Combinatorial calculation of the particle collisions

We can generalize the above approach as follows. Suppose there are N_i particles of ith class and N_j particles of jth class at a given instant. Then the number of possible collisions among ith class and jth class are N_iN_j. When we have collisions among the particles of class i, then the number of possible collisions are $N_i(N_i - 1)/2$.

Table 1. Simulations for collisions among the same size particles

Simulation	N_0	Same size collisions (5)	Same size collisions (4)	Difference
1	5000	6.7155e-05	6.7088e-05	6.7155e-08
2	10000	1.6566e-05	1.6557e-05	8.2828e-09
3	15000	1.4514e-05	1.4509e-05	4.8379e-09
4	20000	1.1776e-05	1.1773e-05	2.9440e-09

We can apply the above combinatorial approach to the fluidized bed as follows. Let $K_{i,j}$ be the collision frequency function of the fluidized bed. Then the number of collisions among the classes i and j per unit time and per unit volume are $N_{i,j} = K_{i,j}N_iN_j$. In case of collisions among the particles in the ith class are $N_{i,j} = K_{i,j}N_i(N_i - 1)/2$. The case of collisions among the same class are not considered during the derivation of the aggregation equation. So here we are showing the effect of this in modelling the aggregation kernel in Table 1. In this simulation we considered 10 classes of particles with an equal number of particles in each class. For details of initial conditions of the simulations and

Discrete Particle Model (DPM) see Narni [9]. We made simulations with an increasing the number of particles.

From the above simulation Table 1, we can observe that when the number of particles is high, i.e. N_0 is large enough, this new correction term has little effect on the aggregation equation. Therefore in industrial level fluidized beds the new correction is negligible, since the number of particles is very high.

6   Conclusions

In the present paper we gave few modelling aspects of aggregation equation and a correction to it. The dimensional analysis of the aggregation equation gives a clear insight into the scaling parameter (kernel) of the equation and its applicability to physical models. The present Discrete Particle Model simulations shows that the new correction for large particle systems has negligible effect.

The present dimensional analysis can be extended to population balance equations with combined processes such as breakage, nucleation, growth, ... etc. The present strategies are very useful for modelling multi particle aggregation (i.e., ternary,...etc.) and multidimensional population balance equations.

Acknowledgments. This work was supported by the Graduiertenkolleg-828, "Micro-Macro-Interactions in Structured Media and Particles Systems", Otto-von-Guericke-University, Magdeburg, Germany. The authors gratefully acknowledge for funding through this Ph. D. program.

Nomenclature

$n(t, u)$	Number density at time t of particle property u	—
t	Time	sec
$N_{i,j}$	Number of collisions between class i and class j	m^3/sec
n_i	Particle concentration of class i	$1/m^3$
n_j	Particle concentration of class j	$1/m^3$
N_i	Number of particles of class i	dimensionless
N_j	Number of particles of class j	dimensionless
$K_{i,j}$	Collision rate function	m^3/sec
V_{bed}	Volume of the bed	m^3
t_{step}	Simulation time steps	sec
u, v	Volume of the particles	m^3

References

[1] Friedlander, S.K.: Smoke, dust and haze. John Wiley & Sons, New York (1977)
[2] Gelbard, F., Seinfeld, J.H.: Numerical solution of the dynamic equation for particulate systems. Journal of Computational Physics 28, 357–375 (1978)
[3] Hoomans, B.P.B.: Granular dynamics of Gas-Solid two-phase flow. PhD thesis, Twente University, The Netherlands (2000)
[4] Hounslow, M.J.: The population balance as a tool for understanding particle rate processes. Kona, 179–193 (1998)

[5] Hu, K.C., Mei, R.: A note on particle collision rate in fluid flows. Phys. Fluids 10, 1028–1030 (1998)

[6] Gantt, J.A., Cameron, I.T., Litster, J.D., Gatzke, E.P.: Determination of coalescence kernels for high-shear granulation using DEM simulations. Powder Technology 170, 53–63 (2006)

[7] Kapur, P.C.: Kinetics of granulation by non-random coalescence mechanism. Chemical Engineering Science 27, 1863–1869 (1972)

[8] Mueller, H.: A calculation procedure for heat, mass and momentum transfer in three dimensional parabolic flow. Int. J. Heat and Mass Transfer 15, 1787 (1922)

[9] Rao, N.N.: Simulations for modelling of population balance equations of particulate processes using discrete particle model. PhD thesis, Otto-von-Guericke University, Magdeburg, Germany (2009)

[10] Peglow, M.: Beitrag zur modellbildung eigenschaftsverteilten dispersen systemen am Beispiel der wirbeschicht-Spruenhagglomeration. PhD thesis, Otto-von-Guericke University, Magdeburg, Germany (2003)

[11] Sastry, K.V.S.: Similarity size disribution of agglomerates during their growth by coalescence in granulation or green pelletization. International Journal of Mineral Processing 2, 187–203 (1975)

[12] Sastry, K.V.S., Fuerstenau, D.W.: Size distribution of agglomerates in coalescing dispersed phase systems. I & EC Fundamentals 9(1), 145–149 (1970)

[13] Smoluchowski, M.V.: Mathematical theory of the kinetics of the coagulation of colloidal solutions. Z. Phys. Chem. 92, 129–168 (1917)

[14] Tan, H.S., Goldschmidt, M.J.V., Boerefijn, R., Hounslow, M.J., Salman, A.D., Kuipers, J.A.M.: Building population balance model for fluidized bed melt granulation: lessons from kinetic theory of granular flow. Powder Technology 142, 103–109 (2004)

Mixed Covering Arrays on Hypergraphs

Yasmeen and Soumen Maity

Indian Institute of Science Education & Research
Pune, India
yasmeensa@students.iiserpune.ac.in, soumen@iiserpune.ac.in

Abstract. Mixed covering arrays are natural generalizations of covering arrays that are motivated by applications in software and network testing. A strength three mixed covering array C is a $k \times n$ array with the cells of row i is filled up with elements from $\mathbb{Z}_{g_i}$ and having the property that in each $3 \times n$ subarray, every 3×1 column appears at least once. In this article, we consider a generalization of strength-3 mixed covering arrays that allows a 3-uniform hypergraph structure which specifies the choices of $3 \times n$ subarrays in C. The number of columns in such array is called its *size*. Given a weighted 3-uniform hypergraph H, a strength three mixed covering array on H with minimum size is called *optimal*. We give upper and lower bounds on the size of strength three mixed covering arrays on 3-uniform hypergraphs based on hypergraph homomorphisms. We construct optimal strength-3 mixed covering arrays on 3-uniform hypertrees, interval, and cycle hypergraphs.

Keywords: Covering arrays, Hypergraphs, Hypergraph homomorphism, Software testing.

1 Introduction

Let n, k, g be positive integers with $k \geq 3$. Three vectors $x, y, z \in \mathbb{Z}_g^n$ are *3-qualitatively independent* if for any triplet $(a, b, c) \in \mathbb{Z}_g \times \mathbb{Z}_g \times \mathbb{Z}_g$, there exists an index $i \in \{1, 2, ..., n\}$ such that $(x(i), y(i), z(i)) = (a, b, c)$. A covering array of strength three, denoted by 3-$CA(n, k, g)$, is an $k \times n$ array C with entries from $\mathbb{Z}_g$ such that any three distinct rows of C are 3-qualitatively independent. The parameter n is called the size of the array. *One of the main problems on covering arrays is to construct a 3-$CA(n, k, g)$ for given parameters (k, g) so that the size n is as small as possible.* The covering array number 3-$CAN(k, g)$ is the smallest n for which a 3-$CA(n, k, g)$ exists, that is

$$3\text{-}CAN(k, g) = min_{n \in \mathbb{N}} \{ n : \exists \ 3\text{-}CA(n, k, g) \}.$$

A 3-$CA(n, k, g)$ of size $n = 3\text{-}CAN(k, g)$ is called *optimal*. An example of a strength three covering array 3-$CA(10, 5, 2)$ is shown below [3]:

J. Mathew et al. (Eds.): ICECCS 2012, CCIS 305, pp. 327–338, 2012.
© Springer-Verlag Berlin Heidelberg 2012

$$
\begin{array}{cccccccccc}
1 & 0 & 1 & 0 & 1 & 0 & 0 & 0 & 1 & 1 \\
1 & 0 & 1 & 0 & 0 & 1 & 0 & 1 & 0 & 1 \\
1 & 0 & 0 & 1 & 0 & 0 & 1 & 1 & 1 & 0 \\
1 & 0 & 0 & 1 & 1 & 1 & 0 & 0 & 0 & 1 \\
1 & 0 & 1 & 0 & 1 & 1 & 1 & 0 & 0 & 0
\end{array}
$$

There is a vast array of literature [9,6,7,2,3,12] on covering arrays, and the problem of determining the minimum size of covering arrays has been studied under many guises over the past thirty years.

An important application of covering arrays is the testing of systems such as software, networks, and circuits [9,4,5]. Consider a new piece of software that has k input parameters, each taking one of g values. It is desirable to test all possible g^k input strings (test cases) for software failure, but if k or g are even reasonably large, this may be infeasible. An effective strategy for achieving certain testing comprehensiveness uses covering arrays of strength t, which guarantees that all possible t-wise interactions of parameter values are tested while minimizing the number of test cases. The test cases are the columns of a covering array t-$CA(n, k, g)$. Relaxing the fixed alphabet-size assumption is a natural extension of covering arrays research, which improve their suitability for applications. In real life, different input parameters in a piece of software or in a network would have different number of possible values. Constructions for mixed covering arrays are given in [8,13]. Another generalization of covering arrays are *strength t (t > 2) covering arrays on hypergraph*. In these arrays, only specified choices of t distinct rows need to be t-qualitatively independent and these choices are recored in hypergraph. As mentioned in [11], this is useful in situations in which some combinations of parameters do not interact; in these cases, we do not insist that these interactions to be tested, which allows reductions in the number of required test cases. Meagher and Stevens [12], and Meagher, Moura, and Zekaoui [11] studied strength two (mixed) covering arrays on graphs. In this article, we consider strength-3 mixed covering arrays on 3-uniform hypergraphs. We now recall the relevant definitions and concepts from the literature. See [1,14,11].

Definition 1. (Voloshin [14]). *Let $V = \{v_1, v_2, ..., v_k\}$ be a finite set, and let $E = \{E_1, E_2, ..., E_m\}$ be a family of subsets of V. The pair $H = (V, E)$ is called a hypergraph with vertex set V also denoted by $V(H)$, and with hyperedge set E also denoted by $E(H)$. If cardinality of every hyperedge of H is equal to r then H is called r-uniform hypergraph. H is called simple hypergraph if it contains no included edges. A complete r-uniform hypergraph containing k vertices, denoted by K_k^r is a hypergraph having every r-subset of set of vertices as a hyperedge. A clique in r-uniform hypergraph is sub-hypergraph which is complete. A maximum clique is the clique of the largest possible size. The clique number $\omega(H)$ of a hypergraph H is the number of vertices in a maximum clique in H.*

The parameters for strength-3 mixed covering arrays on 3-uniform hypergraphs are given by a vertex-weighted hypergraph. A *vertex-weigted hypergraph* is a hypergraph with weight assigned to each vertex. A weighted 3-uniform hypergraph is given by a triple $(V(H), E(H), w_H(.))$ where $V(H)$ is the vertex set,

$E(H)$ is the hyperedge set and $w_H : V(H) \to \mathbb{N}^+$ is the weight function. We assume that the vertices are labeled so that the weights are nondecreasing, that is, $w_H(v_i) \le w_H(v_j)$, for all $1 \le i \le j \le| V(H) |$. When there is no ambiguity on which hypergraph we are using, we can write $w(.)$ in place of $w_H(.)$.

Definition 2. *Let H be a weighted 3-uniform hypergraph with k vertices and weights $g_1 \le g_2 \le ... \le g_k$, and let n be a positive integer. An strength-3 mixed covering array on H, denoted by $3\text{-}CA(n, H, \prod_{i=1}^{k} g_i)$, is an $k \times n$ array with the following properties:*

1. *the cells in row i are filled with elements from a g_i-ary alphabet, which is usually taken to be $\mathbb{Z}_{g_i}$;*
2. *row i corresponds to a vertex $v_i \in V(H)$ with $w_H(v_i) = g_i$;*
3. *triples of rows that correspond to vertices in a hyperedge of H are 3-qualitatively independent.*

Given a weighted 3-uniform hypergraph H with weights $g_1, g_2, ..., g_k$ the *strength-3 mixed covering array number* on H, denoted by $3\text{-}CAN(H, \prod_{i=1}^{k} g_i)$, is the minimum n for which there exists a $3\text{-}CA(n, H, \prod_{i=1}^{k} g_i)$. A $3\text{-}CA(n, H, \prod_{i=1}^{k} g_i)$ of size $n = 3\text{-}CAN(H, \prod_{i=1}^{k} g_i)$ is called *optimal*.

A mixed covering array of strength three, denoted by $3\text{-}CA(n, \prod_{i=1}^{k} g_i)$, is a $3\text{-}CA(n, K_k^3, \prod_{i=1}^{k} g_i)$, where K_k^3 is the complete 3-uniform hypergraph on k vertices with weights g_i, for $1 \le i \le k$. A 3-covering array on a 3-uniform hypergraph H, denoted by $3\text{-}CA(n, H, g)$, is a $3\text{-}CA(n, H, g^k)$ where $k =| V(H) |$.

The generalization considered in this article adds a hypergraph structure to a strength three covering array, obtaining a strength three covering array on hypergraph. We provide constructions for optimal mixed covering arrays with strength three on special classes of hypergraphs. In Section 2, we generalize known results on mixed covering arrays on graphs [11], for mixed covering arrays of strength three on hypergraphs. In Section 3, we define three basic hypergraph operations, and using these operations, we give constructions for optimal mixed covering arrays of strength tree on hypertree, interval, and cycle hypergraphs.

2   Hypergraph Homomorphisms and Strength-3 Mixed Covering Arrays on 3-Uniform Hypergraphs

This section introduces weight-restricted hypergraph homomorphism and the strong chromatic number $\chi_s(H)$ of a hypergraph H. For weighted hypergraphs G and H, if there is a weight restricted hypergraph homomorphism

$$ G \xrightarrow{\ w\ } H, $$

then $\omega(G) \le \omega(H)$ and $\chi_s(G) \le \chi_s(H)$. Based on weight-restricted hypergraph homomorphism, we generalize known results on mixed covering arrays on graphs in [11], for mixed covering arrays of strength three on hypergraphs.

Definition 3. (Weight Restricted 3-uniform Hypergraph Homomorphism). *Let G and H be two 3-uniform hypergraphs, a mapping ϕ from $V(G)$ to $V(H)$ is hypergraph homomorphism if vertices $\phi(x)$, $\phi(y)$, $\phi(z)$ forms a hyperedge in H whenever the vertices x, y, z forms a hyperedge in G. A mapping ϕ from $V(G)$ to $V(H)$ is a weight- restricted hypergraph homomorphism from G to H if ϕ is a hypergraph homomorphism from G to H such that $w_G(v) \leq w_H(\phi(v))$, for all $v \in V(G)$. For weighted hypergraphs G and H, if there exists a weight-restricted hypergraph homomorphism from G to H then we write*

$$G \xrightarrow{\ w\ } H.$$

For $h \geq g$, by *dropping the alphabet size* of a particular row of a strength 3-mixed covering array from h to g we mean one can replace all symbols from $\mathbb{Z}_h \setminus \mathbb{Z}_g$ in the row by arbitrary symbols from $\mathbb{Z}_g$. This operation preserves the 3-qualitatively independent property of the rows of the strength 3-mixed covering array.

Theorem 1. *Let G and H be weighted 3-uniform hypergraphs with weights $g_1, g_2, ..., g_k$ and $h_1, h_2, ..., h_l$, respectively. If there exists a weight-restricted hypergraph homomorphism*

$$\phi : G \xrightarrow{\ w\ } H$$

then $3\text{-}CAN(G, \prod_{i=1}^{k} g_i) \leq 3\text{-}CAN(H, \prod_{j=1}^{l} h_j)$.

Proof. Let $C(H)$ be a $3\text{-}CA(n, H, \prod_{j=1}^{l} h_j)$. One can use $C(H)$ to construct $C(G)$, a $3\text{-}CA(n, G, \prod_{i=1}^{k} g_i)$. Let $i \in \{1, 2, ..., k\}$ be an index of a row of $C(G)$, and let v_i be the corresponding vertex in G. Row i of $C(G)$ is constructed from the row corresponding to $\phi(v_i)$ in $C(H)$ by dropping its alphabet from $w_H(\phi(v_i))$ to $w_G(v_i)$. Now for any hyperedge $\{v_i, v_j, v_m\}$ in G, the triple $\{\phi(v_i), \phi(v_j), \phi(v_m)\}$ is a hyperedge in H and rows of $C(H)$ corresponding to $\phi(v_i)$, $\phi(v_j)$ and $\phi(v_m)$ are 3-qualitatively independent. Since dropping the alphabet size preserves 3-qualitative independence, rows i, j and m of $C(G)$ are 3-qualitatively independent. □

Definition 4. (Berge [1]). *A strong coloring of hypergraph H is a map $\psi : V(H) \to \mathbb{N}$ such that whenever $u, v \in E_i$ for some hyperedge E_i in H, we have that $\psi(u) \neq \psi(v)$. The corresponding strong chromatic number $\chi_s(H)$ is the least number of colors for which H has proper strong coloring.*

The complete weighted 3-uniform hypergraph on n vertices $K_n^3(g_1, ..., g_n)$ is a complete 3-uniform hypergraph on n vertices with weights $g_1, g_2, ..., g_n$ on vertices. For any 3-uniform hypergraph H there exists hypergraph homomorphism

$$K_{\omega(H)}^3 \longrightarrow H \longrightarrow K_{\chi_s(H)}^3$$

These hypergraph homomorphisms can be extended to weight-restricted hypergraph homomorphisms and we get the following lower and upper bounds on $3\text{-}CAN(H, \prod_{i=1}^{n} g_i)$.

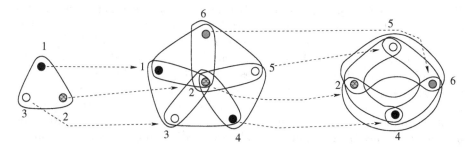

Fig. 1. Illustrating the weight restricted homomorphism: $K^3_{\omega(H)}(g_1, ..., g_{\omega(H)}) \longrightarrow$
$H \longrightarrow K^3_{\chi_s(H)}(g_{k-\chi_s(H)+1}, ..., g_k)$

Corollary 1. *Let H be a weighted 3-uniform hypergraph with k vertices and $g_1 \leq g_2 \leq ... \leq g_k$ be positive weights. Then,*
$$3\text{-}CAN(K^3_{\omega(H)}, \textstyle\prod_{i=1}^{\omega(H)} g_i) \leq 3\text{-}CAN(H, \textstyle\prod_{j=1}^{k} g_j) \leq 3\text{-}CAN(K^3_{\chi_s(H)}, \textstyle\prod_{l=k-\chi_s(H)+1}^{k} g_l).$$

Proof. For given H we know that there exists a hypergraph homomorphism $\phi :$ $K^3_{\omega(H)} \longrightarrow H$. Assign weights $g_1, ..., g_{\omega(H)}$ to $K^3_{\omega(H)}$ so that $w_{K^3_{\omega(H)}}(u) \leq w_{K^3_{\omega(H)}}(v)$ whenever $w_H(\phi(u)) \leq w_H(\phi(v))$. Since the g_i's are ordered by weight, it is clear that ϕ is a weight-restricted hypergraph homomorphism. Similarly there exists a hypergraph homomorphism $\varphi : H \longrightarrow K^3_{\chi_s(H)}$. Assign weights $g_{k-\chi_s(H)+1}, ..., g_k$ to $K_{\chi_s(H)}$ so that $w_{K^3_{\chi_s(H)}}(\varphi(u)) \leq w_{K^3_{\chi_s(H)}}(\varphi(v))$ whenever $w_H(u) \leq w_H(v)$. It is clear that φ is a weight restricted hypergraph homomorphism. $\square$

3   Optimal Mixed Covering Arrays on 3-Uniform Hypergraphs

In this section, we define three basic hypergraph operations, and using these operations, we give constructions for optimal mixed covering arrays of strength three on hypertree, interval and cycle hypergraphs.

A length-n vector with alphabet size g is *balanced* if each symbol occurs $\lfloor n/g \rfloor$ or $\lceil n/g \rceil$ times. A balanced covering array is a covering array in which every row is balanced [11]. For $n \geq g_1g_2$, two length-n vectors x_1 and x_2 with alphabet size g_1 and g_2 are *well balanced* if both vectors are balanced and each pair of alphabets (a, b) occurs $\lfloor n/g_1g_2 \rfloor$ or $\lceil n/g_1g_2 \rceil$ times in (x_1, x_2), where $a \in \mathbb{Z}_{g_1}$ and $b \in \mathbb{Z}_{g_2}$, so we can say that well balanced vectors are always 2-qualitatively independent. A balanced strength-3 covering array is an strength-3 covering array in which every pair of vector is well balanced.

Lemma 1. *For two well balanced length-n vectors x_1 and x_2 on $\mathbb{Z}_{g_1}$ and $\mathbb{Z}_{g_2}$ respectively and for any $g_3 \leq h$, where $h = \lfloor n/g_1g_2 \rfloor$ and $n \geq g_1g_2$, there exists a balanced length-n vector y on $\mathbb{Z}_{g_3}$ such that x_1, x_2 and y are 3-qualitatively independent and pair wise well balanced.*

Proof. We have total $g_1 g_2$ pairs of symbols in $\mathbb{Z}_{g_1} \times \mathbb{Z}_{g_2}$. As each pair of symbols occurs at least h times in (x_1, x_2), so for one pair say (a, b) where $a \in \mathbb{Z}_{g_1}$ and $b \in \mathbb{Z}_{g_2}$ there are at least h positions out of n positions at which (a, b) occurs in (x_1, x_2). Let $y \in \mathbb{Z}_h^n$ be the vector of length-n having h different symbols at positions corresponding to these h positions. We can do this for each pair of symbols in $\mathbb{Z}_{g_1} \times \mathbb{Z}_{g_2}$. Hence each triple of symbols in $\mathbb{Z}_{g_1} \times \mathbb{Z}_{g_2} \times \mathbb{Z}_h$ occurs in (x_1, x_2, y) exactly once. For remaining position if there are, we can assign symbols from $\mathbb{Z}_h$ in any way such that vector y remains balanced. Now for $g_3 \leq h$ we can use the dropping of symbols operation on y to obtain a balanced vector y on $\mathbb{Z}_{g_3}$. □

Observe that given a balanced vector $x_1 \in \mathbb{Z}_{g_1}^n$, one can construct two vectors $x_2 \in \mathbb{Z}_{g_2}^n$ and $x_3 \in \mathbb{Z}_{g_3}^n$ where $g_1 g_2 g_3 \leq n$ such that x_1, x_2 and x_3 are 3-qualitatively independent and pair wise well balanced, by constructing x_2 first such that each pair of symbols occurs near equal number of times in (x_1, x_2) and then constructing x_3 using Lemma 1.

Lemma 2. *Let $x_1 \in \mathbb{Z}_{g_1}^n$, $x_2 \in \mathbb{Z}_{g_2}^n$ and $x_3 \in \mathbb{Z}_{g_3}^n$ be such that x_1 and x_2 are well balanced, and x_1 and x_3 are well balanced, then for any h such that $h g_1 g_2 \leq n$ and $h g_1 g_3 \leq n$, there exists a balanced vector $y \in \mathbb{Z}_h^n$ such that $\{x_1, x_2, y\}$ are 3-qualitatively independent and pair wise well balanced, and $\{x_1, x_3, y\}$ are 3-qualitatively independent and pair wise well balanced.*

Proof. Lemma 2 follows form Lemma 3. □

Lemma 3. *Let $g_1 \leq g_2 \leq g_3$ with $n \geq g_1 g_2 g_3$ and $x_1 \in \mathbb{Z}_{g_1}^n$, $x_2 \in \mathbb{Z}_{g_2}^n$ and $x_3 \in \mathbb{Z}_{g_3}^n$ be three pair wise well balanced vectors. Then for $h = \lfloor n/g_2 g_3 \rfloor$, there exists a balanced vector $y \in \mathbb{Z}_h^n$ such that $\{x_1, x_2, y\}$ are 3-qualitatively independent, $\{x_1, x_3, y\}$ are 3-qualitatively independent, $\{x_2, x_3, y\}$ are 3-qualitatively independent, and $x_1, x_2, x_3,$ and y are pair wise well balanced.*

Proof. Define a length-n vector u_1 corresponding to pair (x_1, x_2) with entries $u_1(i) = (x_1(i), x_2(i))$, $i = 1, 2, ..., n$. We treat each pair of symbol as a new symbol so there are total $g_1 \times g_2$ different symbols appearing in u_1 or we can write $u_1 \in \mathbb{Z}_{g_1 g_2}^n$ so that each symbol appears at least $\lfloor n/g_1 g_2 \rfloor$-times in u_1 making u_1 a balanced vector. Similarly we can define $u_2 \in \mathbb{Z}_{g_1 g_3}^n$ and $u_3 \in \mathbb{Z}_{g_2 g_3}^n$ corresponding to pairs (x_1, x_3) and (x_2, x_3) respectively. Define a tripartite 3-uniform hypergraph H as follows: H has $g_1 g_2$ vertices in the first part, $P \subseteq V(H)$, $g_1 g_3$ vertices in the second part, $Q \subseteq V(H)$ and $g_2 g_3$ vertices in the third part, $R \subseteq V(H)$. Let $P_a = \{i : u_1(i) = a\}$, for $a = 1, 2, ..., g_1 g_2$, be the vertices of P, while $Q_b = \{i : u_2(i) = b\}$, for $b = 1, 2, ..., g_1 g_3$, be the vertices of Q and $R_c = \{i : u_3(i) = c\}$, for $c = 1, 2, ..., g_2 g_3$, be the vertices of R. We also have that $\mid P_a \mid \geq \lfloor n/g_1 g_2 \rfloor$, $\mid Q_b \mid \geq \lfloor n/g_1 g_3 \rfloor$ and $\mid R_c \mid \geq \lfloor n/g_2 g_3 \rfloor$. For each $i = 1, 2, ..., n$ there exists exactly one $P_a \in P$ with $i \in P_a$, exactly one $Q_b \in Q$ with $i \in Q_b$ and exactly one $R_c \in R$ with $i \in R_c$. For each such i, add a hyperedge containing vertices P_a, Q_b and R_c and label it i. Since $\mid P_a \mid \geq \lfloor n/g_1 g_2 \rfloor$, $\mid Q_b \mid \geq \lfloor n/g_1 g_3 \rfloor$ and $\mid R_c \mid \geq \lfloor n/g_2 g_3 \rfloor$, the minimum degree of a vertex in H is at least h. Hence there are h classes of disjoint hyperedge covers. A hyperedge cover is a subset of $E(H)$ such that each vertex in $V(H)$ is contained in at least one hyperedge in that subset. These h disjoint hyperedge covers form a partition T of $[1, n]$, which we use to build a balanced vector $y \in \mathbb{Z}_h^n$. Each hyperedge cover corresponds to a symbol in $\mathbb{Z}_h$ and each hyperedge corresponds to an index from $[1, n]$. For each hyperedge i in a hyperedge cover associated with $l \in \mathbb{Z}_h$, define $y(i) = l$. Fill the remaining positions in the vector y in such a way that y remains balanced. Next, we remain to show that partitions P and T are qualitatively independent. For any $l \in \mathbb{Z}_h$, the class T_l corresponds to a hyperedge cover of H, this means that for any $a \in \mathbb{Z}_{g_1 g_2}$, there exists a hyperedge i in the hyperedge cover incident to P_a. By the definition of T, we have $i \in T_l$. Since the hyperedge labeled i is incident with the vertex corresponding to P_a, we also know that $i \in P_a$. This means that for any $a \in \mathbb{Z}_{g_1 g_2}$, $l \in \mathbb{Z}_h$, there exists an $i \in [1, n]$ such that $i \in P_a$ and $i \in T_l$, or in other words, $u_1(i) = a$ and $y(i) = l$. So, u_1 and y are qualitatively independent and hence x_1, x_2 and y are 3-qualitatively independent and pair wise well balanced. Similarly, we can show that $\{x_1, x_3, y\}$ are 3-qualitatively independent and $\{x_2, x_3, y\}$ are 3-qualitatively independent and pair wise well balanced. $\qquad \square$

3.1 Basic Hypergraph Operations

Let H be a weighted 3-uniform hypergraph with n vertices. Label the vertices $v_1, v_2, ..., v_n$ and for each vertex v_i denote its associated weight by $w_H(v_i)$. Let the *product weight* of H, denoted $PW(H)$, be

$$PW(H) = \max\{w_H(v_i)w_H(v_j)w_H(v_k) : \{v_i, v_j, v_k\} \in E(H)\}.$$

Note that $3\text{-}CAN(H, \prod_{i=1}^n w_H(v_i)) \geq PW(H)$. We now define some hypergraph operations:

1. *Single-vertex One-hyperedge Hooking*
2. *Single-vertex Two-hyperedge Hooking*
3. *Single-vertex Three-hyperedge Hooking.*

A *Single-vertex One-hyperedge Hooking* in a 3-uniform hypergraph H is the operation that inserts a new hyperedge in which two vertices are from $V(H)$ and third one is a new vertex, while a *Single-vertex Two-hyperedge Hooking* is the operation that inserts two new hyperedges say $\{v, v_i, v_j\}$ and $\{v, v_i, v_k\}$ where v is new vertex, v_i, v_j and v_k are from $V(H)$ such that $\{v_i, v_j\} \subset E_s$ and $\{v_i, v_k\} \subset E_t$ for some E_s, E_t in $E(H)$. A *Single-vertex Three-hyperedge Hooking* is the operation that inserts three new hyperedges say $\{v, v_i, v_j\}$, $\{v, v_i, v_k\}$ and $\{v, v_j, v_k\}$ where v is new vertex v_i, v_j and v_k are from $V(H)$ such that $\{v_i, v_j\} \subset E_r$, $\{v_i, v_k\} \subset E_s$ and $\{v_j, v_k\} \subset E_t$ for some E_r, E_s and E_t in $E(H)$.

Proposition 1. (Single-vertex One-hyperedge Hooking). *Let H be a weighted 3-uniform hypergraph with k vertices. Let H' be the weighted 3-uniform hypergraph obtained from H by a single-vertex one-hyperedge hooking of a new vertex v with* *a new hyperedge $\{v, v_1, v_2\}$ and $w(v)$ be such that $w(v)w(v_1)w(v_2) \leq n$. Then, there exists a balanced 3-CA$(n, H, \prod_{i=1}^{k} g_i)$ if and only if there exists a balanced 3-CA$(n, H^{'}, w(v) \prod_{i=1}^{k} g_i)$.*

Proof. If there exists a balanced 3-CA$(n, H^{'}, w(v) \prod_{i=1}^{k} g_i)$ then by deleting the row corresponding to the new vertex v we can obtain a 3-CA$(n, H, \prod_{i=1}^{k} g_i)$. Conversely, let C^H be a balanced 3-CA$(n, H, \prod_{i=1}^{k} g_i)$. C^H can be used to construct a $C^{H^{'}}$, a balanced 3-CA$(n, H^{'}, w(v) \prod_{i=1}^{k} g_i)$. Using Lemma 1, we can build a balanced length-n vector, call it x, corresponding to vertex v such that x is 3-qualitatively independent with the two length-n vectors x_1 and x_2 corresponding to vertices v_1 and v_2 respectively in H. The array $C^{H^{'}}$ is built by appending row x to C^H. Since the only new hyperedge is $\{v, v_1, v_2\}$, and x, x_1, x_2 are 3-qualitatively independent, $C^{H^{'}}$ is a balanced strength-3 mixed covering array on $H^{'}$. $\square$

Proposition 2. (Single-vertex Two-hyperedge Hooking). *Let H be a weighted 3-uniform hypergraph with k vertices. Let $H^{'}$ be the weighted 3-uniform hypergraph obtained from H by a single-vertex Two-hyperedge hooking of a new vertex v with two new hyperedges $\{v, v_1, v_2\}$ and $\{v, v_1, v_3\}$ and $w(v)$ be such that $w(v)w(v_1)w(v_2) \leq n$ and $w(v)w(v_1)w(v_3) \leq n$. Then, there exists a balanced 3-CA$(n, H, \prod_{i=1}^{k} g_i)$ if and only if there exists a balanced 3-CA$(n, H^{'}, w(v) \prod_{i=1}^{k} g_i)$.*

Proof. The proof of the present proposition is similar to that of Proposition 1, except we will make use of Lemma 2 this time rather than Lemma 1. $\square$

Proposition 3. (Single-vertex Three-hyperedge Hooking). *Let H be a weighted 3-uniform hypergraph with k vertices. Let $H^{'}$ be the weighted 3-uniform hypergraph obtained from H by a single-vertex Three-hyperedge hooking of a new vertex v with three new hyperedges $\{v, v_1, v_2\}$, $\{v, v_1, v_3\}$ and $\{v, v_2, v_3\}$, and $w(v)$ be such that $w(v)w(v_1)w(v_2) \leq n$, $w(v)w(v_1)w(v_3) \leq n$ and $w(v)w(v_2)w(v_3) \leq n$. Then, there exists a balanced 3-CA$(n, H, \prod_{i=1}^{k} g_i)$ if and only if there exists a balanced 3-CA$(n, H^{'}, w(v) \prod_{i=1}^{k} g_i)$.*

Proof. The proof of the present proposition follows from that of Proposition 1 and Lemma 3. $\square$

Theorem 2. *Let H be a weighted 3-uniform hypergraph and $H^{'}$ a weighted 3-uniform hypergraph obtained from H via a sequence of single-vertex three-hyperedge hookings, single-vertex two-hyperedge hookings and single-vertex one-hyperedge hookings. Let $v_{k+1}, v_{k+2}, ..., v_l$ be the vertices in $V(H^{'}) \setminus V(H)$ with weights $g_{k+1}, g_{k+2}, ..., g_l$, respectively. If there exists a balanced 3-CA$(n, H, \prod_{i=1}^{k} g_i)$, then there exists a balanced 3-CA$(n, H^{'}, \prod_{i=1}^{l} g_i)$.*

Proof. The result is derived by iterating the previous three propositions. $\square$

3.2 3-Uniform Hypertrees

A *host graph* for a hypergraph is a connected graph on the same vertex set, such that every hyperedge induces a connected subgraph of the host graph [14].

Definition 5. (Voloshin [14]). *A hypergraph $H = (V, E)$ is called a Hypertree if there exists a host tree $T = (V, E')$ such that each hyperedge $E_i \in E$ induces a subtree in T.*

In other words, any hypertree is isomorphic to some family of subtrees of a tree. A 3-uniform hypertree is a hypertree such that each hyperedge in it contains exactly three vertices.

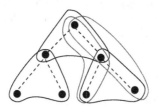

Fig. 2. 3-Uniform Hypertree with a binary tree as host tree

Theorem 3. *Let H be a weighted 3-uniform hypertree with l vertices, having a binary tree as a host tree. If $n = PW(H)$ then there exists a balanced 3-$CA(n, H, \prod_{i=1}^{l} g_i)$ hence, $3\text{-}CAN(H, \prod_{i=1}^{l} g_i) = PW(H)$.*

Proof. We can build H by starting with a hyperedge $\{v_i, v_j, v_k\}$ such that $PW(H) = w(v_i) \times w(v_j) \times w(v_k)$, and by applying successive single-vertex one-hyperedge hooking or single-vertex two-hyperedge hooking operations in the proper order to obtain H. So by Theorem 2, we get $3\text{-}CAN(H, \prod_{i=1}^{l} g_i) = PW(H)$. $\qquad\square$

3.3 3-Uniform Interval Hypergraphs

Definition 6. (Voloshin [14]). *A hypergraph $H = (V, E)$ is called an interval hypergraph if there exists a linear ordering of the vertices $v_1, v_2, ..., v_n$ such that every hyperedge of H induces an interval in this ordering. In other words, the vertices in V can be placed on the real line such that every hyperedge is an interval.*

An interval hypergraph is a hypertree because host graph of an interval hypergraph is a simple path joining the first and last vertices with all the remaining vertices laying on this path in between first and last vertex. A 3-uniform interval hypergraph is an interval hypergraph such that each hyperedge in it contains exactly three vertices.

Fig. 3. 3-Uniform Interval Hypergraph

Theorem 4. *Let H be a weighted 3-uniform interval hypergraph with l vertices, then $3\text{-}CAN(H, \prod_{i=1}^{l} g_i) = PW(H)$.*

Proof. As H is a 3-uniform interval hypergraph, any three vertices of H can form a hyperedge only if they are consecutive three vertices on the host graph of H. We can build H by starting with a hyperedge $\{v_i, v_j, v_k\}$ such that $PW(H) = w(v_i) \times w(v_j) \times w(v_k)$, and then by applying successive single-vertex one-hyperedge hooking operations in the proper order to obtain H. Using Theorem 2, we get $3\text{-}CAN(H, \prod_{i=1}^{l} g_i) = PW(H)$.   □

3.4 3-Uniform Cycle Hypergraph

Definition 7. (Voloshin [14]). *In a hypergraph $H = (V, E)$, an alternating sequence*

$$v_1 E_1 v_2 E_2 ... v_{l-1} E_{l-1} v_l$$

of distinct vertices $v_1, v_2, ..., v_l$ and distinct hyperedges $E_1, E_2, ..., E_{l-1}$ satisfying $v_i, v_{i+1} \in E_i$, $i = 1, 2, ..., l-1$, is called a path connecting the vertices v_1 and v_l, or, equivalently, (v_1, v_l)-path; it is called cycle if $v_1 = v_l$. The value $l-1$ is called the length of the path/cycle respectively.

Definition 8. *A 3-uniform simple cycle in a 3-uniform hypergraph $H = (V, E)$ is a cycle with the property that its distinct vertices can be arranged in a cyclic order such that, any three vertices forms a hyperedge, if only if they are consecutive three in the cyclic ordering of vertices.*

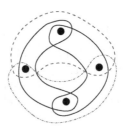

Fig. 4. 3-uniform simple cycle

Theorem 5. *Let H be a weighted 3-uniform simple cycle with l-vertices. Then $3\text{-}CAN(H, \prod_{i=1}^{l} w_H(v_i)) = PW(H)$.*

Proof. Let $\{v_{k-1}, v_k, v_{k+1}\}$ be a hyperedge in H with $w(v_{k-1}) \times w(v_k) \times w(v_{k+1}) = PW(H)$. If $w(v_{k-1}), w(v_k)$ and $w(v_{k+1})$ are three largest weights then we are done. Now we do not assume that $w(v_{k-1}), w(v_k)$ and $w(v_{k+1})$ are three largest weights. We build a balanced mixed 3-$CA(n, H, \prod_{i=1}^{l} w_H(v_i))$ with $n = PW(H)$. Let H_1 be the weighted 3-uniform hypergraph with the single weighted hyperedge $\{v_{k-1}, v_k, v_{k+1}\}$. From Lemma 1, there exists a 3-$CA(n, H_1, w(v_{k-1})w(v_k)w(v_{k+1}))$. For $2 \leq j \leq l-2$, let H_j be the weighted 3-uniform hypergraph obtained from H_{j-1} by hooking a vertex of minimum-weight among the remaining $l - (j+1)$ vertices, at the appropriate position based on the original 3-uniform simple cycle configuration using the single-vertex one-hyperedge hooking or single-vertex two-hyperedge hooking operations as required. If H contains exactly four vertices as shown in Figure 4, then we require single vertex three hyperedge hooking. Note that $H_{l-2} = H$. From Theorem 2, for all $j = 2, ..., l-2$, if $PW(H_j) \leq PW(H)$, then there exists a balanced 3-$CA(PW(H), H_j, \prod_{(u \in H_j)} w_H(u))$. In particular, there exists a balanced 3-$CA(PW(H), H, \prod_{i=1}^{l} w_H(v_i))$. All we need to prove is that $PW(H_j) \leq PW(H)$ for all $j \in \{2, ..., l-2\}$.

To prove this we assume by contradiction that for some $j \in \{2, ..., l-2\}$, $PW(H_j) > PW(H)$. Thus, at some intermediate step, we have a hyperedge $\{u, v, z\}$ in the 3-uniform simple cycle we are building with $w(u) = g, w(v) = h$ and $w(z) = f$ where $ghf > PW(H)$. We know from the definition of $PW(H)$ that the vertices u, v, z are not three consecutive vertices in the original 3-uniform simple cycle. So there must exist at least one vertex between u, v, z in the cyclic arrangement of vertices of the original 3-uniform simple cycle.

Case 1: There is only one vertex in between any two of u, v and z, without loss of generality we can assume there is a vertex in between u and v. Let x be the such vertex with $w(x) = a$, so we have the sequence of vertices in the original cyclic ordering as $v_k, v_{k+1}, ..., u, x, v, ..., v_{k-1}$, i.e, there is a hyperedge $\{u, x, v\}$ in H and vertex x is not in the H_j's we constructed. Since $gah \leq PW(H) < ghf$, we have $a < f$. This is a contradiction since we have hooked the vertices with the smallest weight first.

Case 2: There are more than one vertices in between u and v. Let $x, y, ...$ be the such vertices with $w(x) = a$ and $w(y) = b,...$; so we have the sequence of vertices in the original cyclic ordering as $v_k, v_{k+1}, ..., u, x, y, ..., v, ..., v_{k-1}$ that is there is a hyperedge $\{u, x, y\}$ in H and vertices x, y are not in the H_j we constructed. Since $gab \leq PW(H) < ghf$, that is $ab < hf$ hence we have at least one of the following inequality true.

$$a < h \text{ or } a < f \text{ or } b < h \text{ or } b < f$$

This is a contradiction since we have hooked the vertices with the smallest weight first. This covering array is optimal since $PW(H)$ is a lower bound for the size of a strength three covering array on H.   □

4 Conclusion

In this article, we study strength-3 mixed covering arrays on 3-uniform hypergraphs. We extend results on strength two (mixed) covering arrays on graphs to strength three mixed covering arrays on 3-uniform hypergraphs. We give hypergraph operations that allow us to add vertices to a 3-uniform hypergraph, while preserving the size of a balanced strength-3 covering array on 3-uniform hypergraph. We show that strength-3 mixed covering array number on some special types of 3-uniform hypertree and 3-uniform interval and cycle hypergraph is product weight of that hypergraph. One interesting problem is to study whether the similar result is true for 3-uniform hypergraphs having Helly property.

References

1. Berge, C.: Hypergraphs- Combinatorics of Finite Sets. North-Holland Mathematical Library 45 (1989)
2. Chateauneuf, M.A., Colbourn, C.J., Kreher, D.L.: Covering Arrays of Strength Three. Designs, Codes and Cryptography 16(1), 235–242 (1999)
3. Chateauneuf, M.A., Kreher, D.L.: On the State of Strength-Three Covering Arrays. J. Combin. Designs 10(4), 217–238 (2002)
4. Cohen, D.M., Dalal, S.R., Fredman, M.L., Patton, G.C.: The AETG system: An Approach to Testing Based on Combinatorial Design. IEEE Transaction on Software Engineering 23(7), 437–443 (1997)
5. Cohen, D.M., Dalal, S.R., Parelius, J., Patton, G.C.: The combinatorial design approach to automatic test generation. IEEE Software 13(5), 83–88 (1996)
6. Colbourn, C.J.: Combinatorial aspects of covering arrays. Le Matematiche (Catania) 58, 121–167 (2004)
7. Colbourn, C.J., Dinitz, J.: The CRC Handbook of Combinatorial Designs. CRC Press (1996)
8. Colbourn, C.J., Martirosyan, S.S., Mullen, G.L., Shasha, D.E., Sherwood, G.B., Yucas, J.L.: Products of mixed covering arrays of strength two. J. Combin. Designs 14, 124–138 (2006)
9. Hartman, A.: Software and hardware testing using combinatorial covering suites in Graph Theory. In: Combinatorics and Algorithms: Interdisciplinary Applications, vol. 34, pp. 237–266. Kluwer Academic Publishers (2006)
10. Hartman, A., Raskin, L.: Problems and algorithms for covering arrays. Discrete Mathematics 284, 149–156 (2004)
11. Meagher, K., Moura, L., Zekaoui, L.: Mixed Covering arrays on Graphs. J. Combin. Designs 15, 393–404 (2007)
12. Meagher, K., Stevens, B.: Covering Arrays on Graphs. J. Combin. Theory. Series B 95, 134–151 (2005)
13. Moura, L., Stardom, J., Stevens, B., Williams, A.: Covering arrays with mixed alphabet sizes. J. Combin. Designs 11, 413–432 (2003)
14. Voloshin, V.I.: Introduction to Graph and Hypergraph Theory. Nova Science Publishers, Inc. (2009)

Kalman Filter and Financial Time Series Analysis

M.P. Rajan[1] and Jimson Mathew[2]

[1] School of Mathematics
Indian Institute of Science Education & Research Thiruvananthapuram
CET Campus,Thiruvananthapuram, Kerala, India-695 016
rajanmp@rediffmail.com, rajanmp@iisertvm.ac.in
[2] University of Bristol, United Kingdom
jimson@cs.bris.ac.uk.

Abstract. Kalman filter is one of the novel techniques useful for statistical estimation theory and now widely used in many practical applications. In literature, various algorithms for implementing Kalman filter have been proposed. In this paper, we consider a Fast Kalman Filtering algorithm and applied it to financial time series analysis using ARMA models.

Keywords: Kalman Filter, Fast Kalman, Financial Time Series, ARMA.

1 Introduction

The Kalman filter(cf. [7]) is an indispensable tool in many scientific applications as silicon in the makeup of many electronic systems. It is an estimator used to estimate the state of a linear dynamic system perturbed by Gaussian white noise using measurements that are linear functions of the system state but corrupted by additive Gaussian white noise. The mathematical model used in the derivation of the Kalman filter is a reasonable representation for many problems of practical interest, including control problems as well as estimation problems. Its most immediate applications have been for the control of complex dynamic systems such as continuous manufacturing processes, aircraft, ships, or spacecraft. The Kalman filter model is also used for the analysis of measurement and estimation problems.

The Kalman filter is essentially a set of mathematical equations that implement a predictor-corrector type estimator that is optimal in the sense that it minimizes the estimated error covariance when the observed variable and the noise are jointly Gaussian. Since the time of its inception, the Kalman filter has been the subject of extensive research and application, particularly in the area of autonomous or assisted navigation. This is likely due in large part to advances in digital computing that made the use of the filter practical, but also to the relative simplicity and robust nature of the filter itself. Rarely do the conditions necessary for optimality actually exist, and yet the filter apparently works well for many applications in spite of this situation.

J. Mathew et al. (Eds.): ICECCS 2012, CCIS 305, pp. 339–351, 2012.

An understanding of the Kalman filter is certainly important and useful for time series analysts and researchers. It provides an essential building block for the process used to estimate unknown parameters associated with any linear time series model based on normally distributed disturbances. For trial values of the parameters, it is applied recursively to successive values of a time series, to yield a sequence of one-step ahead predictions. These outputs of the filter are best, linear, unbiased predictors, a consequence being that the associated one-step ahead errors are orthogonal. The Kalman filter conveniently produces these one-step ahead prediction errors, together with their variances. It may be viewed as a device to unravel the dependencies in any time series. As it is always possible to construct this sequence of errors from the original series and vice-versa, both series contain the same information from an estimation perspective. Thus the likelihood associated with the trial parameter values can be evaluated directly from the orthogonal one-step ahead prediction errors. The advantage of this strategy is that explicit inversion of the large variance matrix associated with the original series is avoided. In Econometrics and Statistics, a filter is simply a term used to describe an algorithm that allows recursive estimation of unobserved, time varying parameters, or variables in the system. The basic idea behind the filter is to arrive at a conditional density function of the unobservables using Bayes' Theorem, the functional form of relationship with observables, an equation of motion and assumptions regarding the distribution of error terms. The filter uses the current observation to predict the next periods value of unobservable and then uses the realisation next period to update that forecast.

Fast Kalman filtering (FKF) is developed for statistical calibration of observing systems(cf. [8]). FKF method makes it now possible to apply the theory of optimal linear Kalman filtering to more sophisticated dynamical systems and make real-time use of extremely large time series data set. The key concept behind the FKF is to exploits sparse matrix structures analytically .The statistical calibration based on an optimal linear Kalman Filter is stable under the observability and controllability conditions formulated by Kalman. The stability of calibration can be monitored by predicting or estimating the error covariances $\tilde{P}_t$ or P_t , respectively, from Equations (6).(Prediction error).

We organized the paper in the following manner. Section 2 deals with the basics and derivation of Kalman Filter including the basics of State-Space formulation and a Fast Kalman Filter algorithm. In section 3, we employ kalman filtering technique for financial applications. We consider ARMA model and convert it into a state-space form. The parameters are estimated via maximum likelihood estimation. We supply the numerical result as well.

2   Kalman Filter

In this section , we briefly discuss Kalman filtering and for more details one may refer Hamilton [4]. A key approach applied in Kalman filtering is to express the dynamics in a particular state space form called the state-space representation.

The Kalman Filter is an algorithm for sequentially updating a linear projection for the system. The Kalman filter algorithm provides a way to calculate exact finite-sample forecasts and the exact likelihood function for the Gaussian ARMA processes and various application.

2.1   Assumptions

Let Z_t denote an (n × 1) vector of observed variables at date t. A class of dynamic models for Z_t can be described in terms of a possibly unobserved (r×1) vector X_t known as the state vector. The state-space representation of the dynamics of Z is given by the following system of equations :

$$Z_t = H_t' X_t + v_t \tag{1}$$

$$X_{t+1} = F_t X_t + w_t, \tag{2}$$

where H_t' and F_t are matrices of parameters of dimensions (n × r) and (r×r) respectively. Equation (1) is known as measurement/observation equation and equation (2) is known as State Equation. The (n×1) vector v_t and the (r×1) vector w_t are vector white noise as described below :

$$E(v_t v_\tau') = \begin{cases} R \text{ for } t = \tau; \\ 0 \text{ otherwise.} \end{cases} \tag{3}$$

$$E(w_t w_\tau') = \begin{cases} Q \text{ for } t = \tau; \\ 0 \text{ otherwise.} \end{cases} \tag{4}$$

where Q and R are (r×r) and (n×n) matrices, respectively. The disturbance (white noises) v_t and w_t are assumed to be uncorrelated at all lags

$$E(v_t w_\tau') = 0 \quad for \ all \ t \ and \ \tau. \tag{5}$$

The system of equations (1) to (2) is typically used to describe a finite series of observations $\{Z_1, Z_2, \ldots Z_T\}$ for which the assumptions about the initial value of the state vector X_t are needed. We assume that X_t is uncorrelated with any realizations of v_t and w_t :

$$E(v_t X_t') = E(w_t X_t') = 0 \text{ for } t= 1,2,\ldots\ldots.T \tag{6}$$

2.2   Derivation of the Kalman Filter

We assume that the analyst has supplied with an observed dataset $\{Z_i : i=1,2,\ldots T\}$. One of the ultimate objectives may be to estimate the values of any unknown parameters in the system on the basis of these observations. Further, we assume that the particular numerical values of F, Q, H and R are known with certainty even though it can be estimated from data set. The discussion on Kalman filter here is motivated as an algorithm(the conventional algorithm) for calculating linear least squares forecasts of the state vector on the basis of data observed through data t,

$$\hat{X}_{t+1|t} = E(X_{t+1}|Y_t), \tag{7}$$

where $Y_t = (Z_t, Z_{t-1}, Z_{t-2}........Z_1)$ and $\hat{E}(X_{t+1}|Y_t)$ denotes the linear projection of X_{t+1} on Y_t and a constant. The Kalman filter calculates these forecasts recursively, generating $\hat{X}_{1|0}, \hat{X}_{2|1},......\hat{X}_{T|T-1}$ in succession. Associated with each of these forecasts is a mean squared error *(MSE)* matrix, represented by the following (r×r) matrix :

$$P_{t+1|t} = E[(X_{t+1} - \hat{X}_{t+1|t})(X_{t+1} - \hat{X}_{t+1|t})'] \tag{8}$$

The Kalman algorithm has mainly four components namely, recursion of X_t, forecasting of Z_t, updating the inference about X_t and finally foracsting X_{t+1} and these components will be discussing in the following subsections.

The Recursion. The recursion starts with $\hat{X}_{1|0} := E(X_1)$, the unconditional forecast of X_1 based on observation of Z. with the associated *MSE*

$$P_{1|0} = E\{[X_1 - E(X_1)][X_1 - E(X_1)]'\} \tag{9}$$

To find the unconditional mean of X_t, we take expectations of both sides of equation (2), giving

$$E(X_{t+1}) = F_t.E(X_t) \tag{10}$$

or, since X_t is covariance stationary,

$$(I_r - F_t).E(X_t) = 0 \tag{11}$$

Since unity is not an eigenvalue of F_t, the matrix $(I_r - F_t)$ is nonsingular, and this equation has the unique solution $E(X_t) = 0$. The unconditional variance of X can similarly be found as follows:

$$E(X_{t+1}X'_{t+1}) = E[(F_tX_t + w_t)(X'_tF'_t + w'_t)]$$
$$= F_t.E(X_tX'_t).F'_t + E(w_tw'_t) \tag{12}$$

Letting Σ denote the variance-covariance matrix of X, the above equation implies:

$$\Sigma = F_t\Sigma F'_t + Q \tag{13}$$

whose solution is given by :

$$vec(\Sigma) = [I_{r^2} - (F_t \bigotimes F_t)]^{-1}.vec(Q) \tag{14}$$

Thus, in general, if the eigenvalues of F_t are in unit circle, the Kalman Filter iterations can be started with $\hat{X}_{1|0} = 0$ and $P_{1|0}$ the (r×r) matrix whose elements expressed as a column vector are given by

$$vec(P_{1|0}) = [I_{r^2} - (F_t \bigotimes F_t)]^{-1}.vec(Q) \tag{15}$$

We note that Larger values for the diagonal elements of $P_{1|0}$ register greater uncertainty about the true value of X_t.

Forecasting of Z_t. Given the starting values $\hat{X}_{1|0}$ and $P_{1|0}$, the next step is to compute the values of $\hat{X}_{2|1}$ and $P_{2|1}$. More generally, given $\hat{X}_{t|t-1}$ and $P_{t|t-1}$, the idea is to compute $\hat{X}_{t+1|t}$ and $P_{t+1|t}$ for $t = 2, 3, \ldots\ldots T$. We have

$$\hat{E}(X_t|Y_{t-1}) = \hat{X}_{t|t-1} \tag{16}$$

and the forecasting the value of Z_t as

$$\hat{Z}_{t|t-1} = \hat{E}(Z_t|Y_{t-1}) \tag{17}$$

Recall from Equation (1) that

$$\hat{E}(Z_t|X_t) = H'_t X_t \tag{18}$$

and so from law of iterated projections,

$$\hat{Z}_{t|t-1} = H'_t \hat{X}_{t|t-1} \tag{19}$$

The error of this forecast is :

$$
\begin{aligned}
Z_t - \hat{Z}_{t|t-1} &= H'_t X_t + v_t - H'_t \hat{X}_{t|t-1} \\
&= H'_t(X_t - \hat{X}_{t|t-1}) + v_t
\end{aligned} \tag{20}
$$

with *MSE*

$$
\begin{aligned}
E[(Z_t - \hat{Z}_{t|t-1})(Z_t - \hat{Z}_{t|t-1})'] &= E[H'_t(X_t - \hat{X}_{t|t-1})(X_t - \hat{X}_{t|t-1})' H'_t] \\
&\quad + E[v_t v'_t]
\end{aligned} \tag{21}
$$

Since

$$E[v_t(X_t - \hat{X}_{t|t-1})'] = 0, \tag{22}$$

the above equation can be simplified as:

$$E[(Z_t - \hat{Z}_{t|t-1})(Z_t - \hat{Z}_{t|t-1})'] = H'_t P_{t|t-1} H_t + R \tag{23}$$

Updating the Inference about X_t. Next, the inference about the current value of X_t is updated on the basis of the observation of Z_t to produce

$$\hat{X}_{t|t} = \hat{E}(X_t|Z_t, Y_{t-1}) = \hat{E}(X_t|Y_t) \tag{24}$$

and

$$E[((X_t - \hat{X}_{t|t-1})(Z_t - \hat{Z}_{t|t-1})'] = P_{t|t-1} H_t \tag{25}$$

One can see that

$$\hat{X}_{t|t} = \hat{X}_{t|t-1} + P_{t|t-1} H_t (H'_t P_{t|t-1} H_t + R)^{-1}(Z_t - H'_t \hat{X}_{t|t-1}) \tag{26}$$

The *MSE* associated with this update projection, which is denoted by $P_{t|t}$, can be found as :

$$
\begin{aligned}
P_{t|t} &= E[(X_t - \hat{X}_{t|t})(X_t - \hat{X}'_{t|t})] \\
&= P_{t|t-1} - P_{t|t-1} H_t (H'_t P_{t|t-1} H_t + R)^{-1} H'_t P_{t|t-1}
\end{aligned} \tag{27}
$$

Forecast of X_{t+1}. Our next step is to forecast X_{t+1}

$$\hat{X}_{t+1|t} = \hat{E}(X_{t+1}|Y_t) = F_t.\hat{E}(X_t|Y_t) + \hat{E}(w_t|Y_t) = F_t\hat{X}_{t|t} + 0. \qquad (28)$$

Substituting value of $\hat{X}_{t|t}$ in above equation, we get

$$\hat{X}_{t+1|t} = F_t\hat{X}_{t|t-1} + F_tP_{t|t-1}H_t(H_t'P_{t|t-1}H_t + R)^{-1}(Z_t - H_t'\hat{X}_{t|t-1}) \qquad (29)$$

The coefficient matrix here is known as the *gain matrix* and is denoted as K_t

$$K_t = F_tP_{t|t-1}H_t(H_t'P_{t|t-1}H_t + R)^{-1} \qquad (30)$$

and thus the equation (29) can be rewritten as

$$\hat{X}_{t+1|t} = F_t\hat{X}_{t|t-1} + K_t(Z_t - H_t'\hat{X}_{t|t-1}) \qquad (31)$$

Accordingly, the *MSE* of this forecast can be found from the first equation :

$$P_{t+1|t} = F_tP_{t|t}F_t' + Q \qquad (32)$$

Substituting the value of $P_{t|t}$ from (??) in (33, we get:

$$P_{t+1|t} = F_t[P_{t|t-1} - P_{t|t-1}H_t(H_t'P_{t|t-1}H_t + R)^{-1}H_t'P_{t|t-1}]F_t' + Q \qquad (33)$$

Summary of Kalman Filter. We summarize the above discussion of Kalman filter algorithm in the following.

$$\hat{X}_{t+1|t} = F_t\hat{X}_{t|t-1} + K_t(Z_t - H_t'\hat{X}_{t|t-1}) \qquad (34)$$
$$\hat{X}_{0|-1} = E(X_0) \qquad (35)$$
$$\hat{X}_{1|0} = E(X_1) \qquad (36)$$
$$K_t = F_tP_{t|t-1}H_t(H_t'P_{t|t-1}H_t + R)^{-1} \qquad (37)$$
$$P_{t+1|t} = F_t[P_{t|t-1} - P_{t|t-1}H_t(H_t'P_{t|t-1}H_t + R)^{-1}H_t'P_{t|t-1}]F_t' + Q \qquad (38)$$
$$P_{0|-1} = P_0 \qquad (39)$$
$$\hat{X}_{t|t} = \hat{X}_{t|t-1} + P_{t|t-1}H_t(H_t'P_{t|t-1}H_t + R)^{-1}(Z_t - H_t'\hat{X}_{t|t-1}) \qquad (40)$$
$$P_{t|t} = P_{t|t-1} - P_{t|t-1}H_t(H_t'P_{t|t-1}H_t + R)^{-1}H_t'P_{t|t-1} \qquad (41)$$

Notice that (35) and (40) imply:

$$F_t\hat{X}_{t|t} = K_tZ_t + (F_t - K_tH_t')\hat{X}_{t|t-1} \qquad (42)$$

So (34) is equivalent to

$$\hat{X}_{t+1|t} = F_t\hat{X}_{t|t} \qquad (43)$$

3   Fast Kalman Filter

Fast Kalman filter has been introduced in literature to overcome some of the difficulties posed by Kalman filter algorithm. In Kalman filter, a reliable operation requires continuous fusion of data. Its optimality depends on the use of the error variances and covariances between all measurements and the estimated state and calibration parameters. This large error covariance matrix is obtained by matrix inversion from the respective system of Normal Equations. One may employ traditional numerical techniques to solve this problem. However, these methods can find the solution approximately with high computational cost and it might be thus impossible to do the data fusion in strictly optimal fashion and consequently, the filters stability may become a serious issue. One of the fast algorithms for Kalman filter(FKF) has been developed by Lange in 1999 is an extention of Helmert-Wolf blocking(HWB) method . The algorithm has been used in many practical applications(cf. [8,9,10]). The estimation method make use of Helmert-Wolf formula and C.R. Rao's theory of minimum-norm quadratic unbiased estimation (cf. [9]) and reference therein. The approach is more suitable in the sense that the coefficient matrix is generaly sparse and may often have either a bordered block- or band-diagonal (BBD) structure. If it is band-diagonal it can be transformed into a block-diagonal form. It is first proposed by Hanz Boltz in 1923 who has used it for the inversion of geodetic matrices and later in 1937, Tadeusz Banachiewicz has generalized it and proved its correctness (cf. [2]). The large matrix can thus be most effectively inverted in a block wise manner by using the following analytic inversion formula:

$$\begin{bmatrix} A & B \\ C & D \end{bmatrix}^{-1} = \begin{bmatrix} A^{-1} + A^{-1}B(D - CA^{-1}B)^{-1}CA^{-1} & -A^{-1}B(D - CA^{-1}B)^{-1} \\ -(D - CA^{-1}B)^{-1}CA^{-1} & (D - CA^{-1}B)^{-1} \end{bmatrix} \quad (44)$$

where :

A = a large **block- or band-diagonal (BD)** matrix to be easily inverted and $(D - CA^{-1}B)^{-1}$ = smaller matrix called the **Schur complement** of A.

The computational load of FKF is roughly proportional to $O(N^2)$ where N denotes the number of state parameters to be estimated.

4   Financial Applications of Kalman Filter

In this section, we focus upon the Financial Application of Kalman Filter on ARMA models . ARMA(p,q) model is one of the basic time series model widely used for financial time series analysis. ARMA models can be written in the state-space forms, and the Kalman filter is used to estimate the parameters. It can also be used to estimate time varying parameters in a linear regression and to obtain Maximum likelihood estimates of a state-space model.

4.1   Auto Regressive Moving Average(ARMA) Model

The parameters of ARMA(p,q) models are generally estimated using method of maximum likelihood (cf.[6]) even though there are alternative way to achieve this (cf. [5]). We employ Kalman filtering approach for maximum likelihood estimation. This can be accomplished by rewriting the dynamics of ARMA(p,q) model in state-space form. Here we consider the following ARMA(p,q) process.

$$Z_t = a_1 Z_{t-1} + \cdots + a_p Z_{t-p} + \epsilon_t + b_1 \epsilon_{t-1} + \cdots + b_q \epsilon_{t-q} \tag{45}$$

where ϵ_t is a white noise i.e. a sequence of uncorrelated random variables with zero mean and variance σ^2.

If the original data do not have mean zero mean, we first have to convert to construct $\tilde{Z}_t = Z_t - \sum_{s=1}^{T} \frac{Z_s}{T}$ and then fit the ARMA model to $\tilde{Z}_t$. The model above is rewritten as:

$$Z_t = c + a_1 Z_{t-1} + \cdots + a_m Z_{t-m} + \epsilon_t + b_1 \epsilon_{t-1} + \cdots + b_{m-1} \epsilon_{t-m+1} \tag{46}$$

where $m = \max(p, q+1)$ and some of the coefficients $a_1, \cdots a_m, b_1, \cdots, b_{m-1}$ can be zero.

As it is well known, the ARMA process above is represented as:

$$\text{Measurement equation}: \quad Z_t = H_t' X_t \tag{47}$$

$$\text{Transition Equation}: \quad X_t = F_t X_{t-1} + B\epsilon_t \tag{48}$$

where H_t, F_t and B are defined as :

$$H_t = (1, 0, \cdots, 0) \tag{49}$$

$$F_t = \left(\begin{array}{c|c} \begin{matrix} a_1 \\ \vdots \\ a_{m-1} \\ a_m \end{matrix} & \begin{matrix} I_{m-1} \\ \\ \hline 0 \end{matrix} \end{array} \right) \tag{50}$$

$$B = \begin{pmatrix} 1 \\ b_1 \\ \vdots \\ b_{m-1} \end{pmatrix} \tag{51}$$

Thus,we obtain the state-space form of an ARMA(p,q) model. Next, we address the issue of estimating the parameters using maximum likelihood via Kalman filtering approach.

4.2   Maximum Likelihood Estimation of Parameters of ARMA Model

Kalman filter can be used on the state space model to compute parameters of ARMA(p,q) model. Here, we provide with the equations to achieve Log-likelihood eqation using Kalman Filter. The expected value of X_{t+1} conditioned on the history of observations $(Z_0, \cdots, Z_t)$ is denoted by $\hat{X}_{t+1|t} = $

$E_t[X_{t+1}|(Z_0, \cdots, Z_t); X_0]$. Associated with each $X_{t+1|t}$ forecast given an initial value $\hat{X}_{1|0} = \bar{0}$ is a mean squared error matrix, defined as

$$P_{t+1|t} = E[(X_{t+1} - \hat{X}_{t+1|t})(X_{t+1} - \hat{X}_{t+1|t})'] \tag{52}$$

Using estimates $X_{t|t-1}$, we can compute the innovations

$$g_t = Z_t - H_t' \hat{X}_{t|t-1} \tag{53}$$

The innovation variance, denoted by ω_t, satisfies the following equation

$$\begin{aligned}
\omega_t &= E((Z_t - H_t'\hat{X}_{t|t-1})(Z_t - H_t'\hat{X}_{t|t-1})') \\
&= E((H_t'Xt - H_t'\hat{X}_{t|t-1})(H_t'Xt - H_t'\hat{X}_{t|t-1})') \\
&= H_t' P_{t|t-1} H_t
\end{aligned} \tag{54}$$

Using equation (33) we can get,

$$P_{t+1|t} = F_t[P_{t|t-1} - P_{t|t-1}H_t H_t'/\omega_t]F_t' + BB'\sigma^2) \tag{55}$$

The likelihood function of the observation vector $Z_0, Z_1, \cdots, Z_t$ is given by

$$L = \prod_{t=1}^{T}(2\pi\omega_t)^{-1/2} \exp\left(-\frac{g_t^2}{2\omega_t}\right) \tag{56}$$

Taking logarithms on both sides, dropping the constant 2π and multiplying by 2 we obtain

$$l = -\sum_{t=1}^{T}[ln(\omega_t) + g_t^2/\omega_t] \tag{57}$$

In principle, to find the MLE estimate we maximize (57) with respect to the parameters a_i, b_i and σ^2. However, we can concentrate out the term σ^2 and maximize only with respect to parameters a_i, b_i by following trick. If we initialize the filter with the matrix $\tilde{P}_{1|0} = \sigma^2 P_{1|0}$ then we can optimize first with respect to σ^2. Now, replacing the result in objective function and optimizing the resulting expression with respect to parameters a_i, b_i. The above equation (57) becomes

$$l = -\sum_{t=1}^{T}\left[ln(\sigma^2\omega_t) + \frac{a_t2}{\omega_t\sigma^2}\right] \tag{58}$$

and σ^2 is cancelled out in the evolution equations of $P_{t+1|t}$ and in the projections $\hat{X}_{t+1|t}$. So we can directly optimize (57) to obtain

$$\sigma^2 = \frac{1}{T}\sum_{t=1}^{T}g_t^2/\omega_t \tag{59}$$

Replacing the result into (57) we obtain the concentrated log-likelihood function:

$$l = -\sum_{t=1}^{T}\left[ln(\sigma^2) + ln(\omega_t) + \frac{g_t^2}{\omega_t\sigma^2}\right]$$

$$l = -\left[Tln(\frac{1}{T}) + T + Tln\sum_{t=1}^{T} g_t^2/\omega_t + \sum_{t=1}^{T} ln(\omega_t)\right] \tag{60}$$

or, dropping irrelevant constants,

$$l = -\left[Tln\sum_{t=1}^{T} g_t^2/\omega_t + \sum_{t=1}^{T} ln(\omega_t)\right] \tag{61}$$

we can use the numerical methods to maximize the above equation (61) with respect to a_i, b_i.

4.3   Algorithm and Numerical Results

In this section, we use real time dataset for the computation purpose. We have considered three different datasets of daily stock returns of IBM, Microsoft and Goldman Sachs. We analysed how fast Kalman filter perform agains general Kalman filter algorithm. These are illustrated in Figures 1, 2 and 3. The estimated results are provided in Table 1. We have also compared the computational complexity of fast Kalman algorithm and general Kalman algorithm in Table 2. The result indicates that fast Kalman is efficient than the traditional algorithm. The innovation sequence for GS data is given in Figure 4. The proposed algorithm is given below.

- Use time series containing daily values of stock and calculate their returns.
- Compute the value of p & q to be sent as input for ARMA(p,q) model to be estimated by the program. This can be done using the ACF and PACF properties of the time series and concluding the (p,q) for the model.
- Initialize the parameters of the ARMA(p,q) model by using the regression methods for ARMA initialization [5] and use the initial parameters for further calculations towards the maximization of log-likelihood function.
- Create a function for maximizing(minimizing) the log-likelihood (-log likelihood) (MLE) of the state-space form of ARMA model using Kalman Filter.[2]
- Find out the points/ coefficients at which the log-likelihood maximizes using numerical techniques of maximizing/minimizing.
- Calculate the *number of iterations, number of function calls* and the *total time taken* for the whole procedure.
- Find the required estimated parameters and coefficients $\gamma = (a_1, a_2, \cdots, a_{p-1}, a_p, b_1, b_2, \cdots, b_{q-1}, b_q)$ which maximizes the log likelihood function.
- Compute the σ (volatility) of the model.

Table 1. Estimated AR and MA Coefficients of the fitted models[1]

	GS ARMA(2,2)	**IBM** ARMA(1,1)	**MSFT** ARMA(1,0)
$\underline{a(1)}$	-0.1009	-0.0215	0.0511
$a(1)$	-0.09107	-0.02655	0.05275
Error	(-0.00983)	(0.00505)	(-0.00165)
$\underline{a(2)}$	0.5187	-	-
$a(2)$	0.5158	-	-
Error	(0.0029)	-	-
$\underline{b(1)}$	0.0924	0.0597	-
$b(1)$	0.08448	0.07165	-
Error	(0.00792)	(-0.01195)	-
$\underline{b(2)}$	-0.6005	-	-
$b(2)$	-0.5824	-	-
Error	(-0.0181)	-	-
σ	0.0265	0.0181	0.0244

Table 2. Computational Data[2]

Complexity	**GS**	**IBM**	**MSFT**
No. of iterations	25	9	2
No. of function calls	170	57	14
Time	**GS**	**IBM**	**MSFT**
Fast Kalman(proposed) Method	10.81	4.91	1.09
General Kalman Method	14.55	6.70	1.87
Difference	**(3.74)**	**(1.79)**	**(0.78)**

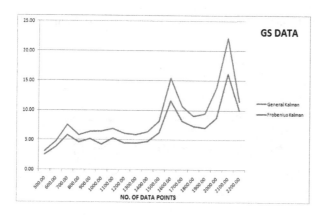

Fig. 1. Time comparison between General & Fast Kalman (GS)

---

[1] * <u>Underline</u> *defines the estimated coefficients and (.) defines the errors.*
[2] * *The time series for analysis are of size appx. 2250.*

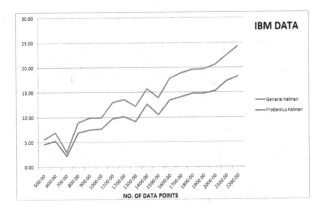

Fig. 2. Time comparison between General & Fast Kalman (IBM)

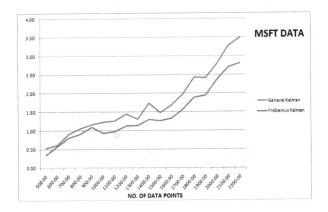

Fig. 3. Time comparison between General & Fast Kalman (MSFT)

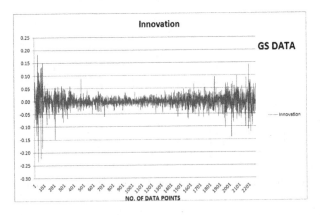

Fig. 4. Residuals (between actual data and estimated values)

5   Conclusion

In this article, we studied the applications of Kalman filter in financial time series analysis. We observe that Kalman filter provides recursive algorithm to estimate the unobserved states from the observed variables which enables us to estimate the parameters of the model from financial time series by converting it into the respective state-space forms. From the application point of view, we take the example of Auto regressive Moving average (ARMA(p,q)) model. The estimated parameters are compared with actual parameters and the errors are calculated which are found to be of order 10^{-4} or less and hence making the estimation procedure through Kalman filter an efficient one.

References

1. Done, W.J.: Use of Fast Kalman Estimation Algorithm for Adaptive System Identification, Acomo Production Company Research Center. Journal of Mathematical Analysis and Applications 6, 886–889 (1981)
2. Grala, J., Markiewicz, K., Styan, G.P.H.: Tadeusz Banachiewicz:18892-1954. Image: The Bulletin of the International Linear Algebra Soceity 25, 24 (2000)
3. Gardner, G., Harvey, A.C., Phillips, G.D.A.: An Algorithm for Exact Maximum Likelihood Estimation of Autoregressive-Moving Average Models by Means of Kalman Filtering. Applied Statistics 29, 311–322 (1980)
4. Hamilton, J.D.: Time Series Analysis. Princeton University Press, New Jersy (1994)
5. Hanna, E.J., McDougal, M.J.: Regression Procedure for ARMA Estimation. Journal of American Statistical Association 83, 490–498 (2001)
6. Harvey, A.C.A., Philips, G.D.A.: Maximum Likelihood Estimation of Regression Models with Autoregressive-Moving Average Disturbances 66, 48–58 (1979)
7. Kalman, R.E.: A new approach to Linear Filtering and Prediction Problems. Transactions of ASME-Journal of Basic Engineering 82, 35–45 (1960)
8. Lang, A.A.: Statistical calibration of observing system. Academic Dissertation, FMI, Helsinki (1999)
9. Lang, A.A.: Simulataneous Statistical Calibration of the GPS Signal Delay Measurements with Metereological Data. Phys. Chem. Earth(A) 26(6-8), 471–473 (2001)
10. Lang, A.A.: Optimal Kalma Filtering for ultra-reliable Tracking. In: Prooceedings of the Symposium on Atmospheric Remote Sensing using Satellite Naviagation Systems, Matera, Italy, October 13-15 (2003)
11. Mung, I.J.: Tutorial on maximum likelihood estimation. Journal of Mathematical Pshycology 34, 5–34 (2001)
12. Penzer, J., Jong, P.: The ARMA model in state space form. Statistics and Probability Letters 70, 2–4 (2004)
13. Pichler, P.: State Space Models and the Kalman Filter, 40461 Vektorautoregressive Methoden (2007)
14. Rao, C.R.: Estimation of Variance and Covaraince Components in Linear Models. Journal of Americal Statistical Association 67(337), 112–115 (1972)
15. Wong, W.-K., Miller, R.B., Shrestha, K.: Maximum Likelihood Estimation of ARMA Model with Error Processes for Replicated Observations. Working Paper No. 0217, Department of Economics, NUS

Fuzzy Properties on Finite Type of Kac-Moody Algebras

A. Uma Maheswari and V. Gayathri

Quaid-E -Millath Government College for Women (Autonomous), Chennai - 600 002, India
umashiva2000@yahoo.com

Abstract. The theory of Kac-Moody algebra was introduced by V.G. Kac & R.V. Moody simultaneously and independently in 1968. The subject Kac-Moody algebras attracted the attention of many Mathematicians because of its various connections and applications to different branches of Mathematics and Mathematical Physics. On the other hand, the concept of fuzzy sets originated in a seminar paper by Lotfi A. Zadeh in 1965. Since then, the subject has developed in leaps and bounds and finds its application in all branches of science, engineering, technology, economics, social and behavioral sciences. In this paper, we introduce the notion of fuzzy sets on Kac-Moody Lie algebras. We define a fuzzy set on the set of simple roots using the Generalized Cartan Matrix elements. Some basic properties like convexity, cardinality, normality are studied. We also compute the α – level sets and strong α – level sets in the case of finite type of Kac-Moody algebras D_l ($l \geq 4$), E_6 , E_7, E_8 , F_4 and G_2.

Keywords: Kac-Moody algebra, root system, finite, Dynkin diagram, fuzzy set, α – level set, cardinality, convex.

1   Preliminaries

1.1   Basic Definitions and Properties of Kac–Moody Algebras

Definition 1[3]: An integer matrix $A = (a_{ij})^n_{i,j\,=\,1}$ is a Generalized Cartan Matrix (abbreviated as GCM) if it satisfies the following conditions:

1. $a_{ii} = 2$ $\forall$ i =1,2,...,n
2. $a_{ij} = 0 \Leftrightarrow a_{ji} = 0$ $\forall$ i, j = 1,2,...,n
3. $a_{ij} \leq 0$ $\forall$ i, j = 1,2,...,n.

Let us denote the index set of A by N ={1,...,n}. A GCM A is said to decomposable if there exist two non-empty subsets I, J $\subset$ N such that I $\cup$ J = N and $a_{ij} = a_{ji} = 0$ $\forall$ i$\in$ I and j$\in$ J. If A is not decomposable, it is said to be indecomposable.

Definition 2[2]: A realization of a matrix $A = (a_{ij})^n_{i,j\,=\,1}$ is a triple (H, Π , Π^v) where l is the rank of A , H is a 2n - l dimensional complex vector space, $\Pi = \{\alpha_1,..., \alpha_n\}$ and $\Pi^v = \{\alpha_1^v,..., \alpha_n^v\}$ are linearly independent subsets of H* and H respectively, satisfying

J. Mathew et al. (Eds.): ICECCS 2012, CCIS 305, pp. 352–363, 2012.

$\alpha_j(\alpha_i^v) = a_{ij}$ for i, j = 1,...,n. Π is called the root basis. Elements of Π are called simple roots. The root lattice generated by Π is $Q = \sum_{i=1}^{n} Z\alpha_i$.

Definition 3[2]: The Kac-Moody algebra g(A) associated with a GCM $A = (a_{ij})_{i,j=1}^{n}$ is the Lie algebra generated by the elements e_i, f_i, $i = 1,2,...,n$ and H with the following defining relations :

$$[h,h'] = 0, \quad h,h' \in H \; ; \; [e_i, f_j] = \delta_{ij}\alpha_i^v$$

$$[h,e_j] = \alpha_j(h)e_j \; ; \; [h,f_j] = -\alpha_j(h)f_j \; , \; i, j \in N$$

$$(ad \; e_i)^{1-a_{ij}} e_j = 0 \; ; \; (ad \; f_i)^{1-a_{ij}} f_j = 0 \quad \forall \; i \neq j, \; i, j \in N$$

The Kac-Moody algebra g(A) has the root space decomposition $g(A) = \bigoplus_{\alpha \in Q} g_\alpha(A)$

where $g_\alpha(A) = \{x \in g(A) / [h,x] = \alpha(h)x, \; for \; all \; h \in H\}$. An element α, $\alpha \neq 0$ in Q is called a root if $g_\alpha \neq 0$. Let $Q = \sum_{i=1}^{n} Z_+\alpha_i$. Q has a partial ordering "$\leq$" defined by $\alpha \leq \beta$ if $\beta - \alpha \in Q_+$, where $\alpha, \beta \in Q$.

Definition 4[2]: Let $\Delta(= \Delta(A))$ denote the set of all roots of g(A) and Δ_+ the set of all positive roots of g(A). We have $\Delta_- = -\Delta_+$ and $\Delta = \Delta_+ \cup \Delta_-$.

Definition 5[2]: To every GCM A is associated a Dynkin diagram S(A) defined as follows: S(A) has n vertices and vertices i and j are connected by max $\{|a_{ij}|, |a_{ji}|\}$ number of lines if a_{ij}. $a_{ji} \leq 4$ and there is an arrow pointing towards i if $|a_{ij}| > 1$. If a_{ij}. $a_{ji} > 4$, i and j are connected by a bold faced edge, equipped with the ordered pair $(|a_{ij}|, |a_{ji}|)$ of integers.

Theorem 6[2]: Let A be a real n x n matrix satisfying (m1), (m2) and (m3).
    (m1) A is indecomposable;
    (m2) $a_{ij} \leq 0$ for $i \neq j$;
    (m3) $a_{ij} = 0$ implies $a_{ji} = 0$

Then one and only one of the following three possibilities holds for both A and ^tA:

(a) det A $\neq 0$; there exists u > 0 such that A u > 0; A v ≥ 0 implies v > 0 or v = 0;
(b) co rank A =1; there exists u > 0 such that A u = 0; A v ≥ 0 implies A v = 0;
(c) there exists u > 0 such that A u < 0; A v ≥ 0, v ≥ 0 imply v = 0.

Then A is of finite, affine or indefinite type iff (a), (b) or (c) is satisfied.

Definition 7[5]: A Kac-Moody algebra g(A) is said to be of finite, affine or indefinite type if the associated GCM A is of finite, affine or indefinite type respectively.

Note 8[5]: We note that for the finite type of Kac-Moody algebra the rank of the GCM A = n. i.e., $l = n$.

For a more detailed study on Kac- Moody algebras, one can refer to [2] and [5].

1.2   Basic Definitions of Fuzzy Sets

Definition 9[6]: A classical (crisp) set is normally defined as a collection of elements or objects $x \in X$ that can be finite, countable or over countable.

Definition 10[6]: If X is a collection of objects denoted generically by x, then a fuzzy set $\tilde{A}$ is defined as $\tilde{A} = \{(x, \mu_{\tilde{A}}(x))/x \in X\}$. $\mu_{\tilde{A}}(x)$ is called the membership function or "grade of membership" of x in $\tilde{A}$ that maps X to the membership space M.

Definition 11[6]: The support of a fuzzy set $\tilde{A}$, $S(\tilde{A})$ is the crisp set of all $x \in X$ such that $\mu_{\tilde{A}}(x) > 0$.

Definition 12[6]: The (crisp) set of elements that belong to the fuzzy set $\tilde{A}$ at least to the degree α is called the α - level set $A_{\alpha} = \{x \in X / \mu_{\tilde{A}}(x) \geq \alpha\}$; $A_{\alpha}' = \{x \in X / \mu_{\tilde{A}}(x) > \alpha\}$ is called "Strong α - level set" or "Strong α - cut".

Definition 13[6]: Let $\tilde{A}$ be a fuzzy set on X. Then the set $\{x \in X / \mu_{\tilde{A}}(x) = 1\}$ is called the core of the fuzzy set $\tilde{A}$. This set is denoted by core ($\tilde{A}$).

Definition 14[6]: A fuzzy set $\tilde{A}$ is said to be normal if $\sup_x \mu_{\tilde{A}}(x) = 1$

Definition 15[6]: The membership function of the complement of a normalized fuzzy set $\tilde{A}$, $\mu_{C\tilde{A}}(x) = 1 - \mu_{\tilde{A}}(x)$ $x \in X$.

Definition 16[6]: For a finite fuzzy set $\tilde{A}$, the cardinality $|\tilde{A}|$ is defined as $|\tilde{A}|$ $= \sum_{x \in X} \mu_{\tilde{A}}(x)$. $\|\tilde{A}\| = \dfrac{|\tilde{A}|}{|X|}$ is called the relative cardinality of $\tilde{A}$.

Definition   17[6]:   A   fuzzy   set $\tilde{A}$ is   convex   if $\mu_{\tilde{A}}(\lambda x_1 + (1 - \lambda)x_2) \geq \min\{\mu_{\tilde{A}}(x_1), \mu_{\tilde{A}}(x_2)\}$, $x_1,\ x_2 \in X,\ \lambda \in [0,1]$.

Definition 18[1]: Let A be a fuzzy set on U and α be a number such that $0 < \alpha \leq 1$. Then by αA we mean a fuzzy set on U, denoted by αA which is such that (αA) $x = \alpha A(x)$ for every x in U. This procedure of associating another fuzzy set with the given fuzzy set A is termed as restricted scalar multiplication.

In our previous paper [4] we introduced the new concept of fuzzy sets on the root systems of Kac- Moody algebras. The fuzzy set on $X = \Pi \times \Pi$ is defined as follows:

$$\mu_{\tilde{A}}(\alpha_i, \alpha_j) = \begin{cases} \dfrac{1}{\max(\ |a_{ij}|, |a_{ji}|\ )} & if\ \ a_{ij} \neq 0 \\ 0 & if\ \ a_{ij} = 0 \end{cases} \quad for\ (\alpha_i, \alpha_j) \in X,\ i, j = 1, 2, ..., l \quad (1)$$

Then $((\alpha_i, \alpha_j),\ \mu_{\tilde{A}}(\alpha_i, \alpha_j))$ forms a fuzzy set on $\Pi \times \Pi$.

Lemma 19[4]: Support of $\tilde{A}$ consists of all (α_i, α_j) such that $a_{ij} \neq 0$, for i, j = 1,...,*l*.

Lemma 20[4]: Core of the fuzzy set $\tilde{A}$ is non – empty if and only if the associated Dynkin diagram contains at least one sub diagram of the form ◯—◯ .

Lemma 21[4]: The fuzzy set $\tilde{A}$ defined by (1) is normal iff a sub diagram of the form ◯—◯ occurs in the Dynkin diagram associated with the GCM A .

2 Some Properties of Fuzzy Set on the Cartesian Product of the Root System $\Pi \times \Pi$ of the Kac-Moody Algebra $g(A)$

In this section we shall study about some more properties of fuzzy sets on the root systems of Kac- Moody algebras.

Lemma 22: A fuzzy set $\tilde{A}$ corresponding to the finite type of Kac-Moody algebras D_l ($l \geq 4$), E_6, E_7, E_8 and F_4(shown in the figures Fig.1 to Fig.5) defined by (1) are convex .

Proof: Consider the finite type of Kac-Moody algebras $D_l (l \geq 4$), E_6, E_7, E_8 and F_4 (shown in the figures Fig.1 to Fig.5). The table showing all possible membership grades for the elements of X and the conditions for checking convexity are listed:

Table 1. Possible membership grades attained by the elements in $\Pi \times \Pi$

$\mu_{\tilde{A}}(\alpha_i, \alpha_j)$	$\mu_{\tilde{A}}(\alpha_k, \alpha_l)$	$\mu_{\tilde{A}}(\ \lambda(\alpha_i, \alpha_j) + (1-\lambda)(\alpha_k, \alpha_l)\ )$	min{ $\mu_{\tilde{A}}(\alpha_i, \alpha_j), \mu_{\tilde{A}}(\alpha_k, \alpha_l)$ }
1	1	1	1
1	0	λ	0
1	1/2	$(\lambda+1)/2$	1/2
0	1	$1-\lambda$	0
0	0	0	0
0	1/2	$1-(\lambda/2)$	0
1/2	1	$1-(\lambda/2)$	1/2
1/2	0	$\lambda/2$	0
1/2	1/2	1/2	1/2

For every $(\alpha_i, \alpha_j), (\alpha_k, \alpha_l) \in X$ & $\lambda \in [0,1]$, we see that the following inequality is satisfied: $\mu_{\tilde{A}}(\lambda(\alpha_i, \alpha_j) + (1-\lambda)(\alpha_k, \alpha_l)\ ) \geq \min\{\ \mu_{\tilde{A}}(\alpha_i, \alpha_j), \mu_{\tilde{A}}(\alpha_k, \alpha_l)\}$. Hence a fuzzy set $\tilde{A}$ corresponding to the finite type of Kac-Moody algebras $D_l (l \geq 4$), E_6, E_7, E_8 and F_4 are convex.

Remark 23: A fuzzy set $\tilde{A}$ corresponding to the classical families of finite type of Kac-Moody algebras A_l, B_l, C_l defined by (1) are convex.

Computation of α – Level Sets

Theorem 24: For the finite type of Kac-Moody algebra D_l($l \geq 4$) associated with the indecomposable GCM A, let $\tilde{A}$ be the fuzzy set defined on $\Pi \times \Pi$ given by (1). Then the α -level sets and strong α -level sets for $\alpha = 1$, 1/2 , ..., 1/k,... are given below :

(i) $A_1 = \{(\alpha_1,\alpha_2),...,(\alpha_{l-2},\alpha_{l-1}),(\alpha_{l-2},\alpha_l),(\alpha_l,\alpha_{l-2}),(\alpha_{l-1},\alpha_{l-2}),...,(\alpha_2,\alpha_1)\}$

(ii) $A_{1/2} = A_1 \cup \{(\alpha_1,\alpha_1),(\alpha_2,\alpha_2),...,(\alpha_l,\alpha_l)\}$

(iii) $A_{1/2}{}' = A_1$ (iv) $A_{1/3}{}' = A_{1/2}$

(v) $|A_1| = 2l-2$, $|A_{1/2}| = 3l-2$, $|A_{1/2}{}'| = 2l-2$. For $k = 3,4,...$, $|A_{1/k}| = |A_{1/k}{}'| = 3l-2$.

Proof: Consider the classical family D_l ($l \geq 4$),

Fig. 1. Dynkin diagram for the finite type of Kac- Moody algebra D_l ($l \geq 4$)

(i) $A_1 = \{(\alpha_i,\alpha_j) \in X / \mu_{\tilde{A}}(\alpha_i,\alpha_j) \geq 1\} = \{(\alpha_i,\alpha_j) \in X / \dfrac{1}{\max(|a_{ij}|,|a_{ji}|)} \geq 1\}$

$= \{(\alpha_1,\alpha_2),...,(\alpha_{l-2},\alpha_{l-1}),(\alpha_{l-2},\alpha_l),(\alpha_l,\alpha_{l-2}),(\alpha_{l-1},\alpha_{l-2}),...,(\alpha_2,\alpha_1)\}$

(ii) $A_{1/2} = A_1 \cup \{(\alpha_1,\alpha_1),(\alpha_2,\alpha_2),...,(\alpha_l,\alpha_l)\}$, $A_{1/3} = A_{1/2}$

From the above relations we have, $A_1 \subset A_{1/2} = A_{1/3} = ... = A_{1/k} = ...$

(iii) $A_{1/2}{}' = \{(\alpha_i,\alpha_j) \in X / \mu_{\tilde{A}}(\alpha_i,\alpha_j) > 1/2\} = \{(\alpha_i,\alpha_j) \in X / \dfrac{1}{\max(|a_{ij}|,|a_{ji}|)} > 1/2\} = A_1$

(iv) $A_{1/3}{}' = A_{1/2}{}'$, $A_{1/4}{}' = A_{1/3}{}'$, $A_1{}' \subset A_{1/2}{}' \subset A_{1/3}{}' = A_{1/4}{}' = ... = A_{1/k}{}' = ...$.

(v) By the above relations, $A_{1/2} = A_{1/3} = ... = A_{1/k} = ... = A_{1/3}{}' = A_{1/4}{}' = ... = A_{1/k}{}'$

$|A_1| = 2l-2$, $|A_{1/2}| = 3l-2$, $|A_{1/2}{}'| = 2l-2$. For $k = 3,4,...$, $|A_{1/k}| = |A_{1/k}{}'| = 3l-2$.

Lemma 25: Let $\tilde{A}$ be the fuzzy set defined on $\Pi \times \Pi$ for the classical algebra D_l($l \geq 4$) given by (1) then $\tilde{A}$ has the following properties:

(a) The cardinality $|\tilde{A}| = (5l-4)/2$ (b) Relative cardinality $\|\tilde{A}\| = (5l-4)/2l^2$

(c) The membership function of the complement of a normalized fuzzy set $\tilde{A}$ corresponding to the classical algebra D_l($l \geq 4$), $\forall$ $(\alpha_i,\alpha_j) \in X$ are listed below :

i. For $i = 2,3,...,(l-1)$ $\mu_{\complement\tilde{A}}(\alpha_{i-1},\alpha_i) = 0$ & $\mu_{\complement\tilde{A}}(\alpha_i,\alpha_{i-1}) = 0$

ii. For $i = 1,2,...,l$ $\mu_{\complement\tilde{A}}(\alpha_i,\alpha_i) = 1/2$

iii. For $i \neq j$, $i \neq j+1$, $j \neq i+1$, if $i = l-2$ then $j \neq l$ & if $i = l$ then $j \neq l-2$

$\mu_{\complement\tilde{A}}(\alpha_i,\alpha_j) = 1$ and $\mu_{\complement\tilde{A}}(\alpha_l,\alpha_{l-2}) = 0$ & $\mu_{\complement\tilde{A}}(\alpha_{l-2},\alpha_l) = 0$.

Proof: (a) By the definition (1), the fuzzy set $\tilde{A}$ corresponding to the classical algebra D_l ($l \geq 4$) contains ($2l-2$) elements in X having membership grade 1, l elements in X having membership grade 1/2 and all the other elements in X having membership grade 0. $|\tilde{A}| = \sum_{x \in X} \mu_{\tilde{A}}(x) = (5l-4)/2$.

(b) $\| \tilde{A} \| = \dfrac{|\tilde{A}|}{|X|} = \dfrac{(5l-4)}{2l^2}$.

(c) Since the fuzzy set $\tilde{A}$ corresponding to the classical algebra $D_l (l \geq 4)$, is normal. For $\forall \ (\alpha_i, \alpha_j) \in X$, the membership function of the complement of a normalized fuzzy set $\tilde{A}$ are listed below :

i. For $i = 2,3,..., (l-1)$ $\mu_{C\tilde{A}}(\alpha_{i-1},\alpha_i) = 1-1 = 0$ & $\mu_{C\tilde{A}}(\alpha_i,\alpha_{i-1}) = 1-1 = 0$

ii. For $i = 1,2,...,l$ $\mu_{C\tilde{A}}(\alpha_i,\alpha_i) = 1-1/2 = 1/2$

iii. For $i \neq j$, $i \neq j+1$, $j \neq i+1$, if $i = l-2$ then $j \neq l$ & if $i = l$ then $j \neq l-2$ implies $\mu_{C\tilde{A}}(\alpha_i,\alpha_j) = 1-0 = 1$ and $\mu_{C\tilde{A}}(\alpha_l,\alpha_{l-2}) = 1-1 = 0$ & $\mu_{C\tilde{A}}(\alpha_{l-2},\alpha_l) = 1-1 = 0$.

Theorem 26: For the finite type of Kac-Moody algebra E_6 associated with the indecomposable GCM A, let $\tilde{A}$ be the fuzzy set defined on $\Pi \times \Pi$ given by (1). Then the α - level sets and strong α - level sets for $\alpha = 1, 1/2, 1/3,...,1/k,...$ are given below:

$(i) A_1 = \{(\alpha_1,\alpha_2),(\alpha_2,\alpha_3),(\alpha_3,\alpha_4),(\alpha_4,\alpha_5),(\alpha_3,\alpha_6),(\alpha_6,\alpha_3),(\alpha_5,\alpha_4),(\alpha_4,\alpha_3),(\alpha_3,\alpha_2),(\alpha_2,\alpha_1)\}$

$(ii) A_{1/2} = A_1 \cup \{(\alpha_1,\alpha_1),(\alpha_2,\alpha_2),...,(\alpha_6,\alpha_6)\}$

(iii) $A_{1/2}{}' = A_1$ (iv) $A_{1/3}{}' = A_{1/2}$

(v) $|A_1| = 10$, $|A_{1/2}| = 16$, $|A_{1/2}{}'| = 10$. For $k = 3,4,...,$ $|A_{1/k}| = |A_{1/k}{}'| = 16$.

Proof: Consider the family E_6,

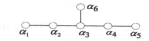

Fig. 2. Dynkin diagram for the finite type of Kac-Moody algebra E_6

(i) $A_1 = \{(\alpha_i,\alpha_j) \in X / \mu_{\tilde{A}}(\alpha_i,\alpha_j) \geq 1\} = \{(\alpha_i,\alpha_j) \in X / \dfrac{1}{\max(|a_{ij}|,|a_{ji}|)} \geq 1\}$

$= \{(\alpha_1,\alpha_2),(\alpha_2,\alpha_3),(\alpha_3,\alpha_4),(\alpha_4,\alpha_5),(\alpha_3,\alpha_6),(\alpha_6,\alpha_3),(\alpha_5,\alpha_4),(\alpha_4,\alpha_3),(\alpha_3,\alpha_2),(\alpha_2,\alpha_1)\}$

(ii) $A_{1/2} = \{(\alpha_i,\alpha_j) \in X / \dfrac{1}{\max(|a_{ij}|,|a_{ji}|)} \geq 1/2\}$ $= A_1 \cup \{(\alpha_1,\alpha_1),(\alpha_2,\alpha_2),...,(\alpha_6,\alpha_6)\}$

$A_{1/3} = \{(\alpha_i,\alpha_j) \in X / \dfrac{1}{\max(|a_{ij}|,|a_{ji}|)} \geq 1/3\} = A_{1/2};$ $A_{1/3} = A_{1/2}$

From the above relations we have, $A_1 \subset A_{1/2} = A_{1/3} = ... = A_{1/k} = ...$

(iii) $A_{1/2}{}' = A_1$ (iv) $A_{1/3}{}' = A_{1/2}$, $A_{1/4}{}' = A_{1/3}$

From the above relations we see that, $A_1{}' \subset A_{1/2}{}' \subset A_{1/3}{}' = A_{1/4}{}' = ... = A_{1/k}{}' = ...$

(v) $A_{1/2} = A_{1/3} = ... = A_{1/k} = ... = A_{1/3}{}' = A_{1/4}{}' = ... = A_{1/k}{}'$

$|A_1| = 10$, $|A_{1/2}| = 16$, $|A_{1/2}{}'| = 10$. For $k = 3,4,...,$ $|A_{1/k}| = |A_{1/k}{}'| = 16$.

Lemma 27: Let $\tilde{A}$ be the fuzzy set defined on $\Pi \times \Pi$ for the finite type of Kac-Moody algebra E_6 given by (1) then $\tilde{A}$ has the following properties:

(a) The cardinality $|\tilde{A}| = 13$ (b) Relative cardinality $\|\tilde{A}\| = 0.36$

(c) The membership function of the complement of a normalized fuzzy set $\tilde{A}$ corresponding to E_6, $\forall$ $(\alpha_i, \alpha_j) \in X$ are listed below :

i. For $i = 2,...,5$ $\mu_{\mathcal{C}\tilde{A}}(\alpha_{i-1}, \alpha_i) = 0$ & $\mu_{\mathcal{C}\tilde{A}}(\alpha_i, \alpha_{i-1}) = 0$

ii. For $i = 1,...,6$ $\mu_{\mathcal{C}\tilde{A}}(\alpha_i, \alpha_i) = 1/2$

iii. For $i \ne j$, $i \ne j+1$, $j \ne i+1$, if $i = 3$ then $j \ne 6$ & if $i = 6$ then $j \ne 3$ implies $\mu_{\mathcal{C}\tilde{A}}(\alpha_i, \alpha_j) = 1$ and $\mu_{\mathcal{C}\tilde{A}}(\alpha_3, \alpha_6) = 0$, $\mu_{\mathcal{C}\tilde{A}}(\alpha_6, \alpha_3) = 0$.

Proof: (a) By the definition (1), the fuzzy set $\tilde{A}$ corresponding to the finite type of Kac-Moody algebra E_6 contains 10 elements in X having membership grade 1, 6 elements in X having membership grade 1/2 and all the other elements in X having membership grade 0. $|\tilde{A}| = \sum_{x \in X} \mu_{\tilde{A}}(x) = 13$.

(b) $\|\tilde{A}\| = |\tilde{A}| / |X| = 0.36$.

(c) Since the fuzzy set $\tilde{A}$ corresponding to the finite type of Kac-Moody algebra E_6 is normal. For $\forall$ $(\alpha_i, \alpha_j) \in X$, the membership function of the complement of a normalized fuzzy set $\tilde{A}$ are listed below :

i. For $i = 2,...,5$ $\mu_{\mathcal{C}\tilde{A}}(\alpha_{i-1}, \alpha_i) = 1 - 1 = 0$ & $\mu_{\mathcal{C}\tilde{A}}(\alpha_i, \alpha_{i-1}) = 1 - 1 = 0$

ii. For $i = 1,...,6$ $\mu_{\mathcal{C}\tilde{A}}(\alpha_i, \alpha_i) = 1 - 1/2 = 1/2$

iii. For $i \ne j$, $i \ne j+1$, $j \ne i+1$, if $i = 3$ then $j \ne 6$ & if $i = 6$ then $j \ne 3$ implies $\mu_{\mathcal{C}\tilde{A}}(\alpha_i, \alpha_j) = 1 - 0 = 1$ and $\mu_{\mathcal{C}\tilde{A}}(\alpha_3, \alpha_6) = 1 - 1 = 0$, $\mu_{\mathcal{C}\tilde{A}}(\alpha_6, \alpha_3) = 1 - 1 = 0$.

Theorem 28: For the finite type of Kac-Moody algebra E_7 associated with the GCM A, let $\tilde{A}$ be the fuzzy set defined on $\Pi \times \Pi$ given by (1). Then the α - level sets and strong α - level sets for $\alpha = 1, 1/2, 1/3,...,1/k,...$ are given below :

(i) $A_1 = \{(\alpha_1, \alpha_2), (\alpha_2, \alpha_3), (\alpha_3, \alpha_4), (\alpha_4, \alpha_5), (\alpha_5, \alpha_6), (\alpha_3, \alpha_7), (\alpha_7, \alpha_3), (\alpha_6, \alpha_5),$
$(\alpha_5, \alpha_4), (\alpha_4, \alpha_3), (\alpha_3, \alpha_2), (\alpha_2, \alpha_1)\}$

(ii) $A_{1/2} = A_1 \cup \{(\alpha_1, \alpha_1), (\alpha_2, \alpha_2),..., (\alpha_7, \alpha_7)\}$

(iii) $A_{1/2}' = A_1$ (iv) $A_{1/3}' = A_{1/2}$

(v) $|A_1| = 12$, $|A_{1/2}| = 19$, $|A_{1/2}'| = 12$. For $k = 3,4,...$, $|A_{1/k}| = |A_{1/k}'| = 19$.

Proof: Consider the family E_7,

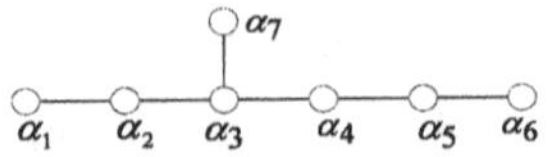

Fig. 3. Dynkin diagram for the finite type of Kac-Moody algebra E_7

(i) $A_1 = \{(\alpha_i, \alpha_j) \in X / \mu_{\tilde{A}}(\alpha_i, \alpha_j) \ge 1\} = \{(\alpha_i, \alpha_j) \in X / \dfrac{1}{\max(|a_{ij}|, |a_{ji}|)} \ge 1\}$

$= \{(\alpha_1, \alpha_2), (\alpha_2, \alpha_3), (\alpha_3, \alpha_4), (\alpha_4, \alpha_5), (\alpha_5, \alpha_6), (\alpha_3, \alpha_7), (\alpha_7, \alpha_3), (\alpha_6, \alpha_5), (\alpha_5, \alpha_4), (\alpha_4, \alpha_3), (\alpha_3, \alpha_2), (\alpha_2, \alpha_1)\}$

(ii) $A_{1/2}$ $= A_1 \cup \{(\alpha_1,\alpha_1),(\alpha_2,\alpha_2),...,(\alpha_7,\alpha_7)\}$; $A_{1/3} = A_{1/2}$

From the above relations we have, $A_1 \subset A_{1/2} = A_{1/3} = ... = A_{1/k} = ...$

(iii) $A_{1/2}' = A_1$ (iv) $A_{1/3}' = A_{1/2}'$, $A_{1/4}' = A_{1/3}'$, $A_1' \subset A_{1/2}' \subset A_{1/3}' = A_{1/4}' = ... = A_{1/k}' = ...$

(v) $A_{1/2} = A_{1/3} = ... = A_{1/k} = ... = A_{1/3}' = A_{1/4}' = ... = A_{1/k}'$

$|A_1| = 12$, $|A_{1/2}| = 19$, $|A_{1/2}'| = 12$. For $k = 3,4,...,$ $|A_{1/k}| = |A_{1/k}'| = 19$.

Lemma 29: Let $\tilde{A}$ be the fuzzy set defined on $\Pi \times \Pi$ for the finite type of Kac - Moody algebra E_7 given by (1) then $\tilde{A}$ has the following properties:
   (a) The cardinality $|\tilde{A}| = 15.5$ (b) Relative cardinality $\|\tilde{A}\| = 0.32$
   (c)The membership function of the complement of a normalized fuzzy set $\tilde{A}$ corresponding to E_7, $\forall$ $(\alpha_i, \alpha_j) \in X$ are listed below:

i. For $i = 2,...,6$ $\mu_{\mathcal{C}\tilde{A}}(\alpha_{i-1},\alpha_i) = 0$ & $\mu_{\mathcal{C}\tilde{A}}(\alpha_i,\alpha_{i-1}) = 0$

ii. For $i = 1,...,7$ $\mu_{\mathcal{C}\tilde{A}}(\alpha_i,\alpha_i) = 1/2$

iii. For $i \neq j$, $i \neq j+1$, $j \neq i+1$, if $i = 3$ then $j \neq 7$ & if $i = 7$ then $j \neq 3$ implies $\mu_{\mathcal{C}\tilde{A}}(\alpha_i,\alpha_j) = 1$ and $\mu_{\mathcal{C}\tilde{A}}(\alpha_i,\alpha_{i-2}) = 0$ & $\mu_{\mathcal{C}\tilde{A}}(\alpha_{i-2},\alpha_i) = 0$.

Proof: (a) By the definition (1), the fuzzy set $\tilde{A}$ corresponding to the finite type of Kac-Moody algebra E_7 contains 12 elements in X having membership grade 1, 7 elements in X having membership grade 1/2 and all the other elements in X having membership grade 0. $|\tilde{A}| = \sum_{x \in X} \mu_{\tilde{A}}(x) = 15.5$.

   (b) $\|\tilde{A}\| = |\tilde{A}|/|X| = 0.32$.

   (c) Since the fuzzy set $\tilde{A}$ corresponding to the finite type of Kac-Moody algebra E_7 is normal. For $\forall$ $(\alpha_i,\alpha_j) \in X$, the membership function of the complement of a normalized fuzzy set $\tilde{A}$ are listed below:
i. For $i = 2,...,6$ $\mu_{\mathcal{C}\tilde{A}}(\alpha_{i-1},\alpha_i) = 1-1 = 0$, $\mu_{\mathcal{C}\tilde{A}}(\alpha_i,\alpha_{i-1}) = 1-1 = 0$
ii. For i $= 1,...,7$ $\mu_{\mathcal{C}\tilde{A}}(\alpha_i,\alpha_i) = 1-1/2 = 1/2$
iii. For $i \neq j$, $i \neq j+1$, $j \neq i+1$, if $i = 3$ then $j \neq 7$ & if $i = 7$ then $j \neq 3$ implies $\mu_{\mathcal{C}\tilde{A}}(\alpha_i,\alpha_j) = 1-0 = 1$ & $\mu_{\mathcal{C}\tilde{A}}(\alpha_3,\alpha_7) = 1-1 = 0$, $\mu_{\mathcal{C}\tilde{A}}(\alpha_7,\alpha_3) = 1-1 = 0$.

Theorem 30: For the finite type of Kac-Moody algebra E_8 associated with the indecomposable GCM A, let $\tilde{A}$ be the fuzzy set defined on $\Pi \times \Pi$ given by (1). Then the α - level sets and strong α - level sets for $\alpha = 1, 1/2, 1/3,...,1/k,...$ are given below:
(i) $A_1 = \{(\alpha_1,\alpha_2),(\alpha_2,\alpha_3),(\alpha_3,\alpha_4),(\alpha_4,\alpha_5),(\alpha_5,\alpha_6),(\alpha_6,\alpha_7),(\alpha_5,\alpha_8),(\alpha_8,\alpha_5),$
$(\alpha_7,\alpha_6),(\alpha_6,\alpha_5),(\alpha_5,\alpha_4),(\alpha_4,\alpha_3),(\alpha_3,\alpha_2),(\alpha_2,\alpha_1)\}$
(ii) $A_{1/2} = A_1 \cup \{(\alpha_1,\alpha_1),(\alpha_2,\alpha_2),..., (\alpha_8,\alpha_8)\}$
(iii) $A_{1/2}' = A_1$ (iv) $A_{1/3}' = A_{1/2}'$
(v) $|A_1| = 14$, $|A_{1/2}| = 22$, $|A_{1/2}'| = 14$. For $k = 3,4,...,$ $|A_{1/k}| = |A_{1/k}'| = 22$.

Proof: Consider the family E_8,

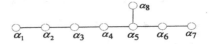

Fig. 4. Dynkin diagram for the finite type of Kac-Moody algebra E_8

(i) $A_1 = \{(\alpha_1,\alpha_2),(\alpha_2,\alpha_3),(\alpha_3,\alpha_4),(\alpha_4,\alpha_5),(\alpha_5,\alpha_6),(\alpha_6,\alpha_7),(\alpha_5,\alpha_8),(\alpha_8,\alpha_5),$
$(\alpha_7,\alpha_6),(\alpha_6,\alpha_5),(\alpha_5,\alpha_4),(\alpha_4,\alpha_3),(\alpha_3,\alpha_2),(\alpha_2,\alpha_1)\}$

(ii) $A_{1/2} = A_1 \cup \{(\alpha_1,\alpha_1),(\alpha_2,\alpha_2),...,(\alpha_8,\alpha_8)\}$; $A_{1/3} = A_{1/2}$

From the above relations we have, $A_1 \subset A_{1/2} = A_{1/3} = ... = A_{1/k} = ...$

(iii) $A_{1/2}' = A_1$ (iv) $A_{1/3}' = A_{1/2}$, $A_{1/4}' = A_{1/3}$, $A_1' \subset A_{1/2}' \subset A_{1/3}' = A_{1/4}' = ... = A_{1/k}' = ...$

(v) $A_{1/2} = A_{1/3} = ... = A_{1/k} = ... = A_{1/3}' = A_{1/4}' = ... = A_{1/k}'$

$|A_1| = 14$, $|A_{1/2}| = 22$, $|A_{1/2}'| = 14$. For $k = 3,4,...$, $|A_{1/k}| = |A_{1/k}'| = 22$.

Lemma 31: Let $\tilde{A}$ be the fuzzy set defined on $\Pi \times \Pi$ for the finite type of Kac-Moody algebra E_8 given by (1) then $\tilde{A}$ has the following properties:

(a) The cardinality $|\tilde{A}| = 18$ (b) Relative cardinality $\|\tilde{A}\| = 0.28$

(c) The membership function of the complement of a normalized fuzzy set $\tilde{A}$ corresponding to E_8, $\forall (\alpha_i,\alpha_j) \in X$ are listed below:

i. For $i = 2,...,7$ $\mu_{\mathcal{C}\tilde{A}}(\alpha_{i-1},\alpha_i) = 0$ & $\mu_{\mathcal{C}\tilde{A}}(\alpha_i,\alpha_{i-1}) = 0$

ii. For $i = 1,...,7$ $\mu_{\mathcal{C}\tilde{A}}(\alpha_i,\alpha_i) = 1/2$

iii. For $i \neq j$, $i \neq j+1$, $j \neq i+1$, if $i = 5$ then $j \neq 8$ & if $i = 8$ then $j \neq 5$ implies $\mu_{\mathcal{C}\tilde{A}}(\alpha_i,\alpha_j) = 1$ & $\mu_{\mathcal{C}\tilde{A}}(\alpha_1,\alpha_{1-2}) = 0$, $\mu_{\mathcal{C}\tilde{A}}(\alpha_{1-2},\alpha_1) = 0$.

Proof: (a) By the definition (1), the fuzzy set $\tilde{A}$ corresponding to the finite type of Kac-Moody algebra E_8 contains 14 elements in X having membership grade 1, 8 elements in X having membership grade 1/2 and all the other elements in X having membership grade 0. $|\tilde{A}| = \sum_{x \in X} \mu_{\tilde{A}}(x) = 18$.

(b) $\|\tilde{A}\| = |\tilde{A}| / |X| = 0.28$.

(c) Since the fuzzy set $\tilde{A}$ corresponding to the finite type of Kac-Moody algebra E_8 is normal. For $\forall (\alpha_i,\alpha_j) \in X$, the membership function of the complement of a normalized fuzzy set $\tilde{A}$ are listed below:

i. For $i = 2,...,7$ ii. For $i = 1,...,7$ $\mu_{\mathcal{C}\tilde{A}}(\alpha_i,\alpha_i) = 1 - 1/2 = 1/2$

iii. For $i \neq j$, $i \neq j+1$, $j \neq i+1$, if $i = 5$ then $j \neq 8$ & if $i = 8$ then $j \neq 5$ implies $\mu_{\mathcal{C}\tilde{A}}(\alpha_i,\alpha_j) = 1 - 0 = 1$ & $\mu_{\mathcal{C}\tilde{A}}(\alpha_5,\alpha_8) = 1 - 1 = 0$, $\mu_{\mathcal{C}\tilde{A}}(\alpha_8,\alpha_5) = 1 - 1 = 0$ ·

Theorem 32: For the finite type of Kac-Moody algebra F_4 associated with the indecomposable GCM A, let $\tilde{A}$ be the fuzzy set defined on $\Pi \times \Pi$ given by (1). Then the α - level sets and strong α - level sets for $\alpha = 1, 1/2, 1/3, ..., 1/k, ...$ are given below:

(i) $A_1 = \{(\alpha_1, \alpha_2), (\alpha_3, \alpha_4), (\alpha_4, \alpha_3), (\alpha_2, \alpha_1)\}$

(ii) $A_{1/2} = A_1 \cup \{(\alpha_1, \alpha_1), (\alpha_2, \alpha_2), (\alpha_3, \alpha_3), (\alpha_4, \alpha_4), (\alpha_2, \alpha_3), (\alpha_3, \alpha_2)\}$

(iii) $A_{1/2}' = A_1$ (iv) $A_{1/3}' = A_{1/2}$

(v) $|A_1| = 4$, $|A_{1/2}| = 10$, $|A_{1/2}'| = 4$. For $k = 3, 4, ...,$ $|A_{1/k}| = |A_{1/k}'| = 10$.

Proof : Consider F_4,

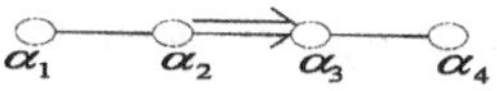

Fig. 5. Dynkin diagram for the finite type of Kac-Moody algebra F_4

(i) $A_1 = \{(\alpha_i, \alpha_j) \in X / \dfrac{1}{\max(|a_{ij}|, |a_{ji}|)} \geq 1\}$ $= \{(\alpha_1, \alpha_2), (\alpha_3, \alpha_4), (\alpha_4, \alpha_3), (\alpha_2, \alpha_1)\}$

(ii) $A_{1/2} = A_1 \cup \{(\alpha_1, \alpha_1), (\alpha_2, \alpha_2), (\alpha_3, \alpha_3), (\alpha_4, \alpha_4), (\alpha_2, \alpha_3), (\alpha_3, \alpha_2)\}$; $A_{1/3} = A_{1/2}$
From the above relations we have, $A_1 \subset A_{1/2} = A_{1/3} = ... = A_{1/k} = ...$

(iii) $A_{1/2}' = A_1$ (iv) $A_{1/3}' = A_{1/2}$, $A_{1/4}' = A_{1/3}$, $A_1 \subset A_{1/2}' \subset A_{1/3}' = A_{1/4}' = ... = A_{1/k}' = ...$

(v) $A_{1/2} = A_{1/3} = ... = A_{1/k} = ... = A_{1/3}' = A_{1/4}' = ... = A_{1/k}'$

$|A_1| = 4$, $|A_{1/2}| = 10$, $|A_{1/2}'| = 4$. For $k = 3, 4, ...,$ $|A_{1/k}| = |A_{1/k}'| = 10$.

Lemma 33: Let $\tilde{A}$ be the fuzzy set defined on $\Pi \times \Pi$ for the finite type of Kac-Moody algebra F_4 given by (1) then $\tilde{A}$ has the following properties:
  (a) The cardinality $|\tilde{A}| = 7$ (b) Relative cardinality $\|\tilde{A}\| = 0.44$
  (c) The membership function of the complement of a normalized fuzzy set $\tilde{A}$ corresponding to F_4, $\forall (\alpha_i, \alpha_j) \in X$ are listed below:

i. For $i = 2, 4$ $\mu_{\mathbb{C}\tilde{A}}(\alpha_{i-1}, \alpha_i) = 0$ & $\mu_{\mathbb{C}\tilde{A}}(\alpha_i, \alpha_{i-1}) = 0$

ii. For $i = 1, ..., 4$ $\mu_{\mathbb{C}\tilde{A}}(\alpha_i, \alpha_i) = 1/2$ iii. For $i \neq j$, $\neq j+1, j \neq i+1$, $\mu_{\mathbb{C}\tilde{A}}(\alpha_i, \alpha_j) = 1$

iv. For $i = 3$, $\mu_{\mathbb{C}\tilde{A}}(\alpha_{i-1}, \alpha_i) = 0$ & $\mu_{\mathbb{C}\tilde{A}}(\alpha_i, \alpha_{i-1}) = 0$.

Proof: (a) The fuzzy set $\tilde{A}$ corresponding to the finite type of Kac-Moody algebra F_4 contains 4 elements in X having membership grade 1, 6 elements in X having membership grade 1/2 and all the other elements in X having membership grade 0. $|\tilde{A}| = \sum_{x \in X} \mu_{\tilde{A}}(x) = 7$.

  (b) $\|\tilde{A}\| = |\tilde{A}| / |X| = 0.44$.

  (c) Since the fuzzy set $\tilde{A}$ corresponding to the finite type of Kac-Moody algebra F_4 is normal. For $\forall (\alpha_i, \alpha_j) \in X$, the membership function of the complement of a normalized fuzzy set $\tilde{A}$ which is denoted as $\mu_{\mathbb{C}\tilde{A}}(x)$ are listed below :

i. For $i = 2,4$ $\mu_{\mathcal{C}\tilde{A}}(\alpha_{i-1}, \alpha_i) = 1 - 1 = 0$ & $\mu_{\mathcal{C}\tilde{A}}(\alpha_i, \alpha_{i-1}) = 1 - 1 = 0$

ii. For $i = 1,...,4$ $\mu_{\mathcal{C}\tilde{A}}(\alpha_i, \alpha_i) = 1 - 1/2 = 1/2$

iii. For $i \neq j$, $i \neq j+1$, $j \neq i+1$, $\mu_{\mathcal{C}\tilde{A}}(\alpha_i, \alpha_j) = 1 - 0 = 1$

iv. For $i = 3$, $\mu_{\mathcal{C}\tilde{A}}(\alpha_{i-1}, \alpha_i) = 1 - 1 = 0$ & $\mu_{\mathcal{C}\tilde{A}}(\alpha_i, \alpha_{i-1}) = 1 - 1 = 0$.

Theorem 34: For the finite type of Kac Moody algebra G_2 associated with the indecomposable GCM A, let $\tilde{A}$ be the fuzzy set defined on $\Pi \times \Pi$ given by (1). Then the α- level sets and strong α- level sets for $\alpha = 1, 1/2, ..., 1/k, ...,$ are given below:

(i) $A_1 = \Phi$ (ii) $A_{1/2} = \{(\alpha_1, \alpha_1), (\alpha_2, \alpha_2)\}$

(iii) $A_{1/3} = \{(\alpha_1, \alpha_1), (\alpha_2, \alpha_2), (\alpha_1, \alpha_2), (\alpha_2, \alpha_1)\} = X$

(iv) $A_{1/2}' = A_1$ (v) $A_{1/3}' = A_{1/2}$ (vi) $A_{1/4}' = A_{1/3}$

(vii) $|A_{1/2}| = 2$, $|A_{1/3}| = 4$, $|A_{1/3}'| = 2$. For $k = 4,5,...,$ $|A_{1/k}| = |A_{1/k}'| = 4$.

Proof : Consider G_2,

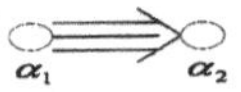

Fig. 6. Dynkin diagram for the finite type of Kac-Moody algebra G_2

(i) $A_1 = \{(\alpha_i, \alpha_j) \in X / \mu_{\tilde{A}}(\alpha_i, \alpha_j) \geq 1\} = \Phi$ (ii) $A_{1/2} = \{(\alpha_1, \alpha_1), (\alpha_2, \alpha_2)\}$

(iii) $A_{1/3} = \{(\alpha_1, \alpha_1), (\alpha_2, \alpha_2), (\alpha_1, \alpha_2), (\alpha_2, \alpha_1)\} = X$. $A_1 \subset A_{1/2} \subset A_{1/3} = ... = A_{1/k}$

(iv) $A_{1/2}' = \{(\alpha_i, \alpha_j) \in X / \mu_{\tilde{A}}(\alpha_i, \alpha_j) > 1/2\} = A_1$ (v) $A_{1/3}' = A_{1/2}$ (vi) $A_{1/4}' = A_{1/3}$

From the above relations we see that, $A_1' \subset A_{1/2}' \subset A_{1/3}' \subset A_{1/4}' = ... = A_{1/k}' = X$

(vii) $A_{1/3} = ... = A_{1/k} = ... = A_{1/4}' = ... = A_{1/k}' = ...$

$|A_{1/2}| = 2$, $|A_{1/3}| = 4$, $|A_{1/3}'| = 2$. For $k = 4,5,...,$ $|A_{1/k}| = |A_{1/k}'| = 4$.

Lemma 35: Let $\tilde{A}$ be the fuzzy set defined on $\Pi \times \Pi$ for the finite type of Kac-Moody algebra G_2 by equation (1) then $\tilde{A}$ has the following properties:
   (a) The cardinality $|\tilde{A}| = 1.66$; (b) Relative cardinality $\|\tilde{A}\| = 0.42$;

Proof: (a) The fuzzy set $\tilde{A}$ corresponding to the finite type of Kac-Moody algebra G_2 fuzzy set $\tilde{A}$ contains 2 elements in X having membership grade 1/2, 2 elements in X having membership grade 1/3. $|\tilde{A}| = 1.66$. (b) $\|\tilde{A}\| = |\tilde{A}|/|X| = 0.42$.

3   Conclusion

We can further compute the level sets for various families of affine, indefinite , hyperbolic, extended hyperbolic and non hyperbolic type of Kac - Moody algebras; Other interesting structural properties on the fuzzy nature of these algebras can also be studied;

Acknowledgment. The first author is grateful to the University Grants Commission, India for sanctioning Major Research Project.

References

[1] Ganesh, M.: Introduction to Fuzzy Sets and Fuzzy Logic. PHI Learning Private Limited, New Delhi (2009)

[2] Kac, V.G.: Infinite Dimensional Lie Algebra, 3rd edn. Cambridge University Press, Cambridge (1990)

[3] Moody, R.V.: A New Class of Lie Algebras. J. Algebra 10, 211–230 (1968)

[4] Uma Maheswari, A., Gayathri, V.: A Fuzzy Approach on the Root Systems of Kac-Moody Algebras. In: Proceedings of the International Conference in Mathematics in Engineering and Businsess Management, March 9-10, vol. 2, pp. 542–549 (2012) ISBN 978-81-8286-015-5

[5] Wan, Z.-X.: Introduction to Kac–Moody Algebra. World Scientific Publishing Co. Pvt. Ltd, Singapore (1991)

[6] Zimmermann, H.J.: Fuzzy Set Theory and its Applications, 4th edn. Kluwer Academic Publishers, London (2001)

Enabling Location Privacy in Pervasive Computing by Fragmenting Location Information of Users

Jeeva Susan Jacob and Preetha K.G.

Rajagiri School of Engineering and Technology, Rajagiri Valley, Cochin, India
jeevasj27@gmail.com
preetha_kg@rajagiritech.ac.in

Abstract. Pervasive Computing integrates communication and computing into real life to enable advanced and uninterrupted services to simplify information manipulation. Location aware computing is a renowned sector of pervasive computing where the systems involved can provide specific services based on the location information of users. The advancements in this area has led to major revolutions in various application areas, especially advertisements. This paper explores the Location Privacy issue in location aware computing. Vertical fragmentation of the stored location information of users has been proposed as a solution for this issue.

Keywords: Pervasive Computing, Location Aware Computing, Location Privacy, Privacy issues.

1   Introduction

Pervasive computing techniques integrate three converging areas namely computing, communications and user interfaces to simplify the lives of users. This is accomplished with the assistance of a handheld user device like a smartphone. Recent advances in technology have paved the way from pervasive computing by enhancing existing technical methods towards ubiquitous computing. One of the significant application area of pervasive computing is Location aware computing. Location aware computing comprises of systems which percepts the location information of the users and changes their behavior or perform specific functions according to the instructions preprogrammed.

Location information describes a user's location over a specific period of time. They can be collected through the following methods.

- Locating Systems: They detect the location of a device or find out the devices existing in a particular location.
- Location Systems: They detect identification of the device and then determine the location information.

J. Mathew et al. (Eds.): ICECCS 2012, CCIS 305, pp. 364–371, 2012.

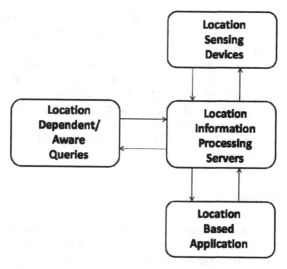

Fig. 1. Block Diagram of Location Aware Computing

Location aware computing consists of location sensing through the implementation of various sensing techniques, processing the location information as per location dependent or location aware queries and then provide them to advanced applications which execute functions based on this processed data. Figure 1 shows the functioning of Location aware computing. Location information can be collected using the methods of proximity location sensing, which determines the nearest known location of a device, or triangulation location sensing, which uses geometrical methods of lateration and angulation. Another method for location sensing which utilizes nonmathematical steps is the scene analysis method. Here the location can be sensed by comparing and analyzing it with the locations of other prominent objects in the scene observed.

2   Location Privacy in Pervasive Computing

The major applications of location aware computing include location based advertising, instantaneous medical services, independent travels, resource discovery, network route selection and the like. The reduction in prices of location sensing hardware and the rise in the widespread use of personal handheld devices are opening up new opportunities in the domain of location aware computing.

One of the major challenges in location aware computing is ensuring privacy by the protection of location information of users. Privacy can be a sensitive issue when it comes to the majority of people. People insist on maintaining the right to determine who should be acquainted with what amount of information about them. Misusing location information can lead to major trouble to people. Exposure of the location information about some public figures might even be a threat to their lives. It can also lead to fraudulent attacks or spam attacks. Who will be responsible if someone changes the value of a coordinate in our location information to frame us for a crime we have not committed? Users might always have the feeling of being exposed or constantly being tracked. This is because location information can uniquely identify us, just like our genetic profile.

Major location privacy attacks comprises of impersonation attacks, spam attacks and constant tracking of user's whereabouts. Laws have been passed in various countries to preserve the right of people to take decisions on the usage of data regarding their whereabouts by the government or other communication service providers. But these do not cover the location tracking of users and using this location information for various applications. Effective methods need to be developed for ensuring location privacy in pervasive computing.

3   Related Work

Numerous methods have been proposed by researchers around the world in this field. Privacy Sensitive Information Diluting Mechanism (PSIUM)[17] ensures privacy in the user location data collected by service providers by sending multiple location information messages along with service requests. Device can select the accurate information from these, but the service providers cannot. Use of mix nodes[1] reorders the packet flow thus confusing attackers since these packets appear to be independent of each other.

LocServ[1] acts as a middleware layer between location sensing devices and location dependent applications. It gives the control of location information to corresponding users. Thus it prevents transfer of location information to other third party agencies which may query the server for location data of users. Another technique is Mist[1], which is specially designed to provide location privacy for users in pervasive computing environments. The GeoPriv[18] model integrates location based services with location privacy preservation measures for effective implementation of location based applications. This is designed for applications dealing with surveying, mapping, tracking and so on.

Majority of these proposals are for preventing privacy attacks by tracking the traffic flow through these systems. In this paper, a proposal for preserving privacy of location information of users by preventing attacks on the servers using fragmentation techniques is discussed. Third party attacks and tracking attacks on the information storage base in the location servers can thus be prevented by this method.

4   Vertical Fragmentation

Fragmentation is a technique commonly used in database management systems to classify data belonging to a single database and store them properly. Usually it is done in the case of distributed systems, where systems distributed among geographically distant regions share resources and applications. Fragmentation and replication of databases is done to ensure durability and redundancy of significant data in distributed systems. Horizontal fragmentation is done to separate tuples or records belonging to a single database into parts. Reconstruction of the original database is much easier in this case.

Vertical fragmentation is done based on different attributes of a database. On each fragment a unique key field is attached so that queries can be executed without rejoining these fragments. In most cases the primary key of the database is replicated in each fragment.

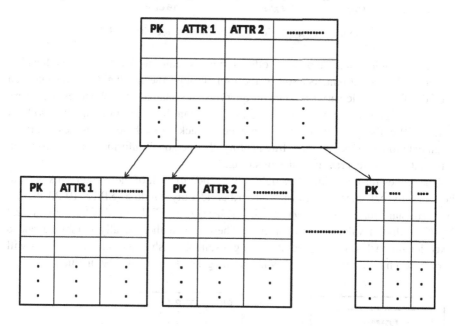

Fig. 2. Vertical fragmentation of Database with PK as the primary key field

Figure 2 shows the vertical fragmentation of a database where each fragment has the primary key field, PK, to interlink them as and when required. By using this method we can fragment the data regarding the users stored in location servers. This separates the location information from the user's personal information. Thus an attack on the data storage cannot effectively reveal a particular user's location. A malicious attacker may never be able to connect the user with the corresponding location information when the details of the user lies in separate fragments of storage.

5   Vertical Fragmentation on Location Information

Location information of users is usually maintained in the location servers using Location Area Identity (LAI). This uniquely identifies the location area (LA) of a particular user at a particular instant of time. This information is stored in the Visitor Location Register (VLR) of the communication network infrastructure used by the user's smartphone or handheld device. LAI comprises of Mobile Country Code (MCC) which identifies the country, Mobile Network Code (MNC) which identifies the mobile network used and the Location Area Code (LAC) which identifies unique locations.

Location Area Identity (LAI)

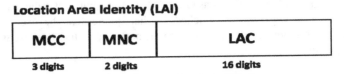

MCC	MNC	LAC
3 digits	2 digits	16 digits

Fig. 3. Vertical Fields of Location Area Identity (LAI)

When a communication or some other service is requested by the user, the location server tracks the current location of the user through the handheld device and then provides the suitable service. Thus on querying the database which stores the user's profile with personal as well as location information the user can be tracked without much difficulty. So, in order to prevent attacks on these databases, vertical fragmentation can be applied fragmenting the personal information and location information into separate modules of storage.

The values for the unique key field to be added to each of these fragments need to be carefully calculated. Here the Subscriber Identity Module (SIM) (19-20 digits), International Mobile Equipment identity (IMEI) (15 or 17 digits), MCC (3 digits) and MNC (2 digits) values are considered. These are mathematically combined and a fixed length hash value is generated using a suitable hash function. These values will vary for each user and thus the new key value generated will also be unique.

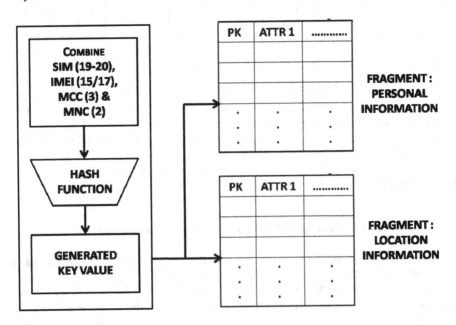

Fig. 4. Vertical Fragmentation to ensure Location Privacy

A key field with unique key values thus calculated is added to each of the database fragments containing personal information and location information respectively. When a service request is sent by the user, the server can calculate the unique key field value and then combine the required records of the fragments to form the original record with both personal and location information. The request is then processed based on the information gathered and the requested service can be provided as per the user's eligibility for receiving that service.

Figure 4 shows the vertical fragmentation of location server databases. In the case of an attack on the server, the attacker cannot identify which tuple in the location information fragment corresponds to the particular entry in the personal information fragment. Also fake requests can be identified since imposter would not be able to fake location information since SIM and IMEI numbers are both considered here. In order to ensure more security to the method the hash function used can be changed periodically.

Thus, this method ensures privacy to the users by preventing attacks on the location server databases by third party or malicious agencies. Unfortunately, it cannot prevent the misuse of location information by the service provider itself, since all the values considered are known to it.

6   Future Enhancements

Although the proposed method can ensure privacy of location information from external attacks, measures need to be adapted to prevent misuse of it by the service provider network. Privacy Laws may prevent it up to a limit, but effective techniques are required. One method would be to provide only an approximate value for the location information instead of the accurate coordinates of the location.

Future work also involves research on methods for preventing other kinds of privacy attacks. Since enhanced pervasive computing or ubiquitous computing techniques does not depend on handheld user devices, other hardware independent measures need to be developed. Advancements in hardware technology and communication standards can accelerate the transformation from Pervasive computing to Ubiquitous computing.

7   Conclusion

As Anthony Burgess says, "To be left alone is the most precious thing one can ask of the modern world". In a world which is evolving into a global apartment building from a global village, ensuring location privacy is a crucial fight against the peeping toms of the communication world. Measures from the world governments, like the Location Privacy Protection Bill (2011) which forces Google and Apple to get user's permission before location information is tracked, reassure the public that all is not lost yet.

This paper discusses the issue of Location Privacy of users in Pervasive computing environments. Vertical Fragmentation technique has been proposed as a solution for the attacks on Location Servers and theft of location information. Its effectiveness in preventing third party attacks and fake users have been analyzed theoretically. Even though internal attacks cannot be prevented, it ensures location privacy for the users of a reliable service provider.

References

1. Bhaskar, P., Ahamed, S.I.: Privacy in Pervasive Computing and Open Issues. In: Second International Conference on Availability, Reliability and Security (ARES 2007) (April 2007)
2. Talukder, N., Ahamed, S.I.: How much Room before you Rely: Balancing Privacy control and Fidelity in the Location-based Pervasive Applications. In: Nineth International Conference on Mobile Data Management Workshops (April 2008)
3. Fischmeister, S., Menkhaus, G., Stumpfl, A.: Location-Detection Strategies in Pervasive Computing Environments. In: Proceedings of the First International Conference on Pervasive Computing and Communications (PerCom 2003) (March 2003)
4. Michael Berman, A., Lewis, S.M., Conto, A.: Location-Aware Computing (November 2008), http://net.educause.edu/ir/library/pdf/DEC0803.pdf
   Kim, Y.J.: Location Aware Computing (November 2002), TermPapers page, http://crystal.uta.edu/~kumar/cse6392/termpapers/YoungPaper.pdf
5. Ardagna, C.A., Cremonini, M., De Capitani di Vimercati, S., Samarati, P.: Location Privacy in Pervasive Computing (November 2008), http://spdp.dti.unimi.it/papers/CL20082.pdf
6. Duckham, M., Kulik, L.: Location Privacy and Location-aware Computing (December 2005), in Papers page http://www.geosensor.net/papers/duckham06.IGIS.pdf
7. Hazas, M., Ward, A.: A High Performance Privacy-Oriented Location System. In: Proceedings of the First IEEE International Conference on Pervasive Computing and Communications, PerCom 2003 (2003)
8. Anthony, D., Kotz, D., Henderson, T.: Privacy in Location-Aware Computing Environments. IEEE Computer Society, IEEE (2007)
9. Ardagna, C.A., Cremonini, M., De Capitani di Vimercati, S., Samarati, P.: An Obfuscation-based Approach for Protecting Location Privacy. IEEE Transactions on Depedable and Secure Computing (2009)
10. Beresford, A.R., Stajano, F.: Location Privacy in Pervasive Computing. Pervasive Computing, IEEE CS and IEEE Communications Society (1), 46–55 (2003)
11. Patterson, C.A., Muntz, R.R., Pancake, C.M.: Challenges in location-aware computing. IEEE Pervasive Computing (June 2003)
12. Yao, L., Lin, C., Kong, X., Xia, F., Wu, G.: A Clustering-Based Location Privacy Protection Scheme for Pervasive Computing, Green Computing and Communications (GreenCom). In: 2010 IEEE/ACM Int'l Conference on & Int'l Conference on Cyber, Physical and Social Computing (CPSCom) (March 2011)
13. Reddy, Y.V.: Pervasive Computing: Implications, Opportunities and Challenges for the Society. In: 2006 1st International Symposium on Pervasive Computing and Applications (January 2007)
14. Jacobsson, M., Niemegeers, I.: Privacy and anonymity in personal networks. In: Third IEEE International Conference on Pervasive Computing and Communications Workshops, PerCom 2005 Workshops (March 2005)
15. Myles, G., Friday, A., Davies, N.: Preserving privacy in environments with location-based applications. IEEE Pervasive Computing (April 2004)

16. Cheng, H.S., Zhang, D., Tan, J.G.: Protection of privacy in pervasive computing environments. In: International Conference on Information Technology: Coding and Computing, ITCC 2005 (May 2005)
17. Morris, J., Peterson, J.: Who's Watching You Now? IEEE Security & Privacy (February 2007)
18. Konings, B., Schaub, F.: Territorial privacy in ubiquitous computing. In: 2011 Eighth International Conference on Wireless On-Demand Network Systems and Services (WONS) (February 2011)
19. Schlott, S., Kargl, F., Weber, M.: Short paper: Random IDs for preserving location privacy. In: First International Conference on Security and Privacy for Emerging Areas in Communications Networks, SecureComm 2005 (March 2006)

R-norm Intuitionistic Fuzzy Information Measures and Its Computational Applications

Rakesh Kumar Bajaj, Tanuj Kumar, and Nitin Gupta

Department of Mathematics, Jaypee University of Information Technology,
Waknaghat, Distt- Solan, H.P., India
{rakesh.bajaj,tanujkhutail,nitinstat}@gmail.com

Abstract. Atanassov introduced the concept of intuitionistic fuzzy sets (IFS), as a generalization of fuzzy sets, which is capable of capturing the information that includes some degree of hesitation. In the present communication, a new R-norm intuitionistic fuzzy entropy and a weighted R-norm intuitionistic fuzzy directed divergence measure have been proposed with their proof of validity. Further, empirical study on the proposed information measures has also been done which explains monotonic nature of the information measures with respect to R and the weight. Computational applications of these information measures in the field of pattern recognition and image thresholding has been proposed with discussion.

Keywords: Intuitionistic fuzzy sets, Fuzzy Information Measure, Directed Divergence Measure, Pattern Recognition, Image Thresholding.

1 Introduction

Intuitionistic fuzzy set (IFS), developed by Atanassov [1] is a controlling tool to deal with vagueness and uncertainty. A prominent characteristic of IFS is that it assigns to each element a membership degree and a non-membership degree with certain amount of hesitation degree, and thus, the IFS constitutes an extension of Zadeh's fuzzy set [17], which only assigns to each element a membership degree. Intuitionistic fuzzy sets can be useful in situations when description of a problem by a (fuzzy) linguistic variable, given in terms of a membership function only, seems insufficient to give best result. Atanassov [2, 4] and many other researchers [6,14] studied different properties of IFSs in decision making problems, particularly in the case of medical diagnosis, sales analysis, new product marketing, financial services, etc.

An intuitionistic fuzzy set (IFS), denoted by $\tilde{A}$, over a finite non empty fixed set X, is well defined and discussed in [1]. We denote $\mathcal{F}(X)$ the set of all the IFSs on X. The following are the few basics of IFSs:

- $\tilde{A}$ is a crisp set iff $\mu_{\tilde{A}}(x_i) = 0$ & $\nu_{\tilde{A}}(x_i) = 1$ or $\mu_{\tilde{A}}(x_i) = 1$ & $\nu_{\tilde{A}}(x_i) = 0$ $\forall x_i \in X$.
- $\tilde{A}$ is sharper than $\tilde{B}$ iff $\mu_{\tilde{A}}(x_i) \leq \mu_{\tilde{B}}(x_i)$ & $\nu_{\tilde{A}}(x_i) \leq \nu_{\tilde{B}}(x_i)$, for $\max\{\mu_{\tilde{B}}(x_i), \nu_{\tilde{B}}(x_i)\} \leq \frac{1}{3}$, $\forall x_i \in X$; and $\mu_{\tilde{A}}(x_i) \geq \mu_{\tilde{B}}(x_i)$ & $\nu_{\tilde{A}}(x_i) \geq \nu_{\tilde{B}}(x_i)$, for $\min\{\mu_{\tilde{B}}(x_i), \nu_{\tilde{B}}(x_i)\} \geq \frac{1}{3}$, $\forall x_i \in X$.
- $\tilde{A}$ is less fuzzy than $\tilde{B}$ iff $\mu_{\tilde{A}}(x_i) \leq \mu_{\tilde{B}}(x_i)$ & $\nu_{\tilde{A}}(x_i) \geq \nu_{\tilde{B}}(x_i)$ for $\mu_{\tilde{B}}(x_i) \leq \nu_{\tilde{B}}(x_i)$ or $\mu_{\tilde{A}}(x_i) \geq \mu_{\tilde{B}}(x_i)$ & $\nu_{\tilde{A}}(x_i) \leq \nu_{\tilde{B}}(x_i)$ for $\mu_{\tilde{B}}(x_i) \geq \nu_{\tilde{B}}(x_i)$ $\forall x_i \in X$.

J. Mathew et al. (Eds.): ICECCS 2012, CCIS 305, pp. 372–380, 2012.

The union operator $\cup$, intersection operator $\cap$ between $\tilde{A}$ & $\tilde{B}$ and the complementary set of $\tilde{A}$ in $\mathcal{F}(X)$ is well known in literature [ref. [10]]. Kaufmann [11] proposed to measure the degree of fuzziness of any fuzzy set A by a metric distance between its membership function and the membership function (characteristic function) of its nearest crisp set. Another way given by Yager [16] was to view the degree of fuzziness in terms of a lack of distinction between the fuzzy set and its complement. Indeed, it is the lack of distinction between sets and their complements that distinguishes fuzzy sets from crisp sets. The less the set differs from its complement, the fuzzier it is. Szimidt and Kacprzyk [13] extended the axioms of De Luca and Termini [7] and proposed the following definition for an entropy measure of intuitionistic fuzzy set $\tilde{A} \in \mathcal{F}(X)$:

- **(IFS1)** : $H(\tilde{A}) = 0$ iff $\tilde{A}$ is a crisp set.
- **(IFS2)** : $H(\tilde{A}) = 1$ iff $\mu_{\tilde{A}}(x_i) = \nu_{\tilde{A}}(x_i); \forall x_i \in X$.
- **(IFS3)** : $H(\tilde{A}) \leq H(\tilde{B})$ iff $\tilde{A}$ is less fuzzy than $\tilde{B}$.
- **(IFS4)** : $H(\tilde{A}) = H(\overline{\tilde{A}})$, where $\overline{\tilde{A}}$ is complement of $\tilde{A}$.

Vlachos & Sergiadis [15] extended the De Luca $-$Termini nonprobability entropy for fuzzy sets, which is given by

$$H_{LT}(\tilde{A}) = -\frac{1}{n \ln 2} \sum_{i=1}^{n} \left[\mu_{\tilde{A}}(x_i) \ln \left(\frac{\mu_{\tilde{A}}(x_i)}{\mu_{\tilde{A}}(x_i) + \nu_{\tilde{A}}(x_i)} \right) + \nu_{\tilde{A}}(x_i) \ln \left(\frac{\nu_{\tilde{A}}(x_i)}{\mu_{\tilde{A}}(x_i) + \nu_{\tilde{A}}(x_i)} \right) - \pi_{\tilde{A}}(x_i) \ln 2 \right].$$

$$(1)$$

Hung & Yang [10] introduced the following axiomatic definition of IFS entropy in a probabilistic setting as a real valued functional $H : \mathcal{F}(X) \to \mathbb{R}^+$:

- **IE1(Sharpness):** $H(\tilde{A}) = 0$ iff $\tilde{A}$ is a crisp set.
- **IE2(Maximality):** $E(\tilde{A})$ assumes a unique maximum if $\mu_{\tilde{A}} = \nu_{\tilde{A}} = \pi_{\tilde{A}} = \frac{1}{3}; \forall x_i$.
- **IE3(Resolution):** $E(\tilde{A}) \leq E(\tilde{B})$ if $\tilde{A}$ is sharper than $\tilde{B}$.
- **IE4(Symmetry):** $E(\tilde{A}^c) = E(\tilde{A})$.

Under the above axioms, Hung and Yang [10] proposed two families of fuzzy entropy of an IFS $\tilde{A}$ given by

$$H_{hc}^{\alpha}(\tilde{A}) = \begin{cases} \frac{1}{(\alpha-1)n} \sum_{i=1}^{n} \left[1 - \left(\mu_{\tilde{A}}^{\alpha}(x_i) + \nu_{\tilde{A}}^{\alpha}(x_i) + \pi_{\tilde{A}}^{\alpha}(x_i) \right) \right]; & \alpha \neq 1(\alpha > 0) \\ -\frac{1}{n} \sum_{i=1}^{n} \left(\mu_{\tilde{A}}(x_i) \log \mu_{\tilde{A}}(x_i) + \nu_{\tilde{A}}(x_i) \log \nu_{\tilde{A}}(x_i) + \pi_{\tilde{A}}(x_i) \log \pi_{\tilde{A}}(x_i) \right); & \alpha = 1, \end{cases}$$

$$(2)$$

$$\text{and} \quad H_r^{\beta}(\tilde{A}) = \frac{1}{1-\beta} \sum_{i=1}^{n} \log \left(\mu_{\tilde{A}}^{\beta}(x_i) + \nu_{\tilde{A}}^{\beta}(x_i) + \pi_{\tilde{A}}^{\beta}(x_i) \right); \quad 0 < \beta < 1. \quad (3)$$

Vlachos & Sergiadis [15] derived the De Luca $-$Termini entropy for IFSs from the concept of intuitionistic fuzzy cross-entropy as

$$I(\tilde{A}, \tilde{B}) = \sum_{i=1}^{n} \left[\mu_{\tilde{A}}(x_i) \ln \left(\frac{\mu_{\tilde{A}}(x_i)}{\frac{1}{2}(\mu_{\tilde{A}}(x_i) + \nu_{\tilde{A}}(x_i))} \right) + \nu_{\tilde{A}}(x_i) \ln \left(\frac{\nu_{\tilde{A}}(x_i)}{\frac{1}{2}(\mu_{\tilde{A}}(x_i) + \nu_{\tilde{A}}(x_i))} \right) \right],$$

which is called the intuitionistic fuzzy cross-entropy between $\tilde{A}$ and $\tilde{B}$.

In the present paper, a new R-norm fuzzy entropy and R-norm weighted fuzzy directed divergence measure for intuitionistic fuzzy sets have been proposed with their proof of validity in section 2 and section 3 respectively. Further, in Section 4, empirical study on the proposed information measures has also been done which explains the monotonic nature of the information measures. Applications of the proposed new information measures in the field of pattern recognition and image thresholding have been presented in section 5.

2   New R-norm Information Measure of IFS

Let $\Delta_n = \{P = (p_1, p_2, \ldots, p_n), p_i \geq 0, i = 1, 2, \ldots, n$ and $\sum_{i=1}^{n} p_i = 1\}$ be the set of all probability distributions associated with a discrete random variable X taking finite values $x_1, x_2, \ldots, x_n$. Boekee and Lubbe [5] defined and studied R-norm information measure of the distribution P for $R \in \mathbb{R}^+$ as given by

$$H_R(P) = \frac{R}{R-1}\left[1 - \left(\sum_{i=1}^{n} p_i^R\right)^{\frac{1}{R}}\right]; \quad R > 0, \ R \neq 1. \tag{4}$$

The measure (4) is a real function from Δ_n to $\mathbb{R}^+$ and is called R-norm information measure. The most important property of this measure is that when $R \to 1$, it approaches to Shannon's entropy and in case $R \to \infty$, $H_R(P) \to (1 - \max p_i); \ i = 1, 2, \ldots, n$. Corresponding to measure (4), we propose the following intuitionistic fuzzy entropy:

$$H_R(\tilde{A}) = \frac{R}{(R-1)}\sum_{i=1}^{n}\frac{1}{n}\left[1 - \left(\left(\mu_{\tilde{A}}^R(x_i) + \nu_{\tilde{A}}^R(x_i) + \pi_{\tilde{A}}^R(x_i)\right)\right)^{\frac{1}{R}}\right]; R > 0, \ R \neq 1. \tag{5}$$

We present following properties for proving validity of above proposed measure:

Property 2.1: Under the condition of **IE3**, we have

$$\left|\mu_{\tilde{A}}(x_i) - \frac{1}{3}\right| + \left|\nu_{\tilde{A}}(x_i) - \frac{1}{3}\right| + \left|\pi_{\tilde{A}}(x_i) - \frac{1}{3}\right| \geq \left|\mu_{\tilde{B}}(x_i) - \frac{1}{3}\right| + \left|\nu_{\tilde{B}}(x_i) - \frac{1}{3}\right| + \left|\pi_{\tilde{B}}(x_i) - \frac{1}{3}\right| \tag{6}$$

$$\text{and } \left(\mu_{\tilde{A}}(x_i) - \frac{1}{3}\right)^2 + \left(\nu_{\tilde{A}}(x_i) - \frac{1}{3}\right)^2 + \left(\pi_{\tilde{A}}(x_i) - \frac{1}{3}\right)^2$$

$$\geq \left(\mu_{\tilde{B}}(x_i) - \frac{1}{3}\right)^2 + \left(\nu_{\tilde{B}}(x_i) - \frac{1}{3}\right)^2 + \left(\pi_{\tilde{B}}(x_i) - \frac{1}{3}\right)^2 \tag{7}$$

Proof: If $\mu_{\tilde{A}}(x_i) \leq \mu_{\tilde{B}}(x_i)$ and $\nu_{\tilde{A}}(x_i) \leq \nu_{\tilde{B}}(x_i)$ with $\max\{\mu_{\tilde{B}}(x_i), \ \nu_{\tilde{B}}(x_i)\} \leq \frac{1}{3}$, then $\mu_{\tilde{A}}(x_i) \leq \mu_{\tilde{B}}(x_i) \leq \frac{1}{3}; \ \nu_{\tilde{A}}(x_i) \leq \nu_{\tilde{B}}(x_i) \leq \frac{1}{3}$ and $\pi_{\tilde{A}}(x_i) \geq \pi_{\tilde{B}}(x_i) \geq \frac{1}{3}$ which implies that equation (6) and (7) hold. Similarly, if $\mu_{\tilde{A}}(x_i) \geq \mu_{\tilde{B}}(x_i)$ and $\nu_{\tilde{A}}(x_i) \geq \nu_{\tilde{B}}(x_i)$ with $\max\{\mu_{\tilde{B}}(x_i), \ \nu_{\tilde{B}}(x_i)\} \geq \frac{1}{3}$ then equation (6) and (7) hold.

Theorem 2.1: Measure (5) is a valid intuitionistic fuzzy information measure.
Proof: In order to prove that the measure (5) is a valid intuitionistic fuzzy information measure, we shall show that four properties (IE1 - IE4) are satisfied.

(IE1)(Sharpness): If $H_R(\tilde{A}) = 0$ then $(\mu_{\tilde{A}}^R(x_i) + \nu_{\tilde{A}}^R(x_i) + \pi_{\tilde{A}}^R(x_i))^{\frac{1}{R}} = 1$. Since $R(\neq 1) > 0$ therefore this is possible only in the following cases:

– Either $\mu_{\tilde{A}}(x_i) = 1$ i.e. $\nu_{\tilde{A}}(x_i) = \pi_{\tilde{A}}(x_i) = 0$ or

– $\nu_{\tilde{A}}(x_i) = 1$ i.e. $\mu_{\tilde{A}}(x_i) = \pi_{\tilde{A}}(x_i) = 0$ or

– $\pi_{\tilde{A}}(x_i) = 1$ i.e. $\nu_{\tilde{A}}(x_i) = \mu_{\tilde{A}}(x_i) = 0$.

In all the cases, $H_R(\tilde{A}) = 0$ implies that $\tilde{A}$ is a crisp set. Conversely, if $\tilde{A}$ be a crisp set i.e., either $\mu_{\tilde{A}}(x_i) = 1$, or $\nu_{\tilde{A}}(x_i) = \pi_{\tilde{A}}(x_i) = 0$ and either $\nu_{\tilde{A}}(x_i) = 1$ or $\mu_{\tilde{A}}(x_i) = \pi_{\tilde{A}}(x_i) = 0$ and either $\pi_{\tilde{A}}(x_i) = 1$ or $\nu_{\tilde{A}}(x_i) = \mu_{\tilde{A}}(x_i) = 0$.

It implies that $\mu_{\tilde{A}}^R(x_i) + \nu_{\tilde{A}}^R(x_i) + \pi_{\tilde{A}}^R(x_i))^{\frac{1}{R}} = 1$ for $R(\neq 1) > 0$, which gives $H_R(\tilde{A}) = 0$. Hence $H_R(\tilde{A}) = 0$ if and only if $\tilde{A}$ is a crisp set.

(IE2)(Maximality): Since $\mu_{\tilde{A}}(x_i) + \nu_{\tilde{A}}(x_i) + \pi_{\tilde{A}}(x_i) = 1$, therefore to obtain the maximum value of the intuitionistic fuzzy entropy, we write $g(\mu_{\tilde{A}}, \nu_{\tilde{A}}, \pi_{\tilde{A}}) = \mu_{\tilde{A}}(x_i) + \nu_{\tilde{A}}(x_i) + \pi_{\tilde{A}}(x) - 1$ and taking the Lagrange's multiplier λ, we consider

$$G(\mu_{\tilde{A}}, \nu_{\tilde{A}}, \pi_{\tilde{A}}) = H_R(\mu_{\tilde{A}}, \nu_{\tilde{A}}, \pi_{\tilde{A}}) + \lambda g(\mu_{\tilde{A}}, \nu_{\tilde{A}}, \pi_{\tilde{A}}). \tag{8}$$

To find the maximum value of $H_R(\tilde{A})$, we differentiate (8) partially with respect to $\mu_{\tilde{A}}$, $\nu_{\tilde{A}}$, $\pi_{\tilde{A}}$ and λ and equating them to zero we get $\mu_{\tilde{A}}(x_i) = \nu_{\tilde{A}}(x_i) = \pi_{\tilde{A}}(x_i) = \frac{1}{3}$. It may be noted that all the first order partial derivatives vanish if and only $\mu_{\tilde{A}}(x_i) = \nu_{\tilde{A}}(x_i) = \pi_{\tilde{A}}(x_i) = \frac{1}{3}$. Hence $H_R(\tilde{A})$ has the stationary point $\mu_{\tilde{A}}(x_i), = \nu_{\tilde{A}}(x_i), = \pi_{\tilde{A}}(x_i) = \frac{1}{3}$. Next, we show that $H_R(\tilde{A})$ is a concave function on the IFS $\tilde{A} \in \mathcal{F}(X)$ by calculating its Hessian at the stationary point. The Hessian of $H_R(\tilde{A})$ is given by

$$\hat{H} = \frac{R \cdot 3^{\frac{1}{R}-1}}{n} \begin{bmatrix} -2 & 1 & 1 \\ 1 & -2 & 1 \\ 1 & 1 & -2 \end{bmatrix}$$

For any $R > 0$, $\hat{H}$ is a negative semi-definite matrix and hence $H_R(\tilde{A})$ is a concave function and has its maximum value at the point $\mu_{\tilde{A}} = \nu_{\tilde{A}} = \pi_{\tilde{A}} = \frac{1}{3}; \forall x_i$.

(IE3)(Resolution): Since $H_R(\tilde{A})$ is a concave function on the IFS $\tilde{A} \in \mathcal{F}(X)$, therefore if $\max\{\mu_{\tilde{A}}(x), \nu_{\tilde{A}}(x)\} \leq \frac{1}{3}$, then $\mu_{\tilde{A}}(x_i) \leq \mu_{\tilde{B}}(x_i)$ and $\nu_{\tilde{A}}(x_i) \leq \nu_{\tilde{B}}(x_i)$ which implies $\pi_{\tilde{A}}(x_i) \geq \pi_{\tilde{B}}(x_i) \geq \frac{1}{3}$. According to the result of property 2.1, we conclude that $H_R(\tilde{A})$ satisfies condition IE3.

Similarly, if $\min\{\mu_{\tilde{A}}(x), \nu_{\tilde{A}}(x)\} \geq \frac{1}{3}$, then $\mu_{\tilde{A}}(x_i) \geq \mu_{\tilde{B}}(x_i)$ and $\nu_{\tilde{A}}(x_i) \geq \nu_{\tilde{B}}(x_i)$. By property 2.1, we again conclude that $H_R(\tilde{A})$ satisfies condition IE3.

(IE4)(Symmetry): It may be noted that from the definition of the complement of intuitionistic fuzzy set, it is clear that $H_R(\bar{\tilde{A}}) = H_R(\tilde{A})$. Hence $H_R(\tilde{A})$ satisfies all the properties of intuitionistic fuzzy entropy and therefore, $H_R(\tilde{A})$ is a valid measure of intuitionistic fuzzy entropy.

3   Weighted R-norm Intuitionistic Fuzzy Directed Divergence Measure

Hooda and Bajaj [9] proposed the following measure of fuzzy directed divergence of fuzzy set A from fuzzy set B for $R > 0$, $R \neq 1$:

$$I_R(A, B) = \frac{R}{R-1} \sum_{i=1}^{n} \left[\left(\mu_A^R(x_i) \mu_B^{1-R}(x_i) + (1 - \mu_A(x_i))^R (1 - \mu_B(x_i))^{1-R} \right)^{\frac{1}{R}} - 1 \right] \quad (9)$$

Analogous to (9), we propose the following measures of fuzzy directed divergence of intuitionistic fuzzy set $\tilde{A}$ from intuitionistic fuzzy set $\tilde{B}$:

$$N_R(\tilde{A}, \tilde{B}) = \frac{R}{R-1} \sum_{i=1}^{n} \left[\left(\begin{matrix} (\mu_{\tilde{A}}(x_i) + \pi_{\tilde{A}}(x_i))^R (\mu_{\tilde{B}}(x_i) + \pi_{\tilde{B}}(x_i))^{1-R} \\ + (1 - (\mu_{\tilde{A}}(x_i) + \pi_{\tilde{A}}(x_i)))^R (1 - (\mu_{\tilde{B}}(x_i) + \pi_{\tilde{B}}(x_i)))^{1-R} \end{matrix} \right)^{\frac{1}{R}} - 1 \right]; \quad (10)$$

where $R > 0$, $R \neq 1$, which is based on the value of membership, and

$$M_R(\tilde{A}, \tilde{B}) = \frac{R}{R-1} \sum_{i=1}^{n} \left[\left(\begin{matrix} (\nu_{\tilde{A}}(x_i) + \pi_{\tilde{A}}(x_i))^R (\nu_{\tilde{B}}(x_i) + \pi_{\tilde{B}}(x_i))^{1-R} \\ + (1 - (\nu_{\tilde{A}}(x_i) + \pi_{\tilde{A}}(x_i)))^R (1 - (\nu_{\tilde{B}}(x_i) + \pi_{\tilde{B}}(x_i)))^{1-R} \end{matrix} \right)^{\frac{1}{R}} - 1 \right]; \quad (11)$$

where $R > 0$, $R \neq 1$, which is based on the value of non-membership.

Also, we propose a new weighted directed divergence measure which is a linear combination of directed divergences (10) and (11), given by

$$I_R^\lambda(\tilde{A}, \tilde{B}) = \lambda M_R(\tilde{A}, \tilde{B}) + (1 - \lambda) N_R(\tilde{A}, \tilde{B}); \text{ where } 0 < \lambda < 1, \ R > 0, \ R \neq 1. \quad (12)$$

Theorem 3.1: (12) is a valid intuitionistic fuzzy directed divergence measure.

Proof: It may be noted that linear combination of two valid intuitionistic fuzzy directed divergence measures is a valid intuitionistic fuzzy directed divergence measure. Therefore, in order to prove that (12) is a valid measure of intuitionistic fuzzy directed divergence, it is sufficient to show that $M_R(A, B) \geq 0$ with equality if $\mu_A(x_i) = \mu_B(x_i)$, $\forall x \in X$, as $M_R(A, B)$ and $N_R(A, B)$ are defined in similar way. Let $\sum_{i=1}^{n} \mu_A(x_i) = s$, $\sum_{i=1}^{n} \mu_B(x_i) = t$, then

$$\sum_{i=1}^{n} \left[\left(\frac{\mu_A(x_i)}{s} \right)^R \left(\frac{\mu_B(x_i)}{t} \right)^{1-R} - 1 \right] \geq 0 \Rightarrow \sum_{i=1}^{n} \mu_A^R(x_i)(\mu_B(x_i))^{1-R} \geq s^R t^{1-R} \quad (13)$$

Similarly, we write $\sum_{i=1}^{n} (1 - \mu_A(x_i))^R (1 - \mu_B(x_i))^{1-R} \geq (n - s)^R (n - t)^{1-R} \quad (14)$

Adding (13) and (14), we get

$$\sum_{i=1}^{n} \mu_A^R(x_i)(\mu_B(x_i))^{1-R} + (1 - \mu_A(x_i))^R (1 - \mu_B(x_i))^{1-R} \geq s^R t^{1-R} + (n - s)^R (n - t)^{1-R} \quad (15)$$

Case 1: $0 < R < 1$

Let $\mu_A^R(x_i)(\mu_B(x_i))^{1-R} + (1 - \mu_A(x_i))^R (1 - \mu_B(x_i))^{1-R} = x_i$, then $x_i < 1$ and $\frac{1}{R} > 1$, implies that, $x_i - 1 > (x_i)^{1/R} - 1$. $\because \frac{R}{R-1} < 0$, $\therefore \sum_{i=1}^{n} x_i - 1 > (x_i)^{\frac{1}{R}} - 1$.

Thus, we have $M_R(A, B) = \frac{R}{R-1}\left[s^R t^{1-R} + (n-s)^R(n-t)^{1-R} - n\right]$.

Further, let $\varphi(s) = \frac{R}{R-1}\left[s^R t^{1-R} + (n-s)^R(n-t)^{1-R} - n\right]$,

then $\varphi'(s) = \frac{R}{R-1}\left[R(s/t)^{R-1} - R((n-s)/(n-t))^{R-1}\right]$,

and $\varphi''(s) = R^2\left[(1/t)(s/t)^{R-2} - (1/n-t)((n-s)/(n-t))^{R-2}\right] > 0$.

This shows that $\varphi(s)$ is a convex function of s whose minimum value arises when $(s/t)(=(n-s)/(n-t)) = 1$ and is equal to zero. Hence, $\varphi(s) > 0$ and vanishes only when $s = t$.

Case 2: $R > 1$ In this case, equation(15) can be written as

$$\left(\sum_{i=1}^{n} \mu_A^R(x_i)(\mu_B(x_i))^{1-R} + (1-\mu_A(x_i))^R(1-\mu_B(x_i))^{1-R}\right)^{1/R}$$
$$\geq \left(s^R t^{1-R} + (n-s)^R(n-t)^{1-R}\right)^{1/R}. \tag{16}$$

Also, $\sum_{i=1}^{n}\left[\left(\mu_A^R(x_i)(\mu_B(x_i))^{1-R} + (1-\mu_A(x_i))^R(1-\mu_B(x_i))^{1-R}\right)^{1/R} - 1\right]$

$$\geq \left(\sum_{i=1}^{n}\mu_A^R(x_i)(\mu_B(x_i))^{1-R} + (1-\mu_A(x_i))^R(1-\mu_B(x_i))^{1-R} - 1\right)^{1/R}. \tag{17}$$

Now (16) and (17) $\Rightarrow M_R(A, B) \geq \frac{R}{R-1}\left(s^R t^{1-R} + (n-s)^R(n-t)^{1-R} - n\right)^{1/R}$.

Let $\varphi(s) = \frac{1}{R-1}\left[s^R t^{1-R} + (n-s)^R(n-t)^{1-R} - n\right]$,

then $\varphi'(s) = \frac{R}{R-1}\left[\left(\frac{s}{t}\right)^{R-1} - \left(\frac{n-s}{n-t}\right)^{R-1}\right]$ \hfill (18)

and $\varphi''(s) = R^2\left[\left(\frac{1}{t}\right)\left(\frac{s}{t}\right)^{R-2} - \left(\frac{1}{n-t}\right)\left(\frac{n-s}{n-t}\right)^{R-2}\right] > 0$. \hfill (19)

$\therefore \varphi(s)$ is a convex function of s whose minimum value arises when $(s/t)(=(n-s)/(n-t)) = 1$ and is equal to zero. Hence, $\varphi(s) > 0$ and vanishes only when $s = t$ i.e. $\forall R \neq 1(> 0)$, $M_R(\tilde{A}, \tilde{B}) \geq 0$ and vanishes only when $\tilde{A} = \tilde{B}$. Thus $M_R(A, B)$ is a valid intuitionistic fuzzy directed divergence measure. Hence, (12) is a valid weighted directed divergence measure for intuitionistic fuzzy sets.

4 Monotonicity of *R*-norm Intuitionistic Fuzzy Information Measures

Let $\tilde{A}_1$, $\tilde{A}_2$ and $\tilde{A}_3$ be any three intuitionistic fuzzy sets over $X = \{x_1, x_2, x_3, x_4\}$.

$$\tilde{A}_1 = \{(x_1, 0.2, 0.5), (x_2, 0.4, 0.4), (x_3, 0.5, 0.2), (x_4, 0.6, 0.3)\}$$
$$\tilde{A}_2 = \{(x_1, 0.3, 0.4), (x_2, 0.2, 0.6), (x_3, 0.5, 0.3), (x_4, 0.6, 0.2)\}$$
$$\tilde{A}_3 = \{(x_1, 0.6, 0.1), (x_2, 0.5, 0.2), (x_3, 0.5, 0.1), (x_4, 0.7, 0.2)\}$$

Considering various values of R, and using (5), we compute and tabulate all the values. On the basis of tabulated data, we plot the figure 1(a).

Let $\tilde{A}$ and $\tilde{B}$ be any pair intuitionistic fuzzy sets over $X = \{x_1, x_2, x_3, x_4\}$.

$$\tilde{A} = \{(x_1, 0.2, 0.5), (x_2, 0.4, 0.4), (x_3, 0.3, 0.4), (x_4, 0.5, 0.3)\}$$

$$\tilde{B} = \{(x_1, 0.3, 0.4), (x_2, 0.2, 0.6), (x_3, 0.5, 0.3), (x_4, 0.6, 0.2)\}$$

Considering various values of R & λ, we compute $I_R^\lambda(\tilde{A}, \tilde{B})$ by using (12) and tabulate them. On the basis of the tabulated data, we plot figure 1 (b).

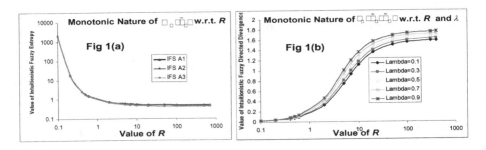

Fig. 1. Monotonic Nature of Intuitionistic Fuzzy Information Measures

On the basis of tabulated values and figures, it may be observed that $H_R(\tilde{A})$ is monotonically decreasing function of R and $I_R^\lambda(\tilde{A}, \tilde{B})$ is monotonically increasing function of R and λ.

5   Computational Applications in Pattern Recognition and Image Thresholding

In the field of pattern recognition, divergence measure describes dissimilarity between pairs of probability distribution which is widely used for the process of statistical inference. It may be noted that the divergence measure and similarity measure are dual concepts. The similarity measure may be defined by a decreasing function of divergence measure, especially when the range of divergence measure is $[0, 1]$. In the present paper, we have proposed the new weighted R-norm intuitionistic fuzzy directed divergence measure, which may be used to define a symmetric directed divergence, denoted by $J_R^\lambda(\tilde{A}, \tilde{B})$, and defined as $J_R^\lambda(\tilde{A}, \tilde{B}) = I_R^\lambda(\tilde{A}, \tilde{B}) + I_R^\lambda(\tilde{A}, \tilde{B})$ Let f be any monotonic decreasing function. Since $0 \le J_R^\lambda(\tilde{A}, \tilde{B}) \le G(R)$; where $G(R)$ is a calculated upper bound of the symmetric divergence measure $J_R^\lambda(\tilde{A}, \tilde{B})$, therefore $f(G(R)) \le f(J_R^\lambda(\tilde{A}, \tilde{B})) \le f(0)$, provided that $f(G(R)) < f(0)$. This implies that the similarity measure between IFSs $\tilde{A}$ and $\tilde{B}$ is given by

$$S_R(\tilde{A}, \tilde{B}) = \frac{f(J_R(\tilde{A}, \tilde{B})) - f(G(R))}{f(0) - f(G(R))}; \quad \text{where } 0 \le S_R(\tilde{A}, \tilde{B}) \le 1. \tag{20}$$

If we choose $f(x) = \frac{1}{1+x}$, then our similarity measure can be defined as follows:

$$S_R(\tilde{A}, \tilde{B}) = \frac{G(R) - J_R(\tilde{A}, \tilde{B})}{(1 + J_R(\tilde{A}, \tilde{B})) \cdot G(R)} \tag{21}$$

On the basis of the proposed similarity measure between two IFSs, the concept of similarity based clustering method (SCM) can be explored and the structure of the considered data set may be studied.

Another application of the theory of intuitionistic fuzzy sets may be found in the field of image thresholding where an image is considered as a intuitionistic fuzzy set. The membership degree of a pixel to the image is in proportion with their gray level, the non-membership degree of the pixel to the image is in inverse proportion with their gray level having a certain amount of hesitation degree. Let us consider the original image as an IFS $\tilde{A}$, the degraded image as an IFS $\tilde{B}$, and the restructured image as an IFS $\tilde{C}$. We try transformed the B to denoised version image as C by an algorithm. Pasha et al., [12] introduced a cost function with the help of fuzzy entropy to choose a threshold value for the denoising the degraded image. In order to accomplish the task, they used Euclidian distance and Kaufmann's entropy. Further, Fatemi [8] used stochastic fuzzy entropy in place of fuzzy entropy and stochastic fuzzy discrimination information for the Euclidean distance. The algorithm first finds the noised pixels then change them with mean of 8 neighbor pixels. The problem is chose a threshold h as unexpected jumping of gray level in the algorithm to find the noised pixels.

Here, it is being suggested that a new weighted (weight $= \lambda$) cost function which includes the intuitionistic fuzzy theory, may be used to find the best threshold by using R-norm intuitionistic fuzzy information measure in place of fuzzy entropy and R-norm intuitionistic fuzzy directed divergence for the Euclidean distance. The basic and fundamental equation for the algorithm is

$$C(\tilde{A}) = H_R(\tilde{C}) + I_R^\lambda(\tilde{A}, \tilde{C}).$$

6   Conclusion

A new R-norm intuitionistic fuzzy entropy and a weighted R-norm intuitionistic fuzzy directed divergence measure have been proposed with their proof of validity. Further, after empirical study on the proposed information measures we find that R-norm fuzzy intuitionistic fuzzy entropy is a decreasing function of R, while the weighted R-norm intuitionistic fuzzy directed divergence measure is increasing function of R as well as the weight λ. The proposed intuitionistic fuzzy information measures have found many applications in the field of pattern recognition and image processing.

Acknowledgments. The authors are thankful to anonymous reviewers for their valuable comments and suggestions.

References

1. Atanassov, K.: Intuitionistic Fuzzy Sets. Fuzzy Sets & Systems 20, 87–96 (1986)
2. Atanassov, K.: More on intuitionistic fuzzy sets. Fuzzy Sets & Systems 33, 37–46 (1989)
3. Atanassov, K.: New operations defined over the intuitionistic fuzzy sets. Fuzzy Sets & Systems 61, 137–142 (1964)
4. Atanassov, K.: Intuitionistic fuzzy sets: theory and applications. Physica-Verlag, Heidelberg (1999)
5. Boekee, D.E., Lubbe, J.C.A.: The R-norm Information Measures. Information and Control 45, 136–155 (1980)
6. De, S.K., Biswas, R., Roy, A.R.: An application of intuitionistic fuzzy sets in medical diagnosis. Fuzzy Sets & Systems 117, 209–213 (2001)
7. De Luca, A., Termini, S.: A definition of a non-probabilistic entropy in the setting of fuzzy sets theory. Information & Control 20, 301–312 (1972)
8. Fatemi, A.: Entropy of Stochastic Intuitionistic Fuzzy Sets. Journal of Applied Sciences 11, 748–751 (2011)
9. Hooda, D.S., Bajaj, R.K.: On Generalized R-norm Information Measures of Fuzzy Information. Journal of the Applied Math., Stat. & Informatics 4, 199–212 (2008)
10. Hung, W.L., Yang, M.S.: Fuzzy Entropy on Intuitionistic Fuzzy Sets. International Journal of Intelligent Systems 21, 443–451 (2006)
11. Kaufmann, A.: Introduction to the Theory of Fuzzy Subsets-Fundamental Theoretical Elements, vol. 1. Academic Press, New York (1975)
12. Pasha, E., Farnoosh, R., Fatemi, A.: Fuzzy entropy as cost function in image processing. In: Proc. of 2nd IMT-GT Regional Conf. on Math., Stat. & Appl., Universiti Sains Malaysia, Penang, pp. 1–8 (2008)
13. Szmidt, E., Kacprzyk, J.: Entropy for intuitionistic fuzzy sets. Fuzzy Sets and System 118(3) (2001)
14. Szmidt, E., Kacprzyk, J.: Group decision making under Intuitionistic fuzzy preference relations. In: Proc. of 7th IPMU Conf., Paris, pp. 172–178 (1998)
15. Vlachos, I.K., Sergiadis, G.D.: Intuitionistic fuzzy information – Applications to pattern recognition. Pattern Recognition Letters 28(2), 197–206 (2007)
16. Yager, R.R.: On the measure of fuziness and negation Part I: Membership in the unit interval. International Journal of General Systems 5, 189–200 (1979)
17. Zadeh, L.A.: Fuzzy Sets. Information and Control 8, 338–353 (1965)

Soft Computing Techniques for Distillation Column Composition Control

Subhadeep Bhattacharjee and Bhaswati Medhi

Department of Electrical Engineering, National Institute of Technology (NIT),
Agartala India-799055
subhadeep_bhattacharjee@yahoo.co.in, bhaswatimtech@gmail.com

Abstract. Distillation columns are important unit operations in chemical process plants. Though many works has been done in this field, but this paper compares some of the artificial intelligence controllers with the conventional controllers which are used to control the composition parameter in the column. The conventional controllers used are PID and PI controllers and the soft computing techniques used are fuzzy logic controllers and neuro-fuzzy controllers. The results obtained by the soft computing techniques are finally compared as to show which controller is best.

Keywords: distillation column, PI/PID control, fuzzy control, neuro fuzzy control.

1 Introduction

The Distillation column [1-3] is the most important separation process in the chemical and petrochemical industries. The difficulties in controlling distillation columns lie in their highly nonlinear characteristics; their multiple inputs multiple outputs (MIMO) structure and the presence of severe disturbances during operation [1]. The nonlinearity of distillation columns is well known. It has been known that the purer the products get, the more nonlinear the system becomes [4]. The interactions occurring between the inputs and the outputs are difficult to identify. These difficulties pose numerous problems challenging control problems and also attract a large number of researchers from different disciplines. Now-a-days, soft computing techniques are in use in a vast manner. Although most columns handle multi-component feeds, many can be approximated by binary (or pseudo binary) mixtures. However, due to the strong cross coupling and significant time delays inherent in the distillation column, the simultaneous control of overhead and bottoms composition using reflux and vapor flow as the control variables is still difficult. A fuzzy logic based scheme or Fuzzy Inference System is a computer paradigm based on fuzzy set theory, fuzzy if-then-rules and fuzzy reasoning [5]. Neural and fuzzy applications have been successfully applied to the chemical engineering processes [6-8] and several control strategies have been reported in the literature for the distillation plant modeling [9-12] and control tasks [13-14]. Recent years have seen a rapid growing number of neuro fuzzy applications, ranging from diagnostics and robotics to control [15-16].

J. Mathew et al. (Eds.): ICECCS 2012, CCIS 305, pp. 381–388, 2012.

2   Distillation Column

In this paper we are considering binary distillation column into account. The mixture is fed onto the feed tray NF (Fig.1), with feed flow rate of F (mol/min) and composition of zF (referred to the more volatile component). The more volatile component leaves the top of the column while the less volatile component leaves the bottom. The overhead vapor is totally condensed in a condenser and flows into the reflux drum with a composition xD and holdup of liquid of MD (moles), and reflux is pumped back to the top tray of the column at rate R, while overhead distillate product is removed at a rate D. At the base of the column, a liquid product stream is removed at a rate B with a composition xB and holdup of MB (moles). Vapor boilup is generated in a reboiler at a rate V.

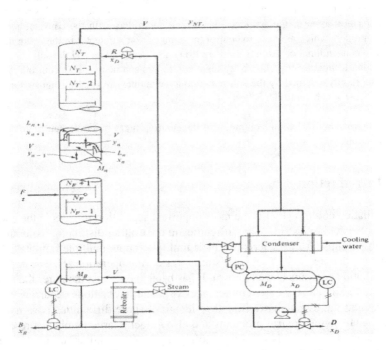

Fig. 1. Schematic of a typical binary distillation column

Both the reflux flow rate R and the vapor boilup V must be regulated to maintain overhead composition xD and bottoms composition xB, which is a precondition for good overall plant performance. Due to inherent complexity in distillation process it is difficult to achieve the simultaneous control of top and bottom composition. To overcome this difficulty a linear model of the Distillation Column has been proposed by Wood and Berry [17]. Mathematically this model can be represented as:

$$\begin{bmatrix} \dfrac{12.8\exp(-s)}{16.7s+1} & \dfrac{-18.9\exp(-3s)}{21s+1} \\ \dfrac{6.6\exp(-7s)}{10.9s+1} & \dfrac{4.9\exp(-3s)}{14.4s+1} \end{bmatrix} \begin{bmatrix} R(s) \\ V(s) \end{bmatrix} + \begin{bmatrix} \dfrac{3.8\exp(-8s)}{14.9s+1} \\ \dfrac{4.9\exp(-3s)}{13.2s+1} \end{bmatrix} \begin{bmatrix} F(s) \\ zF(s) \end{bmatrix}$$

Where

$X_D(s)$ is Distillate composition
$X_B(s)$ is Bottom composition
$R(s)$ is Reflux Flow Rate
$V(s)$ is Vapor Flow Rate
$F(s)$ is Feed Flow Rate
$zF(s)$ is Feed light component

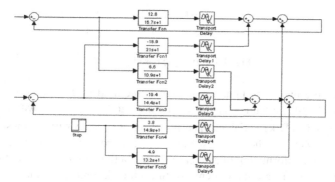

Fig. 2. Transfer function model of binary distillation column

2.1   PI/PID Application to Distillation Column

First of all, in our experiment, PI/PID control of a distillation column is studied. Attempting to use tuning rules such as the well known Ziegler-Nichols rule to adjust the PI/PID controller for each individual loop often leads to deteriorated control performance for the overall system. This is because these tuning rules do not take the interaction between the control loops into consideration [18-19]. Therefore, we have focused more on trial and error method, which is a traditional optimization technique. For the test example, we select the Wood and Berry model [17] because it was derived from an operational distillation column. The simulation results are given in the section 3.

2.2   Fuzzy Inference System Application to Distillation Column

The general form of fuzzy scheme is shown in figure 3. Here Mamdani [20] based FIS has been taken into consideration. The first step in designing a fuzzy based controller [12] is to select the input variables. Here the selected manipulated variables for bottom composition loop are the error in bottom composition ($eB(k)$) and the rate of change of error in bottom composition (eB(k)). Variation in steam flow rate to the reboiler (s(k)) is taken as the output variable for bottom composition controller. The manipulated variables for top composition loop are the error in top composition (eD(k)) and the rate of change of error in top composition (eD(k)). The output variable is the variation in reflux flow rate (R(k)) for top composition controller.

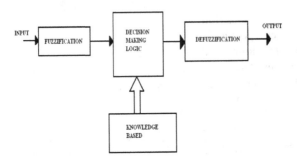

Fig. 3. Fuzzy logic based scheme

After selecting the crisp input variables it is required to decide the linguistic variables. Three linguistic values have been assumed to be associated with each eB(k), eD(k), eB(k) and eD(k) variables. The fuzzy set associated with all these variables are taken as [N, ZO, P], where N is Negative, ZO is Zero and P is Positive.

Next step is the information of rule base for the output composition control for distillation column. There are three members in each $e_B(k)$ and $e_D(k)$ variable fuzzy set and three members each in the $e_B(k)$ and $e_D(k)$ variable fuzzy set. For both the top and bottom composition controllers and same rule base has been developed as shown in the Table 1. After completion of firing of each rule, output will convert to the fuzzy form. To convert this fuzzy output into crisp output, centre of gravity method is used for defuzzification.

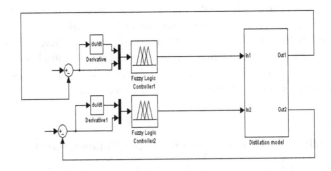

Fig. 4. Simulink model of fuzzy control of distillation column

Table 1. Rule table for top and bottom composition control

		Δe		
		N	ZO	P
e	N	B	S	S
	ZO	B	B	S
	P	B	B	S

2.3   ANFIS Application to Distillation Column

A neural network can model a dynamic plant by means of a nonlinear regression in the discrete time domain. The result is a network, with adjusted weights, which approximates the plant. It is a problem, though, that knowledge is stored in an opaque fashion; the learning results in a (large) set of parameter values, almost impossible to interpret in words. Conversely, a fuzzy rule base consists of readable if-then statements that are almost natural language, but it cannot learn the rules itself. One of the major problems in the not so widespread use of fuzzy control is the difficult of choice and design of membership functions to suit a given problem. A systematic procedure for choosing a type of membership function and ranges of variables in the universe of discourse is still not available. Tuning of the fuzzy controller by trial and error is often necessary to get a satisfactory performance. However, the neural networks have the capability of identification of a system by which the characteristic feature of a system can be extracted from the input output data. This learning capability of the neural network can be combined with the control capabilities of a fuzzy logic system resulting in a Neuro fuzzy inference system. In this sense, an Adaptive-Network based Fuzzy Inference System (ANFIS) is employed.

2.4   Design of the Neuro Fuzzy Controller

The controller block has been implemented as an ANFIS with two inputs, the composition error eD and its integral ieD and two outputs R and V to control the overhead and bottom compositions XD and XB (XB = 1- XD). For each input, the relevant training data has been fed to the ANFIS. These training data are derived from the results of classical PID controller results. The Fuzzy Inference System is generated by considering triangular membership functions for each input. The output membership functions in both the controllers have been considered to be constant type. Training of the FIS is done using Hybrid Optimization method. Based on the training data, the ANFIS automatically generates a Sugeno-type Neuro fuzzy using the above mentioned membership functions and rules as mentioned above.

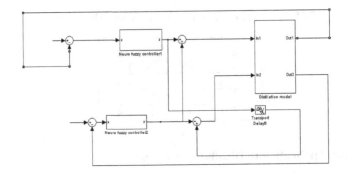

Fig. 5. Simulink model of distillation column with neuro fuzzy controller

3   Results and Discussion

This section presents the simulation results of the PI/ PID, Fuzzy logic, Neuro fuzzy controllers used in the system. Simulation work has been done in Simulink on the environment of Matlab 7.0.5. The results obtained from PI and PID are compared to know the best conventional controller.

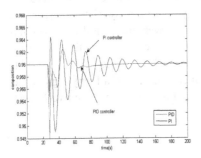

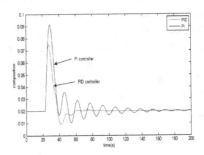

Fig. 6. Comparison of distillate composition using PID/PI controller

Fig. 7. Comparison of bottom composition using PID/PI controller

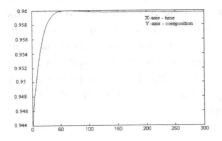

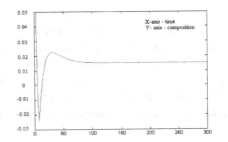

Fig. 8. Distillate composition using fuzzy controller

Fig. 9. Bottoms composition using fuzzy controller

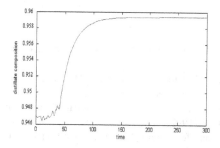
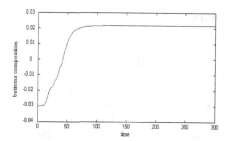

Fig. 10. Distillate composition using neuro fuzzy controller

Fig. 11. Bottoms composition using neuro fuzzy controller

From Fig.6 and Fig.7, we can see that the settling time of PID controller is faster than PI controller. The overshoot of PID controller is less than that of PI controller. Hence, it is clear that the PID controller is the best controller among the conventional controllers and the results obtained from fuzzy and neuro fuzzy controller will be compared with PID controller hereafter.

From Fig. 6-11 it is very much seen that the plots of Neuro fuzzy controller shows better result as compared to other plots, be it in case of settling time or overshoots. Table 2 presents the comparison of controller performance for distillation column.

Table 2. Summary of the result

Composition	Response	PI	PID	Fuzzy logic	Neuro fuzzy
Distillate composition (x_d)	Overshoot	0.9645	0.9625	0.959	0.958
	Settling time	200	82	80	75
Bottoms composition (x_b)	Overshoot	0.09	0.075	0.018	0.015
	Settling time	190	100	90	70

4   Conclusion

We have proposed ANFIS as the better methodology in controlling one of the many parameters in Distillation Column Control, i.e., composition control. This method can be used in controlling other parameters also. The distillation column is used as a test bed to verify the effectiveness of the Neuro Fuzzy controller approach, yielding to stabilization of the composition profiles in the column. The results obtained show the better operability and control of Neuro Fuzzy controller in comparison to the Fuzzy Logic controller and the conventional controller. Other soft computing techniques can be applied to see if better results are obtained.

References

1. Luyben, W.L.: Process Modeling, Simulation and Control for Chemical Engineering. McGraw-Hill (1990)
2. Abdullah, Z., Aziz, N., Ahmad, Z.: Nonlinear Modeling Application in Distillation Column. Chemical Product and Process Modeling 2(3), Article 12 (2007)
3. Skogestad, S.: Dynamics and Control of Distillation Columns – A Critical Review. In: IFAC – Symposium DYCORD+ 1992, Maryland (1992)
4. Luyben, W.L.: Derivation of Transfer Functions for Highly Nonlinear Distillation Columns. Ind. Eng. Chem. Res. 26, 2490–2495 (1987)
5. Psichogios, D., Ungar, L.: Direct and Indirect Model Based Control Using Artificial Neural Networks. Ind. Eng. Chem. Res. 30, 2564–2573 (1991)
6. Godoy Simoes, M., Friedhofer, M.: An Implementation Methodology of a Fuzzy Based Decision Support Algorithm. International Journal of Knowledge- Based Intelligent Engineering Systems 1(4), 267–275 (1997)
7. Bhat, N., McAvoy, T.: Use of Neural nets for Dynamic Modeling and Control of Chemical Process Systems. Computers and Chemical Eng. 14(4), 573–583 (1990)
8. Bulsari, A.: Neural Networks for Chemical Engineers. Elsevier, Amsterdam (1995)
9. Boger, Z.: The Potential of Large Scale Artificial Neural Networks in Distillation Processes Operation. In: Proc. Int. Conference EANN 1999, Warsaw, pp. 117–122 (1991)
10. McAvoy, T., Wang, Y.: Survey of Recent Distillation Control Results. ISA Transactions 25, 5–21 (1986)
11. Gariglio, G., Heidepriem, J., Helget, A.: Identification and Control of a Simulated Distillation Plant Using Connectionist and Evolutionary Techniques. Simulation 63(6), 393–403 (1994)
12. Lee, C.C.: Fuzzy Logic in Control Systems; Fuzzy Logic Controller Part 1 and 2. IEEE Transactions on Systems, Man and Cybernetics 20(2), 404–435 (1990)
13. Riggs, J.: Improve Distillation Column Control. Chem. Eng. Proc. 94(1), 31–47 (1998)
14. Baratti, R., Bacca, G., Servida, A.: Neural Network Modeling of Distillation Columns. Hydrocarbon Processing, 35–38 (1995)
15. Rao, D.H., Gupta, M.: Neuro Fuzzy Controller for Control and Robotics Application. Engineering Applications of Artificial Intelligence 7(5), 479–491 (1994)
16. Li, W., Wu, Z.: Self Organizing Fuzzy Controller Using Neural Networks. In: Proc Int. Conf. on Computers in Engineering, pp. 807–812. ASME, New York (1994)
17. Wood, R.W., Berry, M.W.: Terminal Composition Control of a Binary Distillation Column. Chem. Eng. Science 28, 1707–1717 (1973)
18. Seborg, D., Edgar, T.F., Mellichamp, D.A.: Process Dynamics and Control. John Wiley and Sons, New York (1989)
19. Astrom, K.J., Wittenmark, B.: Adaptive Control. Addition Wesley, Reading (1989)
20. Mamdani, E.H.: Applications of Fuzzy Logic to Approximate Reasoning Using Linguistic Synthesis. IEEE Transactions on Computers 26(12), 1182–1191 (1977)

Efficient Segmentation of Characters in Printed Bengali Texts

Ayan Chaudhury[1] and Ujjwal Bhattacharya[2]

[1] Department of Computer Science and Engineering
University of Calcutta
92, A.P.C.Road, Calcutta 700 009, India
ayanchaudhury.cs@gmail.com

[2] Computer Vision and Pattern Recognition Unit
Indian Statistical Institute
203, B.T.Road, Calcutta 700 108, India
ujjwal@isical.ac.in

Abstract. This paper describes our study of a new and robust approach for character segmentation of printed Bengali text. Like several other Indian scripts, the character set of Bengali consists of basic, modified and conjunct characters. A text line of Bengali has three prominent horizontal zones. Most of its characters appear only in the middle zone while character modifiers or their parts may appear in the upper and lower zones vertically above or below another character. Thus, only vertical segmentation of Bengali texts produces a combinatorially large number of possible shapes making the classification stage intractably difficult. Usually, the problem is tackled by a two-way approach which considers both vertical and horizontal segmentation. Existing approaches for horizontal segmentation of lower zone often fail frequently on old printed Bengali documents due to their typical type-settings. In fact, there is no distinct lower zone in several such Bengali documents. The proposed approach of segmenting modified Bengali characters of the lower zone does not require explicit identification of this lower zone and it is based on the use of a set of empirically designed rules for thinned images of Bengali texts.

Keywords: OCR, lower modifier, segmentation, thinning, recognition.

1 Introduction

Optical Character Recognition (OCR) has been a popular research area for several decades. It has several applications in automation systems like digital library, bank, post-office, language processing etc. Also, OCR is an essential tool of reading aids for the blind.

Robust OCR systems providing acceptable recognition accuracies on printed documents irrespective of the font, style and quality (affected by noise) are now available [1–5] for many scripts such as Roman, Chinese, Japanese, Arabic etc. A detailed survey of its state-of-the-art can be found in [6]. However, there is

J. Mathew et al. (Eds.): ICECCS 2012, CCIS 305, pp. 389–397, 2012.

no such robust OCR methodology for Indian scripts, particularly Bengali, the second most popular language and script of the Indian subcontinent used by 230 million people of India and its neighbouring country Bangladesh. A status report for OCR of Indian scripts including Bengali can be found in [7].

On the other hand, as a part of the well-known 'Million Book Project', the Digital Library of India [8, 9] is now providing free access to a large collection of scanned books of major Indian scripts including Bengali. There are currently 11,009 scanned Bengali books available online at the web site of this digital library. One of the goals of the Digital Library of India (DLI) is to provide support for full text indexing and searching based on OCR technology. However, the existing OCR engines for Bengali may not provide acceptable performance on many of these books. An example scanned page from the book entitled 'Jatak' printed approximately 75 years back is shown in Fig. 1. This is affected by non-uniform skew due to the type-setting methodology used during the said period. We solved this problem by applying skew correction separately on each block of texts (separated by significantly large vertical white space) instead of applying skew correction on the whole page at a time.

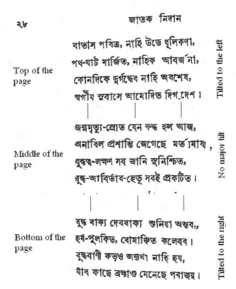

Fig. 1. A page from the book 'Jatak' printed during the period 1920's

A simple OCR software has generally several modules such as image binarization, noise removal, skew correction, text area identification, segmentation of lines, words and characters, character recognition and finally some post-processing for error reduction. The segmentation stage of an OCR algorithm has a major role on the overall accuracy of the OCR system. A few studies of different OCR modules for Bengali documents include [10, 11]. Also, in a recent study [12], a comparison of segmentation driven and recognition driven OCRs for Devanagari, the most popular Indian script, had been presented. Since both

Devanagari and Bangla scripts had evolved from the same ancestor script family Brahmi, they have several commonalities and thus a few initial modules till the segmentation stage of an OCR system of any one of these two scripts should work for building an OCR software of the other script.

In the present work, we studied the performance of existing approaches for segmenting a text line of printed old Bengali documents available in the above DLI into various zones. Also, we proposed a novel robust approach for horizontal segmentation of character modifier appeared in the lower zone of a segmented character.

2 A Few Characteristics of Bengali Script

The script is built around a set of 50 basic characters and the list is shown in Fig. 2. The first eleven in this list are basic vowels and the remaining 39 are basic consonants. Often two or more basic characters merge forming shapes of compound characters. There are more than 200 compound characters in its alphabet several of which occur rarely in a corpus. Shapes of a few compound characters are shown in Fig. 3. Also, all the vowels barring the first one and two among the consonant characters have modified shapes and these are shown in Fig. 4. These modified character shapes can either get attached with or appear adjacent to the shape of a basic or a compound character.

অ	আ	ই	ঈ	উ	ঊ	ঋ	এ	ঐ	ও	ঔ	ক	খ	গ	ঘ	ঙ	চ	ছ
A	AA	I	II	U	UU	.Ra	E	AI	O	AU	ka	kha	ga	gha	nga	ch	chh

জ	ঝ	ঞ	ট	ঠ	ড	ঢ	ণ	ত	থ	দ	ধ	ন	প	ফ	ব	ভ	ম
ja	jha	nya	Ta	Tha	Da	Dha	Na	ta	tha	da	dha	na	pa	pha	ba	bha	ma

য	র	ল	শ	ষ	স	হ	ড়	ঢ়	য়	ৎ	ঁ	ং	ঃ
Ya	Ra	la	sha	Sha	sa	ha	.Da	.Dha	yya	.t	.n	.N	.v

Fig. 2. Bengali Script: Ideal shapes of basic characters

গু(গ + উ)	হু (হ + উ)	রু(র + উ)	ন্ন(ন+ন)	ন্ড(ন+ড)	ক্ষ(ক + ষ)	ষ্ণ (ষ +ঞ)	ন্দ্র(ন+দ+ঋ)
gU(ga + U)	hU(ha + U)	RU(Ra + U)	nna(na + na)	nda(na + da)	kSha(ka + Sha)	Shnya (Sha + nya)	udrra(na+da + rr)

প্ল(প+ল)	ন্ত (ন+ত)	ঙ্গ(ঙ+গ)	ঙ্ক(ঙ+ক)	দ্ব(দ+ব)	ফ্র(ফ+ঋ)	ল্ল(ল+ল)	ষ্ট্র (ষ + ট + ঋ)
pla (pa + la)	nta(na + ta)	ngga(nga+ga)	ngka(nga+ka)	dba(da +ba)	phra(pha+ rr)	lla (la + la)	ShTrra (Sha + Ta + rr)

স্ত(স+ত)	ঞ্জ(ঞ+জ)	শ্ব (শ +ব)	স্ব (স+ব)	ব্ব(ব+ব)	ড্গ(ড +গ)	ষ্ণ(ষ + ণ)	শ্চ(শ+চ)
sta(sa+ta)	nyja (nya + ja)	shba(sha+ba)	mba(ma+ ba)	bba (ba + ba)	.Dga (.Da + ga)	ShNa (Sha + Na)	shcha(sha+ch)

Fig. 3. Shapes of a few compound characters - the constituent basic characters are shown within parentheses for each of them

In a line of printed Bengali text, there are usually three horizontal zones as shown in Fig. 5. The upper and middle zone is separated by the headline or shirorekha. On the other hand, there is no such real line separating the lower zone from the middle one. In the literature [11, 12], an imaginary line, called baseline, had been computed using a small set of handcrafted rules for segmentation of the lower zone.

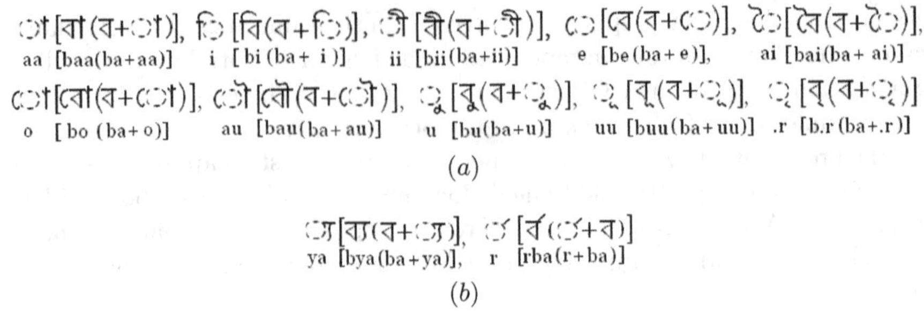

ো [বা (ব+ো)], ি [বি(ব+ি)], ী [বী(ব+ী)], ে [বে(ব+ে)], ৈ [বৈ(ব+ৈ)],
aa [baa(ba+aa)]   i [bi(ba+ i)]   ii [bii(ba+ii)]   e [be(ba+e)],   ai [bai(ba+ ai)]

ো [বো(ব+ো)], ৌ [বৌ(ব+ৌ)], ু [বু(ব+ু)], ূ [বূ(ব+ূ)], ৃ [বৃ(ব+ৃ)]
o [bo (ba+o)]   au [bau(ba+ au)]   u [bu(ba+u)]   uu [buu(ba+ uu)]   .r [b.r (ba+.r)]

(a)

্য [ব্য(ব+্য)], ্র [র্ব (্র+ব)]
ya [bya(ba+ya)],   r [rba(r+ba)]

(b)

Fig. 4. Shapes of Bengali modified characters : (a) modifiers of 10 vowels, (b) modifiers of 2 consonants. Circular shapes show the position of basic or compound character around which the modified character shape appear. Also, instances of their appearances with the basic character 'ba' are shown within respective parentheses.

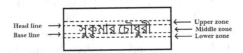

Fig. 5. Different zones of printed Bengali text

Four of the above character modifiers ('aa', 'e', 'o' and 'ya') appear only in the middle zone while another four ('i', 'ii', 'au' and 'ou') of them appear in both the middle and upper zones. There are three character modifiers ('u', 'uu' and '.r') which appear in the lower zone and the remaining character modifier ('r') appears in the upper zone alone.

The goal of the present study is to design a robust method for segmenting a lower character modifier ('u' or 'uu' or '.r') from its associated character situated vertically above it.

3   Existing Approaches to Lower Zone Segmentation

In the approach described in [11], an imaginary line called baseline was computed (as shown in Fig. 6) separating the lower zone of a word from its middle zone. In this strategy, the idea was that three character modifiers ('u', 'uu' and '.r') of the lower zone should appear below the base line. Situations violating this assumption were tackled by considering the relative heights of upper, middle and lower zones of printed Bengali texts. However, these assumptions hold good for a few fonts only and they fail to be satisfied for many old documents available in the said DLI [8].

In Fig. 7, a word from an old book of the above archive shows the possibility of occurrence of the lower modifier at widely different heights of the word.

Additionally, there are books in the said DLI in which the lower modifiers do not appear just vertically below their associated characters but these are disconnected components appearing at the bottom right of the respective main characters. One such situation is shown in Fig. 8.

Fig. 6. Segmentation of lower modifier of printed Bengali text as proposed in [11]

Fig. 7. Variation in vertical position of lower modifiers in a word; dotted lines separate the modifier from its associated characters

Fig. 8. Disconnected lower modifiers appeared at the bottom right of the main character

In another rule based approach described in [12], lower modifier was identified using horizontal run-length distribution of segmented characters. This approach is also seen to be difficult to be used successfully to identify lower character modifiers in various documents of the present archive. As an illustration, a few examples are shown in Fig. 9.

4 Proposed Approach

4.1 Removal of Headline

We first consider horizontal projection profile (hpp) to detect the position of headline in an input line of Bengali text. It corresponds to the maximum value of hpp. Our next goal, as in [11], is to rub off the headline separating the upper zone from the remaining part of the text line. This also results in segmentation of individual characters in a word. However, headline removal from the texts of the said DLI often gets affected due to the variations in their widths and alignment. We tackle this problem as follows. We obtain well-known Canny edge of input text line image and use probabilistic Hough transform(PHT) [13] for detecting the horizontal lines in the Canny edge map. PHT is relatively faster and works well on noisy data compared to the standard Hough Transform(SHT) technique. One of the horizontal edges of each word identified by PHT is nearest to the position of the headline as detected by hpp. If this edge is above the detected headline position, then we move downward and reach at the lower edge of headline. Otherwise, we move upward to identify the upper edge of the headline. Thus, we identify both the edges of the headline and obtain an

Fig. 9. Three characters and the respective horizontal run-length distributions are shown; the two characters at the left are attached with the modifier '.r' appearing in the lower zone but the character at the right is attached with the modifier 'ya' appearing in the middle zone

estimate of the headline width of individual words. Finally, we remove headlines of individual words by turning each black pixel between the two edges of the headline into white. An example of headline removal is shown in Fig. 10.

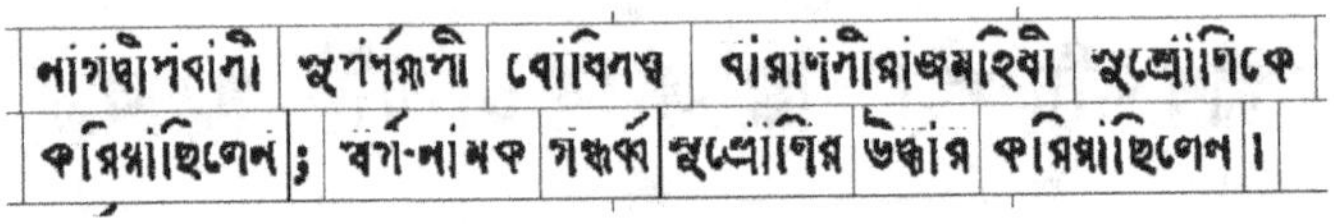

Fig. 10. Headline of each word is removed

Once the headline is removed individual characters or their parts situated below the headline get disconnected and each of them is passed to the next module to identify the possible presence of a lower modifier.

4.2   Detection of Lower Modifier

In the proposed approach for detection of character modifiers present in the lower zone of a text line, we apply thinning technique on each individual component obtained in the above. There are a number of image thinning algorithms available in the literature and we implemented a few of them. We observed that Parker's [14] image skeletonization method produce acceptable results for our character data. Original and skeletal shapes of several characters with a modifier in their lower portions are shown in Fig. 12.

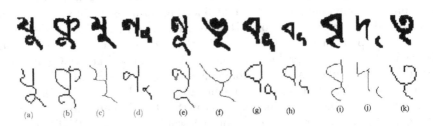

Fig. 11. Variations in the shapes of character modifiers appearing in the lower region; (a) - (d) character modifier 'u', (e) - (h) character modifier 'uu', (i) - (k) character modifier '.r'

Rules for Segmenting Lower Modifier. Based on an extensive study of the shapes of character modifiers and their placements in a text line, we designed the following set of rules for their detection. Let the portion of a text line below the headline be horizontally divided into two equal halves, called *upper* and *lower*. If no part of a character component lie in the region *upper*, then it is a character modifier of the lower zone. Character modifiers satisfying this condition are shown in Figs. 12 (d), (g), (h) and (j). Next, let L denote the bounding box of the part of the skeletal (thinned) image of a character component lying in *lower*. If there is no end pixel (a pixel with only one 8-neighbour) in L, then the corresponding character component does not have a modifier in the lower zone. Such a situation is shown in Fig. 12 (b). In other cases, we trace the skeleton in L as follows and decision towards the presence of a character modifier of the lower zone is taken.

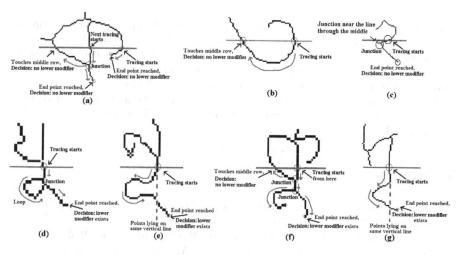

Fig. 12. Different situations of the presence or absence of a character modifier of lower zone. Samples in the first row do not have character modifiers of lower zone while samples of the second row have these modifiers.

Algorithm:
Step 1. Start scanning towards left from upper right corner of T until a black pixel P is found.

Step 2. Start tracing the skeleton from P. If an end pixel (having only one 8-neighbour) is reached before reaching either a junction pixel or the lowest pixel in L, find the next black pixel P on the upper boundary of T and repeat Step 2.

Step 3. If an end pixel is reached at the lower boundary of T or after passing through a point on its lower boundary but before encountering a junction pixel, then go to Step 4. Otherwise go to Step 5.

Step 4. Check whether there are a few pairs of points in the path of tracing such that one point is vertically above the other point of each pair and the number of such pairs is at least one-fifth of the width of T. If so, go to Step 9. Otherwise, go to Step 8.

Step 5. If a junction pixel is reached, check whether a loop is formed and also tracing along the other branch reaches at an end pixel in which case go to Step 9. Otherwise, go to the next Step.

Step 6. If the depth of the junction pixel is less than one-third of the height of T, then trace only along the branch which goes downward and otherwise, traverse along one of the branches which has not been traversed yet and go to Step 3. If no more branch is left, then go to the next step.

Step 7. If there are two end pixels below the last branch visited, then go to Step 9. Otherwise go to Step 8.

Step 8. Decide the non-existence of a character modifier of lower zone and stop.

Step 9. Decide the existence of a character modifier of lower zone and stop.

5   Conclusion

In this paper we have presented a robust approach for detection or segmentation of character modifiers of the lower zone of Bengali texts. The approach has been tested on several pages of 10 old books downloaded from the DLI [8] and obtained significantly good performance. The typesetting in each of these books are not suitable for using previous methods for the same purpose. In future, we shall study recognition of segmented characters (or their segments) so obtained. A planned and successful attempt for automatic reading of Bengali texts available from the DLI will help to achieve the goal of the DLI.

References

1. Mori, S., Suen, C.Y., Yamamoto, K.: Historical review of OCR research andd evelopment. Proc. IEEE 80, 1029–1058 (1992)
2. Nagy, G.: At the frontiers of OCR. Proc. IEEE 80(7), 1093–1100 (1992)
3. Gorman, L.O., Kasturi, R.: Document Image Analysis. IEEE Computer Society Press, Los Alamitos (1995)
4. Trier, O.D., Jain, A.K., Taxt, T.: Feature extraction methods for character recognition – a survey. Pattern Recognition 29, 641–662 (1996)
5. Nagy, G., Seth, S.: Modern optical character recognition. In: Froehlich, F.E., Kent, A. (eds.) The Froehlich/Kent Encyclopedia of Telecommunications, pp. 473–531. Marcel Dekker, New York (1996)
6. Nagy, G.: Twenty years of document image analysis in PAMI. IEEE Transactions on Pattern Analysis and Machine Intelligence 22(1), 38–62 (2000)
7. Pal, U., Chaudhuri, B.B.: Indian script character recognition, a survey. Pattern Recognition 37, 1887–1899 (2004)
8. http://www.new1.dli.ernet.in/ (last accessed on, May 2012)
9. Ambati, V., Balakrishnan, N., Reddy, R., Pratha, L., Jawahar, C.V.: The digital library of India project: process, policies and architecture. In: Int. Conf. on Digital Libraries (ICDL 2006), New Delhi (2006)
10. Chaudhuri, B.B., Pal, U.: Skew angle detection of digitized Indian script documents. IEEE Trans. on Patt. Anal. and Mach. Intell. 19, 182–186 (1997)
11. Chaudhuri, B.B., Pal, U.: A complete printed Bangla OCR system. Pattern Recognition 31, 531–549 (1998)

12. Kompalli, S., Setlur, S., Govindaraju, V.: Design and comparison of segmentation driven and recognition driven Devanagari OCR. In: Proc. of DIAL, pp. 96–102 (2006)
13. Kiryati, N., Eldar, Y., Bruckstein, M.: A probabilistic Hough Transform. Pattern Recognition 24(4), 303–316 (1991)
14. Parker, J.R.: Algorithms for image processing and computer vision, pp. 203–218. John Wiley and Sons, New York (1997)

A GIS Based Approach of Clustering for New Facility Areas in a Digitized Map

J.K. Mandal[1], Anirban Chakraborty[2], and Arun Kumar Chakrabarti[3]

[1] Department of Computer Science & Engineering University of Kalyani
Kalyani, Nadia, W.B., India
jkm.cse@gmail.com
[2] Department of Computer Science
Barrackpore Rastraguru Surendranath College
Barrackpore, Kolkata-700 120, W.B., India
theanirban@rediffmail.com
[3] Department of Computer Science
Barrackpore Rastraguru Surendranath College
Barrackpore, Kolkata-700 120, W.B., India
great.arun786@gmail.com

Abstract. This paper presents the technique for finding the suitable locations for building new facility areas, such as Schools, Cold Storages etc. in the given map. The maps are combination of irregular polygons. Information like the co-ordinate points are being kept in traditional file systems during digitization. As first step of the technique, users need to point the locations of existing facility areas, which is visualized into the map. Through suitable user friendly interfaces, the user entered the number of new facility areas needed. The technique sub-divides the total map into a number of sub-regions, each having equal area. This number of sub-regions is equal to the number of total facility area needed. Finally, the tool finds those sub-regions, which don't contain any facility area and generate the centroid of these sub-regions as the suitable location for the further construction of the facility areas.

Keywords: Facility areas, Raster map, Digitization, irregular polygon, sub-regions, Centroid.

1   Introduction

For a large geographical area like a sub-division of a district (under Indian political system) if it is needed to increase the number of existing facility areas like number of schools or cold-storages etc., at first the most suitable locations for new facility areas should have to be found out. One major aim, while doing this, may be to equally distribute the facility location areas through the whole region under consideration, so that if, for example, new schools are to construct, to spread education among all, then the schools must be evenly distributed through out the whole region under consideration, such that every body could access them. The same thing is also for building new cold-storages or other type of facility locations.

J. Mathew et al. (Eds.): ICECCS 2012, CCIS 305, pp. 398–405, 2012.

Proposed work has been done on a previously digitized map where information are available as stored data. Section 2 of this paper deals with the scheme of the implementation techniques, which starts with sub-division of the map followed by the way of pointing out of the suitable locations for new constructions. The implemented results are given in section 3. Analysis and Comparison and Conclusions are drawn in section 4 and 5 respectively.

2   Scheme

The proposed scheme has been integrated in two phases. The sub-division of the digitized map is made first using proposed algorithm to create the sub regions and then suitable locations for new facility areas are pointed on the digitized map as per requirement of the user. These are discussed in section 2.1 and 2.2 respectively.

2.1   Creation of Sub Regions

This process divides a given polygon, into a number of polygons, each of equal area. When a previously digitized (stored) map is chosen for the work, it can be viewed as an ordered sequence of m-points. If these points are (x_0, y_0), (x_1, y_1), ... (x_n, y_n), then the area of the polygon being constructed with these points is given by $\frac{1}{2}[(x_0y_1-x_1y_0) + (x_1y_2-x_2y_1) + ...+ (x_ny_n+1-x_n+1y_n)]$ or

$$\frac{1}{2} \sum_{i=0}^{n} (x_iy_{i+1} - x_{i+1}y_i)$$

or can be written as (using the determinant form)

$$A = \frac{1}{2}\left\{ \begin{vmatrix} x_0 & x_1 \\ y_0 & y_1 \end{vmatrix} + \begin{vmatrix} x_1 & x_2 \\ y_1 & y_2 \end{vmatrix} + \cdots + \begin{vmatrix} x_{n-2} & x_{n-1} \\ y_{n-2} & y_{n-1} \end{vmatrix} + \begin{vmatrix} x_{n-1} & x_0 \\ y_{n-1} & y_0 \end{vmatrix} \right\}$$

After obtaining the area of a polygon, it is divided into a number equal area polygons using following process.

The user inputs the number of required equal divisions, say D. If A is the area of the original polygon, each sub-polygon will be of area A / D. The scanning is stated from the extreme left point (i.e. the point having minimum x-coordinate value) and moves in a clock-wise rotation. Let at each step we traverse 5 points at a time. Thus, if x_0 is the x-co-ordinate of the starting point, we move to point x_4. At this point we draw a straight line parallel to Y-axis, which intersects the polygon somewhere, as shown in fig. 1.

Fig. 1. Polygon division

Now, we consider the area of the polygon formed at the left side of the line (shown by blue color) and calculate its area. If it is almost equal to the required area (i.e.

A/D), then we save it as one of the sub-polygon by making the line(which has drawn) as permanent, otherwise proceed to the next 5th point and repeat the process again by drawing a fresh straight line.

At each time, while deciding whether the area of a sub-polygon has reached to the required one, some amount of error factor is being associated, this means if the constructed one has an area of ± 10 sq. units (as desired), then it is announced as the final.

While drawing the straight line parallel to Y-axis (point A of fig. 2), we expect that this line will intersect the polygon at B, which has the same x-coordinate value as A but here also n permitted amount of error factor is associated, which for our case is ±5. As practically, more than one point may have their x-values lying in the prescribed range, so the point which has the x-value in the prescribed range and highest y-value is assumed as point B.

A problem may also occurs can be described using fig. 2. While scanning clockwise, during polygon division (shown by arrow-mark), a point B can be found, matching the above criterion, but the line drawn may completely be outside of the map (fig. 2). To overcome this problem while finding A, that point is labeled as B which is the last stored point, matches the above criterion.

Fig. 2. Showing an erroneous situation

2.2 Suitable Location for Facility Area

Locations of existing facility-areas are made input. Mouse clicking onto the points in the map this is done. It is required to input number of additional facility-locations needed and approximated locations for these new constructions are being pointed out. For example, if in a particular digitized area only 2 schools are there and it is required to build 2 more schools then, the process sub-divides the entire area into 4 equal parts using the technique discussed in section 2 B. It then searches for those parts (i.e. polygons generated after division), which does not contain any schools and if it finds two such parts, then the Centroid of each part is located and points it as the suggested location. It will be more clearer from the following Fig. 3.

Fig. 3. Finding the additional suitable locations

In fig. 3, two facility areas already exist, pointed by small black dots. Now the objective is to search for the most suitable locations, where two more facility areas can be built. The proposed locations are pointed by red dots. The Centroid of a polygon (C_x) is obtained by using the formula

$$C_x = \frac{1}{6A} \sum_{i=0}^{n-1} (x_i + x_{i+1})(x_i y_{i+1} - x_{i+1} y_i) \quad C_y = \frac{1}{6A} \sum_{i=0}^{n-1} (y_i + y_{i+1})(x_i y_{i+1} - x_{i+1} y_i)$$

Thus, for the proposed technique, it is necessary to decide frequently whether a point lies inside a polygon or not, because if we find any such polygon which is not containing any existing facility-area, then its Centroid is being pointed out. This (i.e. whether a point existing inside a polygon or not) can be achieved by using the modified Ray-Casting algorithm [8]. The approach has been illustrated as follows.

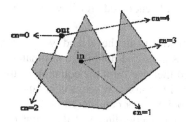

Fig. 4. Ray Casting Algorithm

One simple way of finding whether the point is inside or outside a simple polygon is to test how many times a ray starting from the point intersects the edges of the polygon. If the point in question is not on the boundary of the polygon, the number of intersections is an even number if the point is outside, and it is odd if inside. There are some rules which have to be followed during the design of the algorithm. These are:

- An upward edge includes its starting endpoint, and excludes its final endpoint.
- A downward edge excludes its starting endpoint, and includes its final endpoint.
- Horizontal edges are excluded.
- The edge-ray intersection point must be strictly right of the point P.

The C-like Pseudo code for the above Ray-Casting algorithm, which has been used to implement the software, is as follows:-

```
Initialize as counter = 0, a = x[0], b = y[0];
 for (i=1; i<=N; i++) { c = x[i % N] and d = y[i % N];
   if (tgty > (b < d ? b : d)) { // Downward Crossing.
    if (tgty <= (b > d ? b : d)) { // Upward Crossing.
      if (tgtx <= (a > c ? a : c)) { // Line Extended To
The Right.
        if (b != d) {
          // compute the actual edge-ray intersect x-
coordinate
            xintrsct = (tgty-b) * (c-a) / (d-b) +a;
            if (tgtx <= xintrsct) counter++;}}}}
       set a=c and b=d; }
  if (counter % 2 == 0)return(0); // Outside
    else return(1);// Inside
```

Algorithm 1. Pseudo code for Ray Casting Algorithm

In this algorithm t_gt_x and t_gt_y is the x and y coordinates of the point under consideration. The variables a, b, c, d hold the coordinates of the polygon. All 'Edge Crossing' rules are verified then we form the equation of the line. Our aim is to draw a single straight line parallel to x-axis. When it is parallel to x-axis its y-axis is same as the y coordinate of the point under consideration. Our next task is to find the point of intersection i.e. to find the x coordinate of the point of intersection. We have to check whether this x coordinate is having a value greater than the point under consideration, if so, it is discarded otherwise accepted and the variable counter is incremented by 1. Now when all the points are traversed then it is checked whether count is even or odd. If it is even then it is outside the polygon otherwise it is inside. Since the aim is to check whether a point is inside or outside a polygon, the program return 1 if it is inside the polygon and 0 if outside the polygon.

3   Implementation and Results

The implementation is done using a high level language. Designed pages and its operations are presented and discussed in this section. The implementation has been done by using Net Beans (for Java), based on flat file systems without using any databases. The File-Open window helps the user to choose a Map from any location of the computer (or an online-image), no need to resize the images.

During creation of partition, an input box appears (also showing the area of the original polygon) to input the required number of partitions (Fig. 5) and next screen makes the confirmation regarding the partition (Fig. 6).

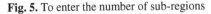

Fig. 5. To enter the number of sub-regions   **Fig. 6.** Showing area of the sub-regions

Snapshots of partitioning are presented in fig. 7A-7C.

Fig. 7. A: Partitioned into 5    **Fig. 7.** B: Partitioned into 3    **Fig. 7.** C: Partitioned into 2

Finally, with these portioned areas, suitable locations for new facility areas are found, as shown in fig. 8A-8B.

Fig. 8. A: Searching one more    **Fig. 8.** B: Searching two more (with 2 existing)

4   Analysis and Comparison

Most important part of this technique is to divide an irregular polygon into a number of sub-polygons each of equal area. Unfortunately, there is no good number of methods for dividing an irregular polygon. Among some available techniques, one is to place the shape onto a grid and then to count the number of cells completely inside the shape (says C) and the number of cells through which the boundary line of the shape passes (say P) as shown in fig. 9. If each cell is of unit area, then the area of the polygon is approximated by the formula, Area = $C + (.5) \times P$.

Fig.9. Finding area of the shape using grid paper

Finally, the original polygon can be divided into a number of sub-polygons of equal area by drawing straight lines onto that grid and each time the number of cells covered are calculated for determining the area of the sub-polygon. Lastly, a sub-polygon is announced to be a final division, when its area becomes equal to the desired one, i.e. a trial-and-error method is incorporated. The comparison of this technique with the proposed one is shown in table 1, which reveals the fact that for the proposed work the time complexity is less as because no need of any extra grids on which to place the polygons, resulting the fact that the area calculation is only on the basis of co-ordinate points, not by calculating the number of grids (or cells); saving calculation time in each step. At the same time more accuracy is being achieved as less approximations are made compared to grid-based method when working with a shape having very curvy boundaries.

Table 1. Comparison of the proposed scheme with the existing grid-based technique

Features	Proposed Technique	Grid-based Technique
Requiring an extra-grid sheet(May be an imaginary)	No such requirement	The shape is required place on a grid sheet
Accuracy	More, as less approximation is done	1. Accuracy is dependent on the size of the grids. Finer the grids, more better the results. 1. At the time of area calculation, approximations are made. Thus if more divisions are made more approximations will be there.
Time taken	Less, because the area is calculated only on the basis of the coordinate points stored.	For area calculation, each time the number of cells lying completely within the shape and those lying on the boundary line has to be calculated separately. Hence the method is quite time consuming. The needed time even increases with more divisions.
Difficulties for very much curvy shaped polygon	No matter how much curvy is the polygon, the method works well.	The accuracy degrades due to more approximations made while dealing with the polygons which are very much curvy in nature

5   Conclusions

Now-a-days, as per Indian scenario, with progression of the Country, new constructions and thus finding the suitable locations for these constructions are frequently being done. In the area of education system, when the terms like "Education for every child", "Education for all, progress for all" etc. are frequently being uttered and to make these words possible new schools are to be constructed in those locations, where no already existing schools are there within few kilometers. Similarly, to progress agriculture and thus to help farmers new cold storages are to built, where no such facilities are available within few kilometers. The effort has been taken to obtain an easy solution for the situations of dividing polygons, by finding the most suitable locations for construction of these new facility-areas.

The advantages of this technique are as follows:

- The existing facility locations are fetched through mouse-clicking, which makes the process less time-consuming and user-friendly.
- Editing is possible to accommodate changing needs.
- The advantage of this work is that it is based on purely flat file systems, no dependence on any particular operating systems and DBMS. But instead if databases are used then, not only the cost would be increased, but as most of the databases are dependent on operating systems, so portability would be lost.
- The digitized map can also be divided into a number of portions, each having same area for future analysis.
- The suitable locations for new constructions are being achieved.

Acknowledgment. The authors express a deep sense of gratitude to the Department of Computer Science, Barrackpore Rastraguru Surendranath College, Kolkata-700 120, India and Department of Computer Science and Engineering, University of Kalyani for providing necessary infrastructural support for the work and to the UGC for financial support under Minor Research Project scheme (sanction number F. PSW-180/11-12 ERO, dated 25.01.2012).

References

1. Chakraborty, A., Mandal, J.K., Chakraborty, A.K.: A File Based GIS anchored Information Retrieval Scheme (FBGISIRS) through Vectorization of Raster Map. International Journal of Advanced Research in Computer Science 2(IV), 132–138 (2011) ISSN No.- 0976-5697
2. Chakraborty, A., Mandal, J.K., Banerjee, N., Patra, P.: A GIS based Interlinked Information Retrieval of a Large Database using KD Tree. In: Proceedings ETCS 2012, UGC Sponsored National Symposium on Emerging Trends in Computer Science (ETCS 2012), January 20-21, pp. 84–89 (2012) ISBN: 978-81-921808-2-3
3. Halder, P., Mandal, J.K., Mal, S.: A GIS Anchored Information Retrieval Scheme (GISIRS) based on Vectorization of Raster Map. International Journal of Engineering Research & Industrial Applications 2(III), 265–280 (2009) ISSN 0974-1518
4. Hongye, B.: Digitization of ancient Maps Based on GIS Technology: The yu ju tu Maps. In: World Library & Information Congress: 75th IFLA General Conference & Council, Milan, Italy, August 23-27 (2009)
5. http://www.maa.org/pubs/Calc_articles/ma063.pdf (accessed March 2012)
6. User manual of TNTmips, the MicroImages Remote Sensing and GIS software
7. http://en.wikipedia.org/wiki/Polygon (accessed March 2012)
8. Hearn, D., Pauline Baker, M.: Computer Graphics, PHI, 2nd edn.

Enhanced Rule Accuracy Algorithm for Validating Software Metrics

Abdul Jabbar and Sarala Subramani

Department of Information Technology, Bharathiar University
Coimbatore, Tamil Nadu, India
Jabbar123p@gmail.com, sriohmau@yahoo.co.in

Abstract. Software metrics has significant scope in the software quality and productivity measurement. In this work the competence of Rule Accuracy Algorithm is enhanced by the Fitness Proportionate selection algorithm that has been previously applied to genetic algorithm as potential recombination of selection, which overcomes the ant colony algorithms multiple path issue. The work analyzes the adequacy of Rule Accuracy Algorithm to validate whether the ant colony algorithm or Fitness Proportionate selection algorithm provide best results. Experimental result shows interesting conclusions on the improvements of the rule accuracy algorithm in Fitness Proportionate selection algorithm. It also opens up new areas of research on the efficiency and effectiveness of software metrics validation.

Keywords: Software metrics quality, Rule Induction, Fitness Proportionate Selection, Rule Accuracy Algorithm, Software Metrics.

1 Introduction

The significance of software metrics in software development is distinguished. With the assistance of metrics the developers gain various improvements in project development. Software metrics plays an important role in maintaining the software quality and cost estimation, which includes readability, maintainability, effort and productivity. Predictive capability of software metrics is attractive, thus it is a significant method to compute the productivity, effort, cost estimation [3], [4] and software quality [21], [22], [28].

The validation process of software metrics is highly difficult and includes complex processes. In related work, properties have been developed for evaluation and validation of software metrics which satisfies metrics attributes. It is not at all identified as an efficient technique, thus Rule Accuracy Algorithm (RAA) based on rule induction technique has been proposed [2]. Rule induction is the common form of knowledge discovery in unsupervised learning systems [25]. RAA can perform well complicated evaluation of software metrics. The experiment was conducted and shows on the given software metrics data set, provides output and the best metrics data were induced by RAA. This work improves the efficiency of RAA using Fitness Proportionate Selection algorithm.

J. Mathew et al. (Eds.): ICECCS 2012, CCIS 305, pp. 406–412, 2012.

2   Related Work

The validation on metric is performed in different ways that are analytical validation and empirical validation [17], [7]. Analytical validation performs the hypothetical basement of metrics as well as the feature coverage associated with the properties. Empirical validation is to identify precision and accuracy of software metrics on metrics relationship and the attitude of different client.

In [19], [20] indicates that metrics validity is to verify based on satisfying defined properties. The authors [7] validate cohesion and information theoretic metrics using properties.

The most related work [10] has show that the described numbers of properties to authenticate the software metrics measures were based only in the context of syntactic features of program. The properties are representing the relation between syntactical deviations throughout the outcome, exactly nine properties were proposed and that perform the validation. Then the responses of validation of these properties are eminent on various complexity measures. It can be concluded that the failure to convince the properties is due to the increased complexity of the metrics. Metrics validation is a complicated process, as the properties might verify to the mathematical precision in all perspectives. As it is a complicated task, it would be desirable to in a few dimensions. This validation approach has been offering hopeful results but the end result is not relevant in all cases. Hence the demand for an alternative to comprehend the absolute unforced validation of metrics in all perception. Therefore RAA approach is proposed to validate metrics adequacy.

Previous work [2] Ant Colony Algorithm performs the optimization of rules. Enhanced view of RAA needs improved selection of rules. Rule induction method is used to evaluate metrics for competence. The rules stimulate the finest data from metrics data set and authenticate the efficiency of the metrics. The rule induction is a significant classification technology [9], [12], [14], [25], [29]. This work evaluates RAA classification algorithm and recognizes the difference in performance. Given a set of metrics data set, its adequacy for validation using RAA is measured. The primary concern is to begin the validation as the resolution of a rule optimization to find the best rule. The earlier versions of work [2], the rules have been formulated based on adequacy criterion, and organized by supervised classification method basis of improved Ant Colony optimization algorithm [18]. To define the ACO, the researchers consider how the almost blind ants found the shortest path to their feeding sources and rear. Basically, the phenomenon is that ants use pheromone for communication. The path of an ant lays a few of pheromone, which are detectable by other ants, along its path. Consequently, the ants can establish the shortest path to the feeding sources and back. Using this methodology the ACO finds the solution for various optimization issues [8], [26], [18], [13], [5], [15], [30], and [31]. ACO finds a partial solution where the problem indicating multiple paths.

3   Fitness Proportionate Selections

ACO finds a partial solution to the problem indicating multiple paths. The improved RAA enhances rule selection method based the metrics adequacy factor. Accordingly,

Fitness proportionate selection performs the rule optimization as an alternative of ACO. Fitness proportionate selection is a genetic operator used in genetic algorithms for selecting potentially useful solutions for recombination [6]. The work [11] and [24] Fitness proportionate selection is used for the optimization. In fitness proportionate selection, since in all selection techniques, the fitness function allocates fitness to possible solutions. This fitness level is used to associate a probability of selection with each individual chromosome. If f_i is the fitness of individual i in the population, its probability of being selected as,

$$p_i = \frac{f_i}{\sum_{j=1}^{N} f_j}$$

Where N is the number of individuals in the population.

4   Rule Accuracy Algorithm

In [2] discusses that rule induction techniques based on the number of approaches has been used to detect the information from data base with good accuracy. The possible if-then-rules are recognized based on adequacy criterion. The rules are recognized, different fuzzy *if- then- rules* are available. The rule is explained in the work linguistic of fuzzy rules [16]. The sample rule is, R1. IF the *rule1* is YES or NO AND THEN adequate. The *if-then rule is* defined as, *rule1, rule2…..rule$_n$* hold and they the structural attributes of metrics, and thus it varies according to the characteristic of metrics. Accordingly the RAA algorithm appears consistent. More precisely, the work focuses on structural metrics, for each rule maximum parameters were considered. The parameters were obtained from a number of available source codes. The significant concern of RAA is how to decide the finest rules from the rule set. The validation of available rules is big issue, so it is optimized by FPS selection method. According to this, metrics adequacy factors distinguish the rules. In addition the adequacy factors reflect on the source code behavior and measuring aspects. The quality coverage may vary considerably depending on the rules clarity. So as to avoid deceptive result, certain criteria are fixed for all rule definition.

Table 1. Henry and Kafura and IFC information flow metrics data set

Si No.	F-in	F-out	P-C	C-L	Henry and Kafura	IFC
1	108	30	19	1133	11893780800	2.31421
2	0	3	1	11	0	0.272727
3	4	0	2	21	0	0.38095238
4	31	12	8	675	93409200	0.50962963
5	15	9	7	134	61053750	1.2537314

RAA algorithm has been defined as the instances are selected according to the rules, which induced the data set and finds the accuracy. According to RAA algorithms with rules, induce the metrics data set in table 1, an individual from a metrics set m= (m_1, m2, m3...m_n) of a metrics set M = (M_1, M2,.M_n) and rule set R=(R1,R2,...Rn). Given rules of the best selection, R, and the preferred metrics, M, indicate a data set, m. RAA approximate the metrics adequacy by each rule metrics values and parameters are examined.   To assist the proposed RAA algorithm's steps are précised as follows.

4.1   Rules Accuracy Algorithm(R, M, a)

Step1: Metrics Set M:={M1, M2,...... Mn}

Step2:   metrics dataset from metrics set

Metrics dataset m=($m_1,m_2,m_3..m_n$).

Step3:   Rules Set R :={R1, R2,..........Rn}

Step4:   check the metrics data M set is satisfied for each Rules R

Step5:   Calculate Accuracy and coverage of metrics.

5   Experimental Results

Software metrics competence prediction using RAA algorithm with rule induction technique is explained with distinct contribution. Initially, metrics data set is selected, followed by program analysis the value has been reflected with result. Information flow metrics has been used to prepare the database [3] which shows in table 1. The metrics are Henry Kafura information flow complexity metrics (1) and IFC (3) [3].

Table 2. IFC, Henry and Kafura Check by optimized the all rules

No	Rules Accuracy Algorithm	IFC	Henry and Kafura
1	RULE 1	Yes	Yes
2	RULE 2	No	Yes
3	RULE 3	Yes	No
4	RULE 4	Yes	Yes
5	RULE 5	Yes	No
6	RULE 6	No	Yes
7	RULE 7	YES	NO
8	RULE 8	YES	NO
	Total	6	4

Accuracy of IFC information flow complexity metrics=75
Accuracy of Henry and Kafura information flow complexity metrics=50

$$\text{H.K Information flow} = \text{length}* (\text{fan-in}*\text{fan-out})^2 \qquad (1)$$

$$\text{IFC} = (\text{Procedure call}) *(\text{fan-in} +\text{fan-out}) / \text{Code length} \qquad (2)$$

RAA examination of conducted by the data set formerly selected best rules from the given set of rules. To be exact, each row in the table 1 that is M1, M2...Mn is validating in the best Rule set R. The rule is defined as to complete each competence intention. However, wide authenticity of rules is possible for merge RRA algorithm. Metric set M1 is validated by rule set R that is, M1 is validated by each rule in Rule set R, and eight rules are obtained after the ACO optimization in previous work. In this work FFP perform the optimization, six rules are obtained. Subsequently, metrics are validated. Table 2 shows the result of the RAA using ant colony algorithm. First column shows optimized rules, even as the second validation result IFC and third validation result of Henry and Kafura in that order. The study has been limited to the induction technique by setting maximum number of rules up to nine using ACO algorithm to ensure the reliability and efficiency. If the metrics M satisfies the rule R, RAA indicate either yes condition or no condition following the evaluation. Table 3 shows the improved result of RAA using fitness proportionate selection. The work shows that the fitness proportionate function is better prediction than ant colony algorithm. Improved algorithm is possible to provide accurate result in empirical research.

Table 3. IFC, Henry and Kafura Check by optimized the all rules

No	Rules Accuracy Algorithm	IFC	Henry and Kafura
1	RULE 1	Yes	Yes
2	RULE 3	Yes	No
3	RULE 4	Yes	Yes
4	RULE 6	No	Yes
5	RULE 7	YES	NO
6	RULE 8	YES	NO
	Total	5	3

Accuracy of IFC information flow complexity metrics=83.33
Accuracy of Henry and Kafura information flow complexity metrics=50

6   Conclusions

Rule Accuracy Algorithm has been devised with the selected rules and data. Fitness proportionate selection algorithm chooses best rules instead of ACO algorithm. It improves the Rule Accuracy Algorithm. RAA investigated by the metrics data set, shows that the IFC is adequate. The processed data set show in table 1, illustrates examples of metrics evaluation. The optimized rules induce the metrics data set. A number of metrics has been scrutinized based on rules induction approach. The Rule Accuracy Algorithm determines that improve the metrics evaluation is found with predictable results. The Rule formation and selection in RAA algorithm offers a suggestion for the future developments. Rules configuration and selection are very beneficial, so the FFP employed and both will improve the adequacy of measurement.

References

1. Jabbar, A., Sarala, S.: Advanced Program complexity measurement. International Journal of Computer Application, 29–33 (2011)
2. Jabbar, A., Sarala, S.: Authenticate Program Complexity Metrics using RAA. In: Das, V.V. (ed.) CIIT 2011. CCIS, vol. 250, pp. 358–362. Springer, Heidelberg (2011)
3. Sarala, S., Abdul Jabbar, P.: Information flow metrics and complexity measurement. In: 3rd IEEE International Conference on Computer Science and Information Technology, pp. 575–578 (2010)
4. Sepasmoghaddam, A., Rashidi, H.: A Novel Method to Measure Comprehensive Complexity of Software Based on the Metrics Statistical Model. In: IEEE UK Sim Fourth European Symposium on Computer Modeling and Simulation, pp. 520–525 (2010)
5. Chan, A., Freitas, A.: A New Classification-Rule Pruning Procedure for an Ant Colony Algorithm. In: Talbi, E.-G., Liardet, P., Collet, P., Lutton, E., Schoenauer, M. (eds.) EA 2005. LNCS, vol. 3871, pp. 25–36. Springer, Heidelberg (2006)
6. Motoki, T.: A new method for modeling the behavior of finite population evolutionary algorithms. In: Evolutionary Computation, pp. 451–489 (2010)
7. Barbara, Pearl: Problem Adopting Metrics from other Disciplines. In: ICSE Workshop on Emerging Trends in Software Metrics, pp. 1–7. ACM (2010)
8. Liu, B., Qin, Z., Rui, Bing, Shao: A hybrid heuristics ant colony system for coordinated multi-target assignment. Information Technology Journal, 156–164 (2009)
9. Cohagan, C., Grzymala-Busse, J.W., Hippe, Z.S.: Mining Inconsistent Data with the Bagged MLEM2 Rule Induction Algorithm. In: 2010 IEEE International Conference on Granular Computing, pp. 115–120 (2010)
10. Elane, Weyuker: Evaluating Software Complexity Measures. IEEE Transactions on Software Engineering, 1357–1365 (1988)
11. Frank, Peitro, Carsten: Theoretical analysis of fitness-proportional selection: Landscapes and Efficiency. In: ACM Genitic and Evolutionary Computation Conference, pp. 835–842 (2009)
12. Shadabi, F., Sharma, D.: Using Knowledge and Rule Induction Methods for Enhancing Clinical Diagnosis: Success Stories. In: International Conference on Future Computer and Communication, pp. 540–542 (2009)
13. Zhang, J., Chen, W.-n., Zhong, J.-h., Tan, X., Li, Y.: Continuous Function Optimization Using Hybrid Ant Colony Approach with Orthogonal Design Scheme. In: Wang, T.-D., Li, X., Chen, S.-H., Wang, X., Abbass, H.A., Iba, H., Chen, G.-L., Yao, X. (eds.) SEAL 2006. LNCS, vol. 4247, pp. 126–133. Springer, Heidelberg (2006)
14. Chaisorn, L., Chua, T.-S.: Story Boundary Detection in News Video Using Global Rule Induction Technique. In: IEEE International Conference on Multimedia and Expo, pp. 2101–2104 (2006)

15. Chen, L., Tu, L.: Parallel Mining for Classification Rules with Ant Colony Algorithm. In: Hao, Y., Liu, J., Wang, Y.-P., Cheung, Y.-m., Yin, H., Jiao, L., Ma, J., Jiao, Y.-C. (eds.) CIS 2005. LNCS (LNAI), vol. 3801, pp. 261–266. Springer, Heidelberg (2005)

16. Mamdani, E.H.: Advances in the linguistic synthesis of fuzzy controllers. Journal of Man-Machine Studies, 669–678 (1976)

17. Mendonça, M.G., Basili, V.R.: Validation of an Approach for Improving Existing Measurement Frameworks. IEEE Transactions on Software Engineering, 484–499 (2000)

18. Dorigo, M., Birattari, M., Stutzle, T.: Ant Colony optimization. IEEE Computational Intelligence Magazine, 28–39 (2006)

19. Mekutha, Ghani, Selamat, Atan.: Complexity metrics for executable business processes. Information Technology Journal, 1317–1326 (2010)

20. Nael, S.: Complexity metrics as predictors of maintainability and inerrability of software components. Journal of Arts and Sciences, 39–50 (2006)

21. Parthasarathy, Anbazhakgan: Analyzing software quality metrics for object oriented technology. Information Technology Journal, 1053–1057 (2006)

22. Bedi, P., Gaur, V.: Trust based quantification of quality in multi agent system. Information Technology Journal, 414–423 (2007)

23. Jiang, Q., Abidi, S.S.R.: From Clusters to Rules: A Hybrid Framework for Generalized Symbolic Rule Induction. In: Yeung, D.S., Liu, Z.-Q., Wang, X.-Z., Yan, H. (eds.) ICMLC 2005. LNCS (LNAI), vol. 3930, pp. 219–228. Springer, Heidelberg (2006)

24. Poli, R.: On fitness Proportionate Selection and Schema theorem in the presence of stochastic effects, Technical Report CSRP-00-2, pp. 1–12 (2000)

25. Tsumoto, S., Hirano, S.: Automated Empirical Selection of Rule Induction Methods based on Recursive Iteration of Resampling Methods and Multiple testing. In: IEEE International Conference on Data Mining Workshops, pp. 835–842 (2010)

26. Ho, S.L., Yang, S., Wong, H.C., Cheng, K.W.E., Ni, G.: An Improved Ant Colony Optimization Algorithm and Its Application to Electromagnetic Devices Designs. IEEE Transactions on Magnetics, 1764–1767 (2005)

27. Tsumoto, S., Hirano, S., Abe, H.: Sampling from Databases for Rule Induction Methods based on Likelihood Ratio Test. In: Proc. 9th IEEE International Conference on Cognitive Informatics, pp. 174–179 (2010)

28. Vinayagasundaram, Srivatsa: Software quality in artificial intelligence system. Information Technology Journal, 835–842 (2007)

29. Bi, Y., McClean, S., Anderson, T.: Combining rough decisions for intelligent text mining using Dempster's rule. In: Artificial Intelligence Review, pp. 191–209. Springer (2006)

30. Zabihinejad, Iranian: Design and optimization of a radial flux direct-drive pm generator using ant colony algorithm. Journal of Applied Science, 2379–2386 (2010)

31. Wang, Z., Feng, B.-q.: Classification Rule Mining with an Improved Ant Colony Algorithm. In: Webb, G.I., Yu, X. (eds.) AI 2004. LNCS (LNAI), vol. 3339, pp. 357–367. Springer, Heidelberg (2004)

Ontology Based Retrieval for Medical Images Using Low Level Feature Extraction

Priyamvada Singh[1], Rohit Rathore[2], Rashmi Chauhan[2],
Rayan Goudar[2], and Sreenivasa Rao[3]

[1] Dept. of Information Technology, Bell Road, Graphic Era University, Dehradun, India
priyakip@gmail.com
[2] Dept. of Computer Science, Bell Road, Graphic Era University, Dehradun, India
{rohitrathor,reshmi06cs,rhgoudar}@gmail.com
[3] CIHL, MSIT Division, IIIT Hyderabad, Gachbowli, Hyderabad, India
prof.srmeda@gmail.com

Abstract. This paper presents a system implemented to evaluate the retrieval efficiency of images when they are semantically indexed using a combination of a web ontology language and the low-level features of the image. Images attached to various entities such as online medical report of a patient are abundant on the web but it is difficult to retrieve accurate information from these entities alone, as using entity names in the search engine gives imprecise results especially where images are concerned. To improve the number of results for medical images we introduced Ontology based image retrieval which is a contextual based images search. Along with the ontologies, we have proposed the use of low level feature extraction to identify a particular image and retrieve it as per user's requirement. The textual and image information is first segregated and relevant data is taken into consideration with the help of semantic filtering based on the contextual exploration (CE), the feature extraction is then applied onto the image and the ontologies are instantiated giving a more relevant result.

Keywords: Semantic web, Ontology, Feature extraction, Image annotation.

1 Introduction

Over the past few years a lot of work has been done in the field of semantic web for enhancing the search; making the information more relevant for the user. But the work is limited to a few domains only. In the area of image retrieval for the medical domain using semantic web a small amount of work has been done so far.

Image retrieval is widely required in the recent decade to express other medium instead of text retrieval. Semantic text retrieval challenges the traditional object based retrieval method, but it cannot satisfy the requirement of an image database [1].

The information retrieval is of great importance in the field of medicine keeping into account the accuracy and the time sensitivity of the results which could in turn affect a doctor's decision for the correct diagnosis. Beside the other textual information semantically stored in a database, image plays a very important role.

J. Mathew et al. (Eds.): ICECCS 2012, CCIS 305, pp. 413–421, 2012.

So image retrieval using semantic web has become an important area of research. By using semantic filtering we can reduce the amount of unnecessary and irrelevant results which we usually get by a normal search.

Semantic Filtering: It is the method of giving text a contextual meaning. This is done by looking at word usage in a document. Words which appear in similar contexts are assumed to have similar meaning and/or relational significance. In this paper we are applying semantic filtering based on contextual exploration method.

Contextual Exploration: In this method the text is filtered on the various information objects such as the indices (spatial features, reference etc.); taking these indices into consideration various rules are defined to infer the result at the extraction time and hence filtering is done efficiently.

All the modern hospital's information systems retrieve such images by querying only the textual details of the patients tagged along with the image which does not ensure an accurate result due to a huge database, so an efficient method must be applied to annotate the anatomy of the images. Another challenge encountered in the image retrieval is the direct translation of informatics methods that are currently applied to the non-imaging data. The image contains rich data that is not explicit like the textual data and not accessible to machine.

The solution of such problems lies in the semantic web, with the help of various ontologies and data annotation technique the unstructured data can be transformed to a machine computable form. For images there are various feature extraction tools which could help in the correct image retrieval along with the annotated textual data.

In the proposed model we will discuss how a system can be constructed which will retrieve both data as well as image of the patient by just querying for it. Previously all the major application in this field took only textual data into consideration because it is free from the low level feature details for the image. It is difficult to deal with image retrieval until it is stored in the text format. So we propose a system where an image is identified with certain unique features (histogram, contour, dominant color etc.) so as to convert it into a text file and hence could be easily accessed just like the normal text.

2   Related Work

Some work has been done on the web retrieval that focuses on image search; a few papers which gave an insightful overview of the different research area are on SEMIA [2] and AIM [3]. For AIM (Annotation and Image Markup), Rubin et al. discussed the various challenges faced by the image retrieval system such as the non-standardization which results in the limited interoperability and another major challenge was the image retrieval on the contextual basis All these issues were resolved with the help of annotating the image to the right context.

AIM is exclusively used for image annotation in a contextual manner with the help of OWL assertions to the appropriate classes, AIM also takes cares of the standardization of the image database e.g. all the annotated images are saved as the AIM XML format and are subsequently transformed to other formats as required by the hospitals or web.

In another paper an online annotation tool for image retrieval named SEMIA (Semantic Medical Image Annotation) was given. The authors have created a hybrid architecture taking into account both the annotation of the image as well as the online search for the various medically relevant information pieces. Kyriazos et al. have used FMA (Foundational Model of Anatomy) as reference ontology for the anatomy; the user can either search for the related images, data or can help in building up the database by annotating the figure manually and submitting the images. They have developed an integrated system of annotation as well as the information retrieval which can act as a knowledge portal between doctors to share similar cases among themselves and enhance their diagnostic approach.

The models discussed deals either only with image annotation or with adding image-description to the image. No work has been done separately onto the image and the data associated with it. So we are proposing a model which will help in extracting the information based on the image properties and not the data alone that has been tagged along with the images.

3 Proposed Model

In our proposed model we are taking into consideration the data along with the image which is there on the web in an unstructured format and converting it into the semantic form. The architecture is divided into three main modules:

a) Semantic Search engine module
b) Image database module
c) Low Level Features Extraction

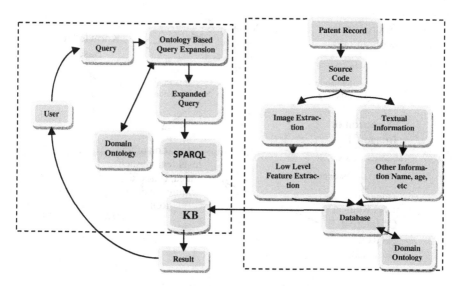

Fig. 1. Overall System Architecture

3.1   Semantic Search Engine

The user enters the query on the database which is converted to the RDF (Resource Description Framework) format with the help of SPARQL.RDF is a directed, labeled graph data format for representing information in the Web. This specification defines the syntax and semantics of the SPARQL query language for RDF [8]. **SPARQL** can be used to express queries across diverse data sources, whether the data is stored natively as RDF or viewed as RDF via middleware.

SPARQL contains capabilities for querying required and optional graph patterns along with their conjunctions and disjunctions [4]. SPARQL also supports extensible value testing and constraining queries by source RDF graph.

The results of SPARQL queries can be results sets or RDF graphs. Once the query is converted to the RDF format, it is then expanded based on the ontology.

Query expansion in general is a method of supplementing the query with additional data to make the retrieved data more meaningful as most of the times the initial query that the user fires is incomplete and inadequate. In the above model the ontology based query expansion takes place in which the words entered by the user is matched with all the concepts of the domain ontology and are further expanded as required. Once the query is expanded, it is stored in the knowledge base which is further used in retrieving the results.

Domain Ontology: It models a specific area of interest which is the part of a bigger domain [10] here we are interested in the medical domain and specifically the images related to the patients record hence we have developed a domain ontology for various image concepts related to the respective patient(see fig 2).

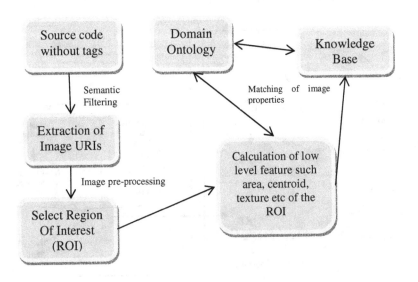

Fig. 2. Image Extraction Module

3.2   Image Database

The image database contains the reference image collection such as the Magnetic Resonance Imaging (MRI), Computer Tomography (CT) and other anatomical modes. The images are selected such that they have different color, texture and shape. All the images are first converted to jpeg format with a particular size and resolution. Image database module works in the following format:

1) First the patient records are taken as raw data which will be in the form of XML files consisting of both image and text. We need to first extract the image and text separately [9]. For that, first the xml page is converted to text with no tags by viewing it as a source code.

2) For the extraction of relevant data from the web pages we use semantic filtering. Semantic filtering is applied with the help of semantic annotation and domain ontology, it involves two steps:

   a)   First we have to design a filter specifying which information object are of interest.

   b)   Second we will use an inference engine to determine which information object can pass through the filter, thus ensuring if the result is correct or not.

The above steps help to determine which text data is relevant by making the use of appropriate object addresses for further analysis. With the help of the semantic filtering which is based on the Contextual Exploration (CE) [6] the images and text is segregated into two different files. Contextual exploration works on the basis of *hierarchical indices* which consist of the systematic collection of data associated with it, *exploration rule* which are to be defined and used at the extraction time, these rule help to refine the result on the contextual level and creating the *database*.

Once the relevant data is extracted the images are also separated by extracting the URI's of the images present in the text, certain geometrical selectors are used as ROI markers to select the required Region Of Interest (ROI) in the image. The next challenge for us was to convert the image in such a format so that it can be used to populate the ontology too. For this purpose we have used the feature extraction tool which extracts the various components such as color distribution, color histogram, contour and texture as well as the other physical features such as the shapes to identify which organ we are dealing with. The calculation of the a few features such as the area, centroid, eccentricity, are shown below in the next section [11]. Once all these features are extracted they can be saved in a separate text file and finally can be used to instantiate the ontology just as a normal text file. The combination of both the file (one associated with the textual description and the other is the image description) will make a database which will in turn be used to create the knowledge base.

3.3   Low Level Features Extraction

Low level feature with the combination of semantic descriptors plays a helpful role in the medical domain for example if the doctor enters a query such as "round tumor", here "tumor" will act as a semantic descriptor connecting it to the ontology whereas the keyword "round" is a textual annotation of a low level feature. In our ontology we are storing the numeric value of the low level feature to retrieve the most relevant results. Various properties needs to be calculated for the image such as the area of the specific region of the image, its centroid, eccentricity, texture, color and contour shape [12] etc.

Area and Centroid - The standard image moment calculation will provide us with the first three features, *area, x centroid,* and *y centroid*. The moments *Mij* are calculated by,

$$M_{i,j} = \sum_x \sum_y x^i y^j I(x,y)$$

Where x and y are pixel locations and, I, is a binary region. The area is calculated by $M00$, and the centroid is $(xc, yc) = (M10/M00, M01/M00)$.

Eccentricity - The eccentricity of a region can be calculated using the central moments μ_{pq},

$$\mu_{pq} = \sum_x \sum_y (x - x_c)^p (y - y_c)^q I(x,y)$$

Color Features Extraction: Commonly color feature is extracted using the histogram technique. The color histogram describes the distribution of different colors of the image in a simple and computationally efficient manner. Other color feature extraction techniques are region histogram, color coherence vector, color moments, correlation histogram etc.

Dominant Color: Dominant color can be identify by calculating the distance between the color descriptors F_1 and F_2 (Tsechpenakis et al.,2002):

$$D^2(F_1, F_2) = \sum_{i=1}^{N1} P^2_{i1} + \sum_{j=1}^{N2} P^2_{2j} - \sum_{i=1}^{N1} \sum_{j=2}^{N2} 2a_{1i,2j} P_{1i,} P_{2j}$$

F = Dominant Color
C = Color
P = Percentage
$a_{k,l}$ = Similarity Coefficient between two colors c_k and c_l

The image retrieval on the basis of color histogram can be done on the following two basis:
1. Global Color Histogram
2. Block Color Histogram

Global Color Histogram: Once the color feature is extracted with the help of the above calculations, the image is retrieved only taking into account the frequency of the color. The direction or the spatial features are not considered.

Block Color Histogram: In this image retrieval method the whole image is divided into nxn matrix and each block is considered at a time. Specific weight of each block is calculated after applying the color space conversion and color quantization.

It is observed [15] that blocked color histogram gives more specific result as compared to the global color histogram method because it takes care of the spatial feature so as to return more accurate results.

Texture Features Extraction: Some of the texture extraction methods are co-occurrence matrix, wavelet decomposition, Fourier filters, etc. In the co-occurrence matrix method, the image is first converted to grey scale and according to the four co-occurrence matrices each texture parameter such as capacity, entropy moment of inertia and relevance is calculated. The comparison of all these parameters are done with the corresponding other images with the help of Euclidean function:

$$D = \sum_{i=1}^{n} (A_i - B_i)^2$$

Contour Shape: Contour shape is determined by calculating the distance between Global Curve Parameter and Curvature Scale Space (CSS) peak associated with the object (Tsechpenakis et al.,2002):

$$M = 0.4 \text{ X } E + 0.3 \text{ X } C + M_{css}$$

$$M_{css} = \sum_i ((x_{peak}[i] - x_{peak}[j])^2 + (y_{peak}[i] - y_{peak}[j])^2) + \sum_j (y_{peak}[i])^2$$

E = Absolute value of eccentricity
C = Absolute value of circularity
M_{css} = Distance between the CSS matching peaks x_{peak} and the missing peak height y_{peak}.

Knowledge Base: Knowledge base is a formalized form of a database which contains the instantiated ontologies. This knowledge base deals directly with the expanded query and extracts the result directly as per the query requirement. The database consists of the data in the form of RDF triples (subject-object-predicate) which can be used directly to extract the relevant web page with the given URI.

4   Implementation

On the implementation level we have developed an ontology exclusively defining the properties of the images for a patient record along with some basic information. We have identified certain concepts in the form of features such as the contextual features, physical features and spatial features. In the contextual Feature class, the concepts are classified on the basis of their relevance e.g. independent features which are independent of the image properties such as the patient's name, doctor's specialty.

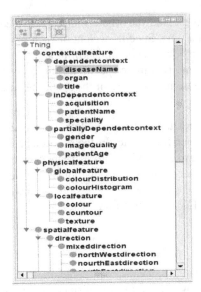
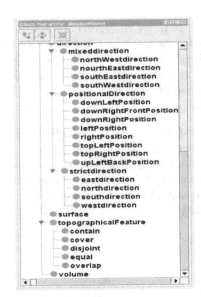

Fig. 3. Example Domain Ontology of the image features

Similarly there are dependent features as well. The physical subclass will cover the local and global features of the image such as the color distribution and the color histogram which will help to identify the image exclusively.

Finally there is the spatial feature class which deals with the directional aspect of the content of the image such as the location of the organ to be detected, the direction are further divided into the mixed and strict directions.

When we instantiate our ontology it should work in the following manner, if a doctor queries about certain patients who had a history of "brain tumor of second stage", then the result will be displayed with the various images and their respective details on the basis of the concepts identified in the ontology such as the texture, color histogram, contours etc. which gives a better and precise result as compared to the previous results which were based on the data tagged along the image and not the image properties itself.

5   Future Scope

In the above proposed model we have used the database where we are storing the information about the image as well as the textual details of the patient, but this will take a lot of time to create that database of the already existing records. Even for a single hospital it is a very tedious task to create a database of all the patient's details, so in future a system can be designed which could search for the patient's record from the web and automatically keep on updating the database. This task can be started on a smaller level such as only for a single hospital or so and can slowly be applied to the whole web making it semantic for medical domain with the help of the ontologies.

6   Conclusion

The vision of the architecture proposed is to create a semantic web structure of the whole medical images available on web thus making medical image data structured and contextually accessible. Up till now all the images were retrieved just on the basis of the text associated with them and not by their own characteristics. The most important module which is introduced in this paper is the concept of extraction of low level features and storing them in a text format hence making it easier to access. The approach proposed in this paper will benefit the medical community to a large extent as large collections of medical images can be indexed and retrieved semantically.

References

1. Khalid, Y.I.A., Noah, S.A.: A Framework for Integrating DBpedia in a Multi- Modality Ontology News Image Retrieval System. In: 2011 International Conference on Semantic Technology and Information Retrieval, Putrajaya, Malaysia, June 28-29, pp. 145–149 (2011)
2. Kyriazos, G.K., Gerostathopoulos, I.T., Kolias, V.D., Stoitsis, J.S.: A semantically-aided approach for online annotation and retrieval of medical images. In: 33rd Annual International Conference of the IEEE EMBS, Boston, Massachusetts USA, August 30-September 3, pp. 2372–2375 (2011)
3. Rubin, D.L., Mongkolwat, P., Kleper, V., Supekar, K., Channin, D.S.: Medical Imaging on the Semantic Web: Annotation and Image Markup. In: Association for the Advancement of Artificial Intelligence (2007)
4. http://www.w3.org/TR/rdf-sparql-query
5. http://code.google.com/apis/searchappliance/ documentation/50/help_gsa/serve_query_expansion.html
6. Shahnaz, B.: Multimedia Medical Application by Filimage System Automatic Extraction of Images Associated to Textual Comments, pp. 2974–2978. IEEE (2006)
7. http://www.who.int/classifications/icd/en/
8. Grobe, M.: RDF, Jena, SparQL and the "Semantic Web". Indiana University, Indianapolis, 1.317.278.6891
9. Radhouani, S., Joohweelim, Chevallet, J.-P., Falquet, G.: Combining textual and visual ontologies to solve medical multi model queries, pp. 1853–1856. IEEE (2006)
10. Ahmad, W., Muhammad Shahzad Faisal, C.: Context based Image Search. IEEE 978-1-4577-0657-8/11, pp. 67–70 (2011)
11. Chávez-Aragón, A., Starostenko, O.: Ontological shape-description, a new method for visual information retrieval. In: Proceedings of the 14th International Conference on Electronics, Communications and Computers (CONIELECOMP 2004). IEEE (2004)
12. Kim, E., Huang, X., Rodney Long, G.T.L., Antani, S.: A Hierarchical SVG Image Abstraction Layer for Medical Imaging
13. Yuea, J., Li, Z., Liu, L., Fub, Z.: Content-based image retrieval using color and texture fused features. Mathematical and Computer Modelling 54, 1121–1 (2011)

Ontology Based Automatic Query Expansion for Semantic Information Retrieval in Sports Domain

Rashmi Chauhan[1,*], Rayan Goudar[1], Rohit Rathore[1], Priyamvada Singh[1], and Sreenivasa Rao[2]

[1] Dept. of Computer Science & Information Technology,
Graphic Era University, Bell Road, Dehradun, India
{rashmi06cs,rhgoudar,rohitrathor,priyakip}@gmail.com
[2] CIHL, MSIT Division, IIIT Hyderabad, Gachbowli, Hyderabad, India
prof.srmeda@gmail.com

Abstract. Ontology is the basis of representing semantic information for a particular domain. In this paper, a conceptual model for ontology based semantic information retrieval, is proposed and we constructed ontology for sports domain. Query expansion techniques have been broadly applied in information retrieval. We have proposed a semantic Query Expansion technique that includes a mathematical model to compute semantic similarity between concepts and an algorithm for query expansion, based on domain ontology. Our approach is different from others in that, 1). It utilizes the query concepts as well as the synonyms of these concepts to perform Query Expansion. 2). the new terms are added only if consisting of a similarity within a threshold 3).The method is being applied in semantic search. The proposed model provides the most relevant information according to user query.

Keywords: query expansion, semantic similarity, ontology, OWL, semantic web, RDF, semantic annotation, SPARQL, WordNet.

1 Introduction

The World Wide Web is a rich source of information, which can be used by different individuals in many ways (as per their requirements). The various search engines are being used to facilitate this service. The current search engines are based on keyword-based search for information retrieval. The search engines like Google, Yahoo, and MSN etc are quite useful for obtaining information from the World Wide Web; but users still have problems to incur the most relevant information for their web queries. Generally, they do not get the appropriate information, they wish. These search engines perform the search on the basis of Keyword- matching. To overcome the problems of the keyword-based search the concept based search is used i.e. semantic search.

Semantic search attempts for improving search accuracy by understanding searcher intent and the contextual meaning of terms as they appear in the searchable data store, to come up with more relevant results. The Semantic Web, as described by the W3C,

---

* Corresponding author.

J. Mathew et al. (Eds.): ICECCS 2012, CCIS 305, pp. 422–433, 2012.

is a Web of data – the collection of Semantic Web technologies, e.g. ROF, OWL, SKOS, SPARQL, provides an environment where software application can query the data, make inferences using vocabularies (ontology) [1]. Ontology is a conceptualization of a domain into a human understandable, machine readable format consisting of entities, attributes, relationships and axioms [2]. It is used as a standard knowledge representation for the semantic web. The use of Ontology to overcome the limitations of keyword-based search has been put forward as one of the motivations of the semantic web [3]. It is constructed by using some standard ontology languages as OWL and frameworks as RDF. We have constructed the ontology for Sports which should be presented in a structured way; so that the relevant information can be given. Ontology can be used to reformulate a query on the basis of related concepts to the user query. This is known as query expansion.

Query expansion is a technique that refines the user initial query and expands it to obtain the relevant results. We have proposed an algorithm for query expansion that follows a mathematical model for calculating semantic similarity between concepts.

2   Related Work

In [4], a review of ontology based query expansion is performed. J. Bhogal et. al. presented the work related to query expansion, following various techniques such as relevancy feedback, corpus dependent and ontology based query expansion. There are two types of techniques for automatic query expansion i.e. global techniques and local techniques. Xu and Croft (2000) compared both the techniques. The performance of global techniques is not good enough because to analyse the total set to find relationships among words, much time is consumed. The local techniques are based on the most frequently occurring terms in the top ranked documents. J. Bhogal et. al. has presented case studies related to query expansion based on WordNet and domain ontology[4]. The domain-independent ontologies such as WordNet consist of some problems because they have a broad coverage, conflicting terms within the ontology can be difficult to figure out. Fu, G. et. al. (2005) proposed query expansion techniques based on both a domain and a geographical ontology. A query is expanded by origin of its geographical footprint. In this work, spatial terms such as place names are modeled in the geographical ontology and non-spatial terms such as 'near' are encoded in tourism domain ontology. Through the experiments it was clear that this improved searches. In 1993, Gruber defined the Ontology which is an explicit specification of a conceptualization, using the conceptualization form of $\langle D, R \rangle$, where D denotes the domain and R denotes a relation set of the concepts of D. Ontology is also used in case-based information retrieval, geographic information retrieval system and bio-engineering for gene annotation.

In [10] Ahmad Maziz Esa et. al. pointed out various problems in searching process through internet. In this paper, they proposed Nutch (which is a web search engine based on the open source information retrieval system called Lucene) with Zenith expansion architecture to add semantic capabilities to it through a plug-in Zenith.

"Subject Relationship Predicate" and semantic indexing are used for query expansion. In this relationship is limited to only two types: positive and negative.

In [9] an ontological approach for semantic-aware learning object retrieval is presented. In this paper an automatic ontology-based query expansion algorithm for aggregating user intent based on seed query, and "ambiguity removal" procedure for correcting improper user query terms are proposed. In [11] corn plant ontology is constructed by FCA (Format Concept Analysis); in which a concept lattice has to be constructed.

To overcome the limitations we have constructed domain ontology; in which a number of relationships between concepts can be defined.J. Bhogal et. al. mentioned the three success factors in using an ontology to expand a query as knowledge model quality, knowledge model familiarity and navigability of knowledge model [4].

In [12] a context based query expansion is proposed based on user interest.; in which weight calculation is being performed for concept similarity; while there can be other factors too. In [13] a conjunctive query expansion approach and an algorithm for query expansion is proposed which is based on some inference rules. Clustering and indexing is being used for similarity. The inference rules are limited to the conjunction of query.

With the purpose to overcome the shortcomings of existing information retrieval, we have presented architecture for semantic information retrieval. Through the literature survey we found that various techniques and algorithms for query expansion have been proposed at conceptual level but from the perspective of implementation, not much work has been done. The techniques like relevancy feedback, fuzzy rules based, corpus based, cluster based have some limitations as most of the techniques are based on WordNet, relationships of words that are limited, follow limited axioms to find related terms etc. Keeping in view all these things we are motivated to propose an algorithm for query expansion. We implemented one module i.e. Sports Domain-ontology construction and the other modules will be implemented in future.

3   Proposed Architecture

The proposed architecture of semantic information retrieval (figure1) is comprised of four modules: Domain ontology construction, User Interface, query handling and semantic search engine. The system provides ontology-based information retrieval for semantic web. The proposed model can solve most of the problems of keyword-based search and with query expansion the degree of relevancy of results can be enhanced.

3.1   Ontology Construction Module

Ontology is a conceptualization of a domain into a human understandable, machine readable format consisting of entities, attributes, relationships, and axioms [2]. Ontology includes a knowledge base that consists of the various entities of a particular domain, and relations among them that how the terms are related to one another, in a structured form. OWL (web ontology language) is the standard language for ontology construction. We have constructed ontology for sports domain (figure2) using Protégé tool. Ontology can be used to reformulate a query on the basis of related concepts to the user query. This is known as query expansion.

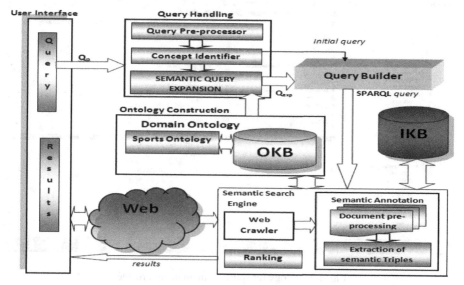

Fig. 1. Architecture of semantic Information retrieval system using query expansion based on domain ontology

3.2 User Interface Module

The user interface module interacts with user which consists of two parts, query and results. The user puts a query to the search engine which is firstly handled by Query handling module and then other modules. User gets the results according to query.

3.3 Query Handling Module

Query Pre-processing

Query pre-processing is the process of identifying the information contained in the initial query, entered by the user including the determination of query structure and length. The first chore of query handling module is to identify the syntax and length of the seed query.

Concept Identifier

Concept identifier extracts the semantic keywords as concepts by removing raw words from seed query. These concepts are then expanded by semantic query expansion to obtain a set of expanded concepts.

Semantic Query Expansion

The concepts identified by concept identifier are expanded to find new concepts related to seed concepts so that the semantic information retrieval can be attained. The method of Semantic Query Expansion is described in section 4.

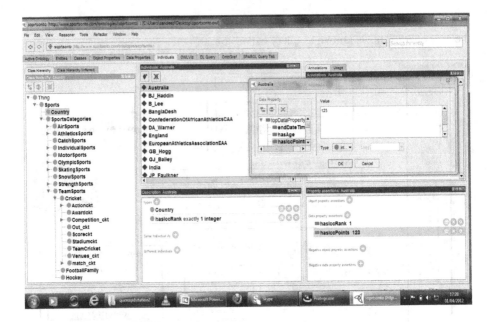

Fig. 2. Ontology for sports domain using Protégé

3.4   Semantic Search Engine Module

Web Crawler.
The web crawler collects the documents from the web. The documents that are not semantically annotated are annotated.

Semantic Annotation
In Semantic annotation, the two main operations are performed:

Document Pre-processing
The web crawler collects the documents from the web. These documents are pre-processed by segmenting into words. In this paper, sports ontology is used to indicate word segment so that domain-specific words can be identified.

Extraction of Semantic Triples
RDF [6] (resource description framework) statements are depicted in the form of triples, indicating as subject, predicate, and object. The words obtained by word segmentation are the words which have same or have high similarity with concepts, properties and instances in ontology and refer to the sentence subject, predicate and object. After semantic annotation, RDF triples are formed that are stored in IKB.

IKB (Information Knowledge Base)
The semantically annotated documents (collected by web crawler and by semantic annotation) are stored in form of RDF triples in IKB.

Ranking

The results are ranked according to their relativity based on domain (sports) ontology.

Query Builder

The concepts identified in seed query and the expanded concepts will be submitted to a query builder that constructs the query using a query language SPARQL [6] [8]. SPARQL takes a query in the form of concepts, and returns the information regarding to query in the form of RDF graph. It is used with a protocol (such as HTTP and SOAP). The SPARQL query language is based on matching graph patterns. The graph pattern is a triple pattern, which is like an RDF triple, but with the possibility of a variable instead of an RDF term in the subject, predicate or object positions. To find a graph pattern, triples are aggregated. This SPARQL query is used to search from the Information Knowledge Base.

3.5   Algorithm of Proposed System

Step 1. User enters any query through user interface module.

Step 2. The query is handled and expanded by query handler module.

Step 3. The initial and expanded query is built in the form of triples using query builder and SPARQL: a standard query language.

Step 4. The SPARQL query is used to search for the semantic information from semantic search engine. The semantic search engine matches the query in IKB (Information knowledge Base).

Step 5. The RDF triples are returned regarding to the query.

Step 6. The relevant documents are ranked based on their relativity using sports domain ontology.

Step 7. The ranked results are then sent to user as final results.

4   Semantic Query Expansion

Query expansion is the process of expanding the seed query by obtaining new terms, related to initial query; so that the most relevant information can be accomplished. Mostly, the existing QE algorithms are based on phrases matching, axioms, fuzzy rules, synsets of WordNet ontology etc; we identified some limitations in those algorithms as we discussed earlier. This motivated us to propose a new approach for QE. In our approach we are expanding the query in two rounds figure3.

4.1   First Query Expansion

The first Query expansion is accomplished using a machine-readable dictionary WordNet [6]. The output of first expansion is a set of synonyms Qsyn for all concepts of seed query extracted for a particular domain (sports). These expanded concepts are then combined with seed query concepts Qp for further expansion.

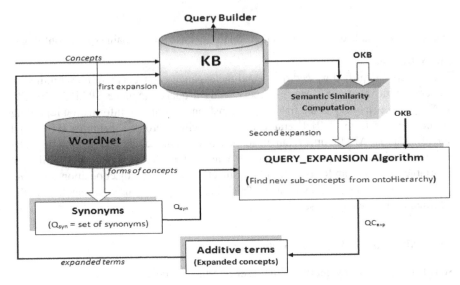

Fig. 3. Query expansion based on ontology and WordNet

4.2   Semantic Similarity Computation

Ontology essentially represents a particular domain in conceptual form. The various terms are viewed as concepts and set up in a hierarchical manner. This hierarchy contains the three main aspects *classes, properties, and individuals.* These concepts can also form ontology individually. Each concept is related to each other in ontology hierarchy; so semantic query expansion can be performed through the concepts that senses with query keywords. To accomplish this intent, semantic similarity computation between concepts is needed. This paper puts forward a *mathematical model* to compute semantic similarity. It considers three main aspects: semantic distance between concepts, Layer factor, and degree of upper concepts.

- **Semantic Distance factor between concepts**

The semantic distance between concepts pertains to the minimum number of edges between two concepts in ontology hierarchy. As the distance increases the similarity decreases as the concepts are set up according to their relativity. Therefore, semantic similarity is inversely proportional to semantic distance.

$$\text{SSim} \propto \frac{1}{dist(c1,c2)} \text{ and thus, Sim1} = \quad p * \frac{1}{1+a*dist(c1,c2)} \tag{1}$$

Where p and a are the coefficients and dist (c1, c2) is the semantic distance between concepts c1 and c2.

The chart in figure 4 dictates the variation of distance factor according to a; where $y = \frac{1}{1+ax}$ and x = dist (c1, c2). As shown in figure 4 the semantic distance factor decreases with increasing distance. At a = 0.5 and x = 6, y < 0.3 and SSm can also be <

0.4 and at a=2, x = 6, y >> 0.3 but at a=0.4, y ≥ 0.3 and SSm will be around 0.4. Thus the value of a can be considered 0.4.

- **Layer Factor**

Since ontology represents the concepts in a hierarchical manner; in which the concepts are set up at different layers. The concept ci at any layer will have the greater similarity with a concept c, than a concept cj, located at down layer. As we consider the farther layer, the similarity is decreased. So, the difference between layers of two concepts can be considered as the influence factor on semantic similarity. Hence,
The semantic similarity is inversely proportional to the difference of the layer of the concepts.

$$SSim \; \propto \; \frac{1}{|L(c1)- L(c2)|} \quad \text{and thus, } Sim2 \;\; = q * \frac{1}{|L(c1)- L(c2)|+1} \qquad (2)$$

Where, L(c) denotes the layer of a concept c.

- **Degree of Upper concepts**

As, we discussed earlier, ontology represents the hierarchy of concepts. A concept c can have more than 1 successor. If two concepts c1 and c2 consist of same parent concept c than c1 and c2 will have the greater similarity than other concepts having different parent concept. So, the degree of upper concepts can be considered an influence factor on semantic similarity.

Upper Concept, Down Concept.
For concepts c1 and c2 in the same ontology,
If L (c1) < L (c2) then c1 is upper concept
Otherwise c1 is down concept.
Hence, similarity is directly proportional to degree of common upper concepts.

$$SSim \; \propto \; no. \, of \, common \, upper \, concepts$$

$$Sim3 \quad = r * \frac{|comupc(c1,c2)|}{|upc(c1,c2)|} \;, \qquad (3), \qquad \text{Where comupc} = \; \boldsymbol{upc(c1) \cap upc(c2)}$$

$$upc \, (c1, c2) = \boldsymbol{upc(c1) \cup upc(c2)}$$

upc denotes the upper concept and comupc as common upper concepts.

$\frac{|comupc(c1,c2)|}{|upc(c1,c2)|}$, is known as the degree of upper concepts.

$$So, \, SSim \, (c1, c2) = \sum \boldsymbol{Sim \, (c1, c2)} \qquad (4)$$

The similarity between two concepts cannot exceed 1; where both of the concepts are same and it cannot be less than 0 where, the concepts are located at ∞ distance. The value of SSm should be between [0, 1]. So, the coefficients p, q and r cannot exceed 1. Hence, the values of these coefficients should be small enough such that $\sum(\boldsymbol{p,q,r}) = \boldsymbol{1}$ i.e. p + q + r = 1

Thus semantic similarity between concepts c1 and c2:

$$SSim(c1, c2) = \begin{cases} 0 & \text{if } dist(c1,c2) = \infty \\ SSm & \text{otherwise} \\ 1 & \text{if } dist(c1,c2) = 0 \text{ or } c1 = c2 \end{cases}$$

$$Where, \quad SSm = p * \frac{1}{1 + a * dist(c1,c2)} + q * \frac{1}{[|L(c1) - L(c2)| + 1]} + r * \frac{|comupc(c1,c2)|}{|upc(c1,c2)|}$$

$$and \quad p + q + r = 1$$

4.3   Second Query Expansion

Consider a set of seed query concepts as Qp, a set of synonyms obtained by first query expansion Qsyn and a set of query concepts Qcp for second query expansion. The following algorithm provides a set of expanded concepts Qcexp.

QUERY_EXPANSION Algorithm.

Input: $Q_{cp} = Q_p \cup Q_{syn}$ where $Q_p = \{c_1, c_2..............................c_n\}$

Output: Q_{exp} (expanded concepts)

Step1: $Q_{cp} = Q_{syn} \cup Q_p \, \Xi \, \{qc_1, qc_2,.........qc_m\}$

Step2: Check whether the qc_i is class, property or individual by concept matching and ascertain the position of qc_i in ontology to find the similarity between concepts.

Step3: for each query concept qc_i do

If qc_i is a class then find

CC= $\{c_1, c_2..............c_j\}$, where SSim $(c_j, qc_i) \geq 0.4$ and ꓬ $c_j \in$ classname related to qc_i
CI = getIndividual (qc_i) CI is a set of concept individuals associated with
qc_i CP = getProperty (qc_i) and go to step4;

Otherwise, if qci is an individual then find

CI= $\{I_1, I_2.............I_j\}$,where SSim $(I_j, qc_i) \geq 0.4$ and ꓬ Ij $\in$ Individuals related with qc_i

CP = getproperty (qc_i) CP is a set of concept properties associated with qc_i

CC = getClass (qc_i) and go to step4;

Otherwise, if qci is a property then find

CP= $\{P_1, P_2.......P_j\}$,where SSim $(P_j, qc_i) \geq 0.4$ and ꓬ $P_j \in$ Individuals related with qc_i

CC = getClass (qc_i) CC is a set of concept classes associated with qc_i

CI = getIndividual (qc_i) and go to step4;

End for

Step4: Find the set of expanded concepts.

$$Qc_{exp} = Qc_{exp} \cup CC \cup CP \cup CI$$

Hence, Qc_{exp} will consist of a set of all possible expanded concepts. These concepts are stored in OB-IKB (ontology-based information knowledge base); which are then used for semantic information retrieval.

Example: Consider small example ontology class hierarchy in figure 5. Let any user enters a query like *"who are the participants of test-match?"* This query will firstly be handled by query handler module. In query pre-processing, the structure, type and length etc of query are checked. Concept identifier extracts the concepts from this query as {participants, test-match} by removing raw keywords.

Q_p = {participants, test-match} and Q_p is expanded by the semantic expansion part.

First expansion: Qsyn = {players, match-game}.

Second expansion: For second expansion the algorithm is followed as:
Qcp = {players, participants, test-match, match-game}
For each qci = players, participants, test-match, and match game check whether its class or property or individual in ontology hierarchy
In figure 5 players and test-match are matched in class hierarchy. Ascertain the position and Qcp = {test-match, players, participants, match game}
For qc1 = test-match, find similarity with all other concepts as:
SSm(test_match, cricket)   = 0.77 > 0.4
Hence, CC = {cricket, bat_and_ball_Sport, players, events ....} and CI = { }
CP = {haslocation, date, year, team, average, doc1, doc2.....} and find QCexp
Similarly, for qc_2 =   CC = {test_match, event, spot, action, cricket ...}
    CI = {Sachin, Yuvraj.....},CP = {isparticipant in, playerwith,..........} and find QC_{exp}
Hence, QC_{exp} = {test-match, players, cricket, bat_and_ball_Sport, event, sports.........}
The expanded query will be generated for these expanded concepts.
In this way, the seed query will be expanded using the proposed algorithm. The initial query and the expanded query will be built in form of triples using SPARQL query builder. These queries are used for information retrieval by IKB.
The results obtained by the proposed approach are more relevant to user query.

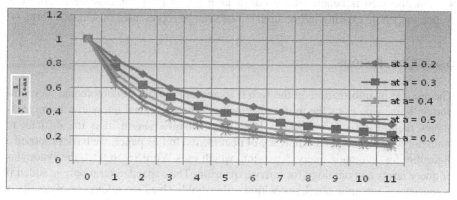

Fig. 4. Variation of semantic distance factor affecting via a $y = \frac{1}{1+ax}$ and x = dist (c1, c2)

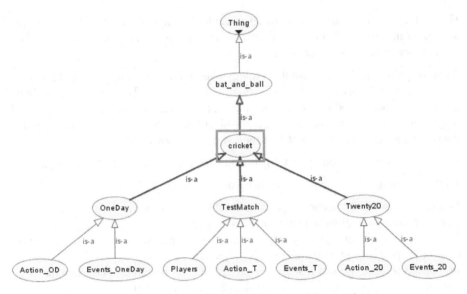

Fig. 5. A part of sports domain ontology as cricket ontology

Hence, the proposed system differs from existing techniques and provides advantages:

- The information of a particular domain that exists currently on the Web is mostly semi-structured or unstructured; it can be represented in a structured form by constructing domain ontology.
- Our Query Expansion method differs from others in that, 1). It utilizes the query concepts as well as the synonyms to perform Query Expansion. 2). the new terms are added only if consisting of a similarity within a threshold.
- Using query expansion, only the relevant documents will have top rank. The existing algorithms and techniques are applied mostly for keyword-based search but we applied query expansion for semantic search.
- Semantic web is beneficial for information retrieval. It contains the future view that the whole existing information on web can be represented in semantic form.

5   Conclusions

The amount of information on the web is increasing rapidly. It is more and more important to provide users the useful information. The traditional query expansion techniques require users to consider a large piece of information. It is too difficult to maintain the precision and efficiency of documents. In this paper, we have proposed a algorithm for automatic query expansion, which uses WordNet to fetch the synonyms of query terms and ontology for query expansion. The expanded term only is added if it consists of semantic similarity within a threshold. Thus, the relevancy of results is improved; as the expanded terms are closely related to seed query concepts.

Acknowledgment. This work has been done in the College Graphic Era University, Dehradun; where the authors1, 3, 4 are currently the students of MTech (CSE/IT). The authors would like to express their sincere thanks to Dr. Rayan Goudar for his support rendered during the work of this module.

References

[1] Yang, C.-Y., Lin, H.-Y.: An Automated Semantic Annotation based-on WordNet Ontology

[2] Fensel, D., van Harmelen, F., Horrocks, I., McGuinness, D.L., Patel Schneider, P.F.: OIL: An Ontology Infrastructure for the Semantic Web. IEEE Intelligent Systems 16(2), 38–45 (2001)

[3] Castells, P., Fernandez, M., Vallet, D.: An adaptation of the vector-space model for ontology-based information retrieval. IEEE Transactions on Knowledge and Data Engineering 19(2), 261–272 (2007)

[4] Bhogal, J., Macfarlane, A., Smith, P.: A review of ontology based query expansion. Elsevier ScienceDirect, 0306-4573 (2006)

[5] Selvaretnam, B., Belkhatir, M.: Natural language technology and query expansion issues, state-of-the-art and perspectives. Springer Science+Business Media, LLC (2011)

[6] Miller, G.: WordNet: A Lexical Database for English. Communications of the ACM 38(11), 39–41 (1995)

[7] Latiri, C., Haddad, H., Hamrouni, T.: Towards an effective automatic query expansion process using an association rule mining approach. Springer Science+Business Media, LLC (2011)

[8] Zhai, J., Zhou, K.: Semantic Retrieval for Sports Information Based on Ontology and SPARQL. IEEE (2010) 978-0-7695-4132-7/10

[9] Lee, M.-C., Tsai, K.H., Wang, T.: A practical ontology query expansion algorithm for semantic-aware learning objects retrieval. Elsevier ScienceDirect, 0360-1315 (2006)

[10] Esa, A.M., Taib, S.M., Hong, N.T.: Prototype of Semantic Search Engine Using Ontology. In: IEEE Conference (2010)

[11] Qi, H., Zhang, L., Gao, Y.: Semantic Retrieval System Based on Corn Ontology. IEEE (2010) 978-0-7695-4139-6/10

[12] Najmeh Ahmadianm, M., Nematbakhsh, A., Vahdat-Nejad, H.: A Context Aware Approach to Semantic Query Expansion. IEEE (2011) 978-1-4577-0314-0/11

[13] Pahal, N., Gulati, P., Gupta, P.: Ontology Driven Conjunctive Query Expansion based on Mining User Logs. In: International Conference on Methods and Models in Computer Science (2009)

Privacy Preserving Technique in Data Mining by Using Chinease Remainder Theorem

P. Rajesh[1], G. Narasimha[2], and Ch. Rupa[3]

[1,3] Department of CSE,
VVIT, Nambur, Andhra Pradesh, India
[2] Department of CSE,
JNTUH, Hyderabad, Andhra Pradesh, India
{rajesh.pleti,narasimha06,rupamtech}@gmail.com

Abstract. Human users are facing lots of problems, when they are transmitting sensitive data and confidential data. The sensitive data is intended to share between only authorized persons, not for all. The collection and sharing of person specific sensitive data for data mining raise serious concerns about the privacy of individuals. Privacy preserving data mining also concentrate on sensitive knowledge pattern that can be exposed when mining the data. Therefore, researchers, for a long time period, have been investigating paths to provide privacy for sensitive data in data mining analysis task process. So, many techniques have been introduced on privacy preserving issues in data mining by using equivalence operator. In this paper we proposed an effective and efficient approach based on Chinease remainder theorem (CRT) that uses congruence operator to provide privacy of sensitive data in the transformation of extracted data mining analysis from servers to clients.

Keywords: Privacy preserving, data mining, chinease remainder theorem (CRT), Equivalence, Congruence.

1 Introduction

Privacy issues in data mining have attracted significant interest in the past decade with its vast domain of applications [1].The collection, analysis and sharing of person specific data for publication or data mining raise serious concerns about the privacy of individuals [2]. Privacy preserving data mining also concentrate on sensitive knowledge patterns that can be exposed when mining the data [3, 4]. There are many challenges that require further investigation both from a theoretical and practical point of view. New privacy preserving data mining techniques as personal data is protected (individual privacy), knowledge extracted is private (corporate privacy) [15, 16]. Privacy preserving data mining has become prevalent for sharing sensitive data [17, 18]. It has become increasingly popular because it allows sharing of privacy sensitive data analysis purposes.

Consider a scenario in which two or more parties owning confidential data wish to share among different parties without revealing any unnecessary information [8].

J. Mathew et al. (Eds.): ICECCS 2012, CCIS 305, pp. 434–442, 2012.

There we have to provide the security of sensitive information. Then we have to transform this information to data mining system without revealing its data bases to any other party and perform privacy preserving data mining to extract the hidden patterns in databases. In this process of knowledge sharing, most recent techniques utilizes equivalence relation operator (one-one) means userid, password (pwd) to protect privileges of individual databases and provide the privacy of sensitive information, which is intended to share among people [19, 26]. The process of one – one (equivalence) is shown in Fig 1.

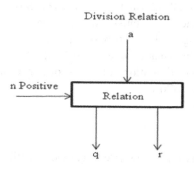

Fig. 1. Equivalence Operation

In this paper we proposed a new privacy preserving technique based on chinease remainder theorem to share   sensitive information in secure manner between clients and server. The paper is structured in the following way. In section 2 describes related literature and the problems in existing technique. Section 3 consists of the proposed technique by using chinease remainder theorem. Solutions to the problems in existing technique are to be explained in section4. Section 5 explains different case studies of CRT based approach.

2   Privacy Preserving Techniques in Data Mining

Privacy preserving data mining is natural tradeoff between privacy quantification and data utility. Identification of problems related to all aspects of privacy and security issues in data mining are significantly growing in real time environment applications. Two current main categories are Perturbation method and the secure computation method to perform data mining tasks without compromising privacy. Yet both have some disadvantages, in the first one reduced accuracy and increased overhead for the second. Some methods published in privacy data mining based on the orientation of the above two categories are as follows.

(1) Differential privacy [7]
(2) Randomization [1,2, 5]
(3) Synthetic data [7]
(4) Secure multiparty computation [10]

(5) Smart cards[9, 13]
(6) Summarization[1, 2]
(7) Cryptography [6,11,12]
(8) K-anonymization [14]

Now days, often companies collect personal information for costumers target product recommendations. Again those companies are giving data or distributing data to other companies for another purpose. In that perspective of social networking, numbers of peoples are sharing their personal information through orkut, face book and scraps and their personal information is misused. Most traditional data mining techniques [1] analyze the data statistically in terms of aggregation and secure mechanisms are also using in this distribution process. Even though, unauthorized parties are accessing the information due to the limitations which are existed in the existing systems. The main limitations of one – one mechanism (user id, pwd) of equivalence relation are as follows.

2.1   Limitations

(1) In that perspective any one is possible to get view profile of other persons. There may be a chance to misuse.
(2) Trusted third parties are not concentrating on privacy of individual information.
(3) Transmitting the secure data by using one-one mechanism. In this, there is more chance to vulnerability of attacker to get sensitive data by minimum scope of boundary as one –one.

3   Proposed Technique Using CRT

In privacy preserving data mining proposing novel technique that allows extract knowledge while trying to protect the privacy of individual users. Privacy preserving data mining has become increasingly popular because it allows sharing of privacy sensitive data analysis purposes. The collection, analysis and sharing of person specific data for publication or data mining raise serious concerns about the privacy of individuals. There are many challenges that require further investigation both from a theoretical and practical point of view. New privacy preserving data mining techniques provide personal data is protected (individual privacy) and knowledge extracted is to be shared in secret (corporate privacy) manner. The proposed privacy preserving technique uses modulo arithmetic operator (Congruence) for achieving these objectives. The congruence operator (many – one) is same as the equality operator. It is not one-one, infinite members map to one member [26, 24]. For example: 2 mod10=2, 12 mod 10=2, 22 mod10=2. In cryptography, we often used the concept of congruence instead of equality [20]. The operation of congruence modulo operator is shown in fig 2.

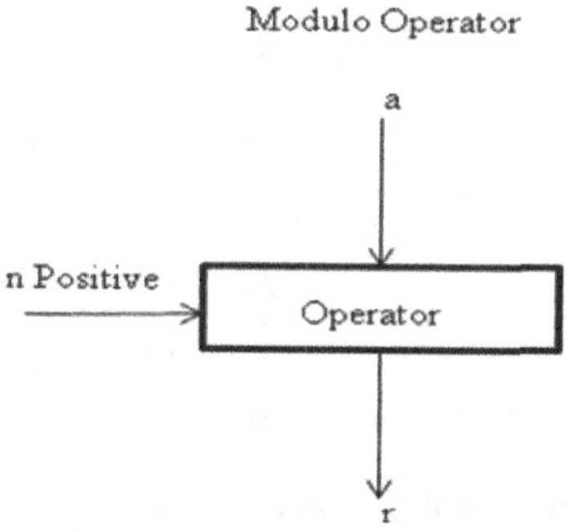

Modulo Operator

Fig. 2. Congruence modulo operation

3.1 Objectives

The main goals of proposed privacy preserving technique are as follows

(1) Privacy preserving data mining can offer guarantees about the level of achieved data quality and utility for demanding applications.
(2) It describes knowledge discovery of data with the integration of data mining techniques in terms of privacy preserving manner.
(3) Publishing and sharing various forms of data without disclosing sensitive personal information while preserving mining results.

3.2 Chinease Remainder Theorem

Let the numbers $n_1 n_2 n_3 \ldots\ldots n_k$ be positive integers which are relatively prime in pairs, i.e. gcd $(n_i, n_j) = 1$ when $i \neq j$. Furthermore, let $n = n_1 n_2 n_3 \ldots\ldots n_k$ and let $x_1, x_2, \ldots x_k$ be integers. Then the system of congruence's

$$x \equiv x_1 \bmod n_1$$

$$x \equiv x_2 \bmod n_2$$

$$x \equiv x_3 \bmod n_3$$

$$\ldots\ldots\ldots\ldots\ldots\ldots\ldots$$

$$x \equiv x_k \bmod n_k$$

has a simultaneous solution x to all the congruence's and any two solutions are congruent to one another modulo n. Furthermore, there exists exactly one solution x between 0 and n-1[21, 22, 23]. The evaluation of CRT [27] is shown in Fig 3.

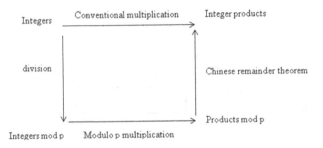

Fig. 3. Integer multiplication by mod p transformation

Furthermore, all the solutions X to this system are congruent modulo the product of $n = n_1 n_2 \ldots n_k$. The solution of system equations satisfies $x \equiv y \bmod n_i$ for $1 \leq i \leq k$, if and only if $x \equiv y \bmod n$. Some times, the simultaneous congruence can be solved even if the n_i's are not pair wise co prime. A solution X to the system of equations exists if and only if $x_i = x_j (\bmod \gcd(n_i, n_j))$ for all i and j. All solutions X are then congruent modulo the least common multiple of n_i.

3.3   Privacy Preserving by CRT

The efficiency of privacy preserving information methods is based on the two factors i.e. level of security and the amount of computation. The proposed method will improves the efficiency of the system by using congruence modulo arithmetic operator and CRT on client server distributed architecture. The architecture of the proposed method is shown in Fig 4. This system architecture consists of curator (admin) in between client and server. It holds two modules. Module 1 describes the secure information transmission between group of clients and curator. Module 2 consists of the proliferation between curator and server. In module 1, identities of the requested clients are verified by the curator with in a session of time period and then assign various prime numbers $(n_1 n_2 n_3 \ldots n_k)$ to clients' requests which are relatively co prime to each other. In module 2, Curator will transfer assigned prime numbers to server. Server computes 'n' by the product of prime numbers.

$$n = n_1 * n_2 * \ldots \ldots * n_k$$

By using Chinese remainder theorem (CRT) unique solution 'x' will be finding out by the server and transfer to curator along with secret information. Now, Curator will transfers the information in secure manner by using unique solution 'x' and corresponding assigned prime numbers $(n_1 n_2 n_3 \ldots n_k)$ to the respective clients.

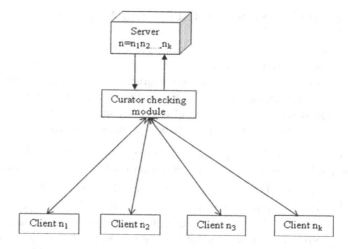

Fig. 4. Architecture of sharing secret information using CRT

3.4   CRT Using K- Threshold System for Secret Sharing

How many people are allowed by the system to access the services or to share secret sensitive information at particular instance of time? The Chinese remainder theorem can also be used in secret sharing, which consists of distributing a set of shares among a group people who, all together (but no one alone), can recover a certain secret from the given set of shares. Each of the shares is represented in congruence and the solution of the system of congruence's using the CRT is the secret to be recovered. Secret sharing of privacy preserving data mining using CRT uses, along with the CRT, special sequence of integers that guarantee the impossibility of the secret from set of shares with less than a certain cardinality[22, 25].

4   Solutions to the Limitations of the Existing System by CRT

Chinese remainder theorem (CRT) based privacy preserving data mining uses congruence authentication approach provides many to one authentication to share the sensitive information. Here there is no chance of masquerading the information of clients by intruders because it may changes information based on particular instance of time stamp requests. Hence it helps to reduce the limitations of the existing privacy preserving approaches by the following way.

(1) To protect the privacy of individual sharing personal information in secret transmission using chinese remainder theorem.
(2) No interference with the third party.
(3) Dynamically generating random number helps to improve the security to provide services to various clients.
(4) Using many – one mechanism (of congruence) by Chinease   remainder theorem, we have enhanced the confusion complexity (more scope of confusion boundary) to attacker and their by reducing chance to vulnerability of attacks, provide more security to sensitive extracted knowledge from data mining..

5   Different Case Studies of CRT Based Approach

All the previous authentication methods based on equality and the equality provide one to one authentication (client produces, server checks). In Chinese remainder theorem based privacy preserving data mining, congruence authentication approach provides many to one authentication to share the sensitive information. Here there is no chance of masquerading the information of clients by intruders why because it may changes information based on particular instance of time stamp requests.

The mapping f: $Z/u \rightarrow Z/u_1 * Z/u_2 * \ldots \ldots Z/u_k$ (chinease remainder theorem in principal ideal domains) is an isomorphism implies that f is homomorphism, injective and surjective. The mapping f is homomorphism means that $f(a + b) = f(a) * f(b)$. The operations of domain of f can be changed into the operations of co domain (quotient ring) of f, i.e. the protocols provide by service station can be changed into protocols at substations. Like HTTP is a protocol used mainly to access data on the World Wide Web. Functions of HTTP are like FTP and SMTP. But there is no separate control connection, unlike SMTP; the HTTP messages are not destined to be read by humans. They are read and interpreted by the HTTP.

The mapping f is injective means that $f(a) = f(b) \rightarrow a = b$ i.e. there is no duplicate congruent modulo of quotient ring. All elements in domain are uniquely mapped to elements in co domain. So there is no chance of duplication or masking to access the sensitive information in privacy preserving data mining. The mapping f is onto means that for every element b in co domain there exist one element a in domain such that f (a) = b i.e. a group of persons is possible to access the services provided by the server in a secure secret sharing manner. Such type of solution is always unique.

The mapping $f: (Z * +) \rightarrow (\emptyset(n) + *)$ is an isomorphism. Hence $n(A) > n(B)$, then there is no chance to vulnerable the contents of information based on the highest number (n) computed by server site and that should be dynamically changed. If any unauthorized want to get access information at sever site, he must has to know all the identities of clients $n_1 n_2 n_3 \ldots n_k$. Why because, the unique solution provided by the sever can be calculated based on the factorization of $n_1 n_2 n_3 \ldots n_k$ and $x = (x_1 r_1 s_1 + x_2 r_2 s_2 + \ldots x_k r_k s_k) \bmod n$ where $r_i = \frac{n}{n_i}$ and $s_i \equiv r_i^{-1} \bmod n_i$ for $i = 1,2 \ldots \ldots k$. Means that s_i is the inverse of r_i under modulo n_i for $i = 1,2 \ldots k$ and the largest number n should be dynamically changed based on particular period of time requests. If anything is going wrong to hijack the information at server site, that information is monitor by curator. He will be acts as a continuous checking module.

If any vulnerable person attack on client's site and suppose he has already get the identity of one client n_1, then it is not sufficient to access the secret information at server site. He has to know about all the identities of other systems $n_1 n_2 n_3 \ldots n_k$. Which are so far located from client1 and these $n_1 n_2 n_3 \ldots n_k$ depends on dynamically changed number (n). If attacker selects a random number at client site to hijack the information, that may not be a prime factor of largest number 'n' selected by group of persons at particular instance of time requests.

How the largest number (n) dynamically changed without disturbing existed sensitive secret sharing information person's in communication channel? It provides services in terms of isolation to group persons in session time. Under the assumption of information security algorithm is an analytically insoluble mathematical problem that can be defined as a function applied to' x' to give 'y'; y=f(x). The function is made to be so complex that reversing it is impossible, like trying to unmix different coloured paints in pot. The time complexity of CRT with given a set of congruence's $a_i \bmod p_i$ for $1 \leq i \leq r$ is $O(r^2)$ [27]..

6   Conclusion

A major social concern of data mining is the issue of privacy and data security. Privacy preserving data mining emerged in response to two equally important needs: data analysis in order to deliver better services and ensuring the privacy rights of the data owners. No party should learn anything more than its prescribed output. Privacy preserving issues are exacerbated now the World Wide Web makes it easy for new data to be automatically collected and added to databases. Privacy preserving data mining faces a lot of challenges to ensure the privacy of individual and secret sharing of sensitive information. We expect that computer scientists, policy experts and counterterrorism experts will continue to work with social scientists, lawyers, companies and consumers to take responsibility in building solution to ensure data privacy protection and security. In this way, we may continue to reap the benefits of data mining in terms of time and money savings and the discovery of new knowledge. Privacy preserving data mining field is expected to flourish.

References

1. Agrawal, R., Srikant, R.: Privacy preserving data mining. In: Proceedings of ACM SIGMOD Conference on Management of Data, Dallas, TX (2000)
2. Agrawal, R., Srikant, R., Eufimieuski, A.: Information Sharing across Private Databaes. In: Proceedings of ACMSIGMOD International Conference on Management of Data, pp. 86–97 (2003)
3. Castano, S., Fugini, M., Martella, G., Samarati, P.: Database Security. Addison Weley (1995)
4. Derosa, M.: Data mining and Data analysis for counterterrorisim. Center for Strategic and International studies (2005)
5. Adam, N.R., Workman, J.C.: Security control methods for Statistical databases. A Comparative Study on ACM Computing Surveys, 515–556 (2006)
6. Sharma, A., Ojha, V.: Implementation of Crypotography for Privacy Preserving Data mining. In: Proceedings of ITDMS, vol. 2(3) (2010)
7. Dwork, C.: A Firm Foundation for private Data analysis. Proceedings of Microsoft Research

8. Dinur, L., Nissm, K.: Revealing Information while Preserving Privacy. In: Proceedings of 22nd ACM Symposium on Principles of Database Systems, pp. 202–210 (2003)

9. Andrew, Y.: Making Privacy Preserving Data mining Protical with Smartcards (2009)

10. Canetti, R.: Security and Composition of multiparty Cryptographic Protocols. Journal of Cryptology, 143–202 (2000)

11. Canetti, R., Ishai, Y., Kumar, R., Reiter, M.K.: Selective Private Function Evaluation with Application to Private Statistics. In: 20th PODC, pp. 243–304 (2001)

12. Lindell, Y., Pintas, B.: Preserving Data mining. Journal of Cryptology, 177–206 (2002)

13. Witreman, M.: Advances in Smartcard Security. Information Security Bultrin, 11–22 (2002)

14. Bayardo, R.J., Agrawal, R.: Data Privacy through Optimal k-anonymization. In: Proceedings of the 21st International Conference on Data Engineering (ICDE 2005), pp. 217–228 (2005)

15. http://Research.microsoft.com/en-us/projects/Databaseprivacy

16. http://www.nextgenerationdatabasesystems.com

17. http://www.privacypreservingbiblography.com

18. Recent IEEE Papers on Privacy Preserving datamining

19. Kolitz, N.: A Course in Number theory and Cryptography. In: Proceedings of Springer-verlay (1994)

20. Nuiven, I., Herbertz, S., Zuckerman: An Introduction to theory of Numbers. Wiley Easteren limited

21. Edpegg, J.R.: Chinese Remainder theorem. In: Wolfram Demonstrations Project (2007)

22. Ding, C., Dingwipei, Salomaa, A.: CRT Applications in Computing, Cooling, Cryptography, pp. 1–213. World Scientific Publishing (1996)

23. Weisstein, Eric, W.: CRT, from mathworld

24. Asmuth, L.A., Bloom, J.: A modular approach to key safe guarding. IEEE Transactions on Information Theory (1983)

25. Ifrene, S.: General Packett sharing Based on the CRT with Applications in E-voting. ENTCS, pp. 67–84 (2007)

26. Behrouz, Forouzan, A.: Cryptography and network security, pp. 20–40. The MC Graw Hill Companies (2005)

27. Borodin, A.B., Muhro, I.: For an interesting collection of papers that deal with evaluation, interpolation and modular arithmetic See. The Computational Complexity of Algebraic and Numeric Problems

Author Index